CHEMICAL SENSES
Volume 3

CHEMICAL SENSES

Series

Volume 1: Receptor Events and Transduction in Taste and Olfaction, *edited by Joseph G. Brand, John H. Teeter, Robert H. Cagan, and Morley R. Kare*

Volume 2: Irritation, *edited by Barry G. Green, J. Russell Mason, and Morley R. Kare*

Volume 3: Genetics of Perception and Communication, *edited by Charles J. Wysocki and Morley R. Kare*

Volume 4: Appetite and Nutrition, *edited by Mark I. Friedman and Morley R. Kare*

CHEMICAL SENSES

Volume 3
Genetics of Perception and
Communications

EDITED BY
CHARLES J. WYSOCKI
MORLEY R. KARE
Monell Chemical Senses Center
Philadelphia, Pennsylvania

Marcel Dekker, Inc.   New York · Basel · Hong Kong

Library of Congress Cataloging-in-Publication Data

Genetics of perception and communication.

(Chemical senses; v. 3)
Based on the International Conference on Genetics and Immunology of Chemosensation and Chemical Communication held at the Monell Center in Philadelphia, Nov. 8-10, 1988.
Includes bibliographical references and index.
1. Chemical senses----Congresses. 2. Chemoreceptors----Congresses. 3. Molecular genetics----Congresses. I. Wysocki, Charles J. II. Kare, Morley Richard. III. International Conference on Genetics and Immunology of Chemosensation and Chemical Communication (1988: Monell Center) IV. Series: Chemical senses (New York, N.Y.); v. 3. [DNLM: 1. Animal Communication-congresses. 2. Chemoreceptors--physiology--congresses. 3. Smell--genetics--congresses. 4. Taste--genetics--congresses.
WL 702 G515 1988 v. 3]
QP455.G46   1991   591.1'826   90--15728
ISBN 0--8247--8370--0 (alk. paper)

This book is printed on acid-free paper.

Copyright © 1991 by MARCEL DEKKER, INC. All Rights Reserved

Neither this book nor any part may be reproduced or transmitted in any form or by any means, electronic or mechanical, including photocopying, microfilming, and recording, or by any information storage and retrieval system, without permission in writing from the publisher.

MARCEL DEKKER, INC.
270 Madison Avenue, New York, New York 10016

Current printing (last digit):
10 9 8 7 6 5 4 3 2 1

PRINTED IN THE UNITED STATES OF AMERICA

In Memoriam

MORLEY RICHARD KARE
March 7, 1922 - July 30, 1990

Most of us dream that we will do something that will make a difference; something that will be important not just for ourselves but for many people. These dreams often prove elusive as we compromise with and submit to the short-term and daily demands of our world.

Morley Kare made a difference. He dreamed of a productive interaction between basic and applied research in chemosensory function, nutrition, and appetite. This dream was realized. Morley's unlimited effort, enthusiasm, guidance, and self-sacrifice made it possible.

Occasionally one of us has a vision, a vision of creating something far greater and more enduring than ourselves. These visions usually fade and eventually disappear. We say it was not meant to be.

Morley Kare envisioned an institute to which investigators from many disciplines would come to pursue an integrated research program. Applied studies on nutrition would complement basic experiments on chemosensory mechanisms. Biochemical investigations of digestive processes would lead to applied aspects of appetite. Morley made this institute a reality.

IN MEMORIAM

The Monell Chemical Senses Center is the instantiation of Morley Kare's dream and vision; without him, this institution would not exist. His hard-won success now provides the opportunity for many present and future chemosensory and nutritional scientists to realize their dreams and visions. Their contributions will be more productive and far-reaching, their chances better, because of Morley R. Kare.

Bruce P. Halpern
Department of Psychology
Section of Neurobiology and Behavior
Cornell University
Ithaca, New York

Series Introduction

The Monell Chemical Senses Center celebrated its twentieth anniversary in 1988. Founded as a multidisciplinary organization dedicated to research in all aspects of the chemical senses, the Monell Center is today the only such institution of its kind and has become the focal point for chemosensory research. While the center has always had strong programs of research in many of the traditional areas of chemosensory science, it has also nurtured and helped to expand the scope of less traditional approaches to the study of the chemical senses. With this in mind, the center commemorated its twentieth year by hosting four international conferences on specialized topics within the framework of the chemical senses.

The proceedings of the four conferences are published by Marcel Dekker, Inc. as the series *Chemical Senses*. The first volume, *Receptor Events and Transduction in Taste and Olfaction,* contains the proceedings of the first international symposium that highlighted recent advances in understanding of receptor mechanisms in taste and olfaction. The second volume, *Irritation*, contains the proceedings of the second international symposium, dedicated exclusively to the developing field of common chemical sensitivity. The impact of genetics on the chemical senses was the topic of the third conference, which is covered in the third volume, *Genetics of Perception and Communication.* The proceedings of the final conference which focused on recent research into the interdependence of the chemical senses and nutrition, appear in the fourth volume, *Appeitite and Nutrition.*

The editors of these volumes are confident that the next twenty years will be even more fruitful than the past twenty and that as our research disciplines mature, this anniversary year of the Monell Chemical Senses Center will be recognized as a major turning point in our understanding of the mechanisms, processes, and functions of the chemical senses.

Joseph G. Brand, Mark I. Friedman, Barry G. Green,
Charles J. Wysocki, and Morley R. Kare

Preface

Until recently, only modest progress had been made in the application of the techniques of behavioral and molecular genetics to an understanding of chemosensory function. For example, bacterial chemotaxis was genetically dissected during the 1970s (research in this area continues today). Furthermore, studies of chemosensory variation among inbred strains of fruit flies and mice were also undertaken in the same decade, and by that time a few pedigrees of selective deficits in human olfaction had been published. Already, the genetic approach was providing powerful tools that could be applied to foster a greater understanding of the chemical senses. More recently, progress has occurred in the genetic dissection of chemoreception in other single-cell organisms and invertebrates. Work on *drosophila* has proceeded to the point of identifying olfactory-specific genes. In vertebrates the isolation of bitter-taste variants among strains of mice and production of congenic lines now offer a mammalian model of taste function worthy of intensified study. Identification of distinct variants in mouse olfactory bulb neuroanatomy will provide valuable models for studies of central processing of olfaction.

To explore and discuss these and other research strategies, the International Conference on Genetics and Immunology of Chemosensation and

Chemical Communication was held at the Monell Center in Philadelphia. The conference grouped investigators from many disciplines, including anatomy, biochemistry, physiology, behavior, and clinical practice. Initially a unifying theme was not evident to many of the participants. Soon, however, a theme did emerge—the utility of genetic methods and strategies in understanding the mechanisms of chemoreception and chemical communication. By the end of the conference, contributors had discussed issues ranging from unicellular chemoreception to human taste and olfaction. Along the way, conference participants were exposed to genetic approaches in invertebrate chemoreception, resource utilization and pheromone communication systems, mammalian models for studies of the mechanisms of taste and olfaction, genetic regulation of individual-specific "odor-prints" in mice and rats, and molecular genetic approaches to isolating and identifying a putative hamster pheromone. It was hoped that these outwardly disparate lines of research could be combined to foster communication among investigators interested in applying a genetic approach to understanding mechanisms of the chemical senses and to convey relevant and substantive information to the readers of this volume, a summary of the conference proceedings. We hope that its contents are as interesting and exciting to the reader as the three-day conference was to participants.

This volume, the third in the *Chemical Senses* series, is intended for anyone interested in unique approaches to an understanding of the chemical senses, including those with research interests in molecular mechanisms of transduction, pheromonal modulation of behavior, animal models of taste and smell, and human chemoreception. We hope that this volume will be of use to those who wish to adopt similar genetic strategies in their areas of expertise. It is also intended to summarize the status of genetic strategies as they are applied in the chemical senses. Here, however, the informed geneticist will undoubtedly note voids that can be filled. Perhaps this volume may stimulate further research from these individuals.

In addition to the contributors in this volume, the input and assistance of numerous others should be acknowledged. These include Dr. Gordon Shepherd (Yale University), Dr. Sol Katz (University of Pennsylvania), Dr. Bruce Waldman (Harvard University), Dr. Walter Nance (Medical College of Virginia), Dr. Frank Margolis (Roche Institute of Molecular Biology), Dr. Sally H. Zigmond (University of Pennsylvania), Dr. Laurie Tomkins (Temple University), Dr. Joseph Brand (Monell Center), Dr. D. Bernard Amos (Duke University), and Dr. Gary K. Beauchamp (Monell Center). Special recognition is extended to those who worked behind the scenes to insure the success of the conference: Janice Blescia, Jodi Carr, Sharron Hoffman, Linda Robinson, Tony Tann, and Linda Wysocki. Funding for the

PREFACE

conference was provided by the National Science Foundation and the National Institute of Neurological and Communicative Disorders and Stroke, National Institutes of Health.

Charles J. Wysocki
Morley R. Kare

Contents

Series Introduction   *v*

Preface   *vii*

Contributors   *xv*

PART I   Genetic and Molecular Genetic Approaches

1. Current Methods in Mouse Genetics   1

   Benjamin A. Taylor

2. Genetic and Immunologic Probes of Signal Transduction in Olfaction   13

   Richard C. Bruch

3. Induction of Odorant-Evoked Current Transients In Ovo by RNA Isolated from the Olfactory Mucosa   25

   Thomas V. Getchell

PART II Invertebrate Inquiries

4. Molecular Genetics of Sensory Signaling in
   Bacterial Chemotaxis — 29
   John S. Parkinson

5. Chemoreception in *Paramecium*: A Genetic Approach — 47
   Judith Van Houten

6. Evolution of Pheromonal Specificity in
   Insect Chemoreceptors — 61
   Richard W. Mankin

7. Olfaction in *Drosophila* — 79
   Obaid Siddiqi

8. A New Homoeotic Mutant with Defects in the
   Drosophila Olfactory System — 97
   Richard Ayer, Paula Monte, and John Carlson

9. Genetics of a Moth Pheromone System — 109
   Wendell Roelofs and Thomas J. Glover

10. Genetic Alteration of the Multiple Taste Receptor
    Sites for Sugars in *Drosophila* — 125
    Teiichi Tanimura

11. A Possible Mechanism of the High Differential
    Sensitivity of Taste in *Drosophila* — 137
    Ichiro Shimada

12. Chemical Cues from Conspecifics and Resource
    Response Variation in *Drosophila* — 147
    Ary A. Hoffmann

CONTENTS

PART III Genetics in Chemical Communication

13. A Hamster Macromolecular Pheromone Belongs to a Family of Transport and Odorant-Binding Proteins — 169

    Foteos Macrides and Alan G. Singer

14. Excretion of Transplantation Antigens as Signals of Genetic Individuality — 187

    Bruce Roser, Richard E. Brown, and Prim B. Singh

15. Chemosensory Identity and Immune Function in Mice — 211

    Kunio Yamazaki, Gary K. Beauchamp, Judith Bard, Edward A. Boyse, and Lewis Thomas

PART IV Mouse Model Systems

16. The Genetics of Bitterness, Sweetness, and Saltiness in Strains of Mice — 227

    Ian E. Lush

17. Congenic Lines Differing in Ability to Taste Sucrose Octaacetate — 243

    Glayde Whitney, David B. Harder, Kimberley S. Gannon, and John C. Maggio

18. Taste Preference and Taste Bud Prevalence Among Inbred Mice — 263

    Inglis J. Miller, Jr.

19. Taste Receptor Mechanisms Influenced by a Gene on Chromosome 4 in Mice — 267

    Yuzo Ninomiya, Noritaka Sako, Hideo Katsukawa, and Masaya Funakoshi

20. Linkage Studies of Genes for Salivary Proline-Rich Proteins and Bitter Taste in Mouse and Human — 279

    Edwin A. Azen

| 21. | Genetics and the Neurobiology of Olfactory Bulb Circuits | 291 |

Charles A. Greer

PART V   Human Analyses

| 22. | A Clinical Manifestation and Its Implications | 317 |

H. N. Wright

| 23. | A Graphic History of Specific Anosmia | 331 |

John E. Amoore and Shelton Steinle

| 24. | Individual Differences in Human Olfaction | 353 |

Charles J. Wysocki and Gary K. Beauchamp

Index *375*

Contributors

John E. Amoore Olfacto-Labs, El Cerrito, California

Richard Ayer Biology Department, Yale University, New Haven, Connecticut

Edwin A. Azen Department of Medicine and Medical Genetics, University of Wisconsin-Madison, Madison, Wisconsin

Judith Bard Department of Microbiology and Immunology, Health Sciences Center, University of Arizona, Tucson, Arizona

Gary K. Beauchamp Monell Chemical Senses Center, Philadelphia, Pennsylvania

Edward A Boyse Department of Microbiology and Immunology, Health Sciences Center, University of Arizona, Tucson, Arizona

Richard E. Brown Dalhousie University, Halifax, Nova Scotia, Canada

Richard C. Bruch Department of Neurobiology and Physiology, Northwestern University, Evanston, Illinois

John Carlson Department of Biology, Yale University, New Haven, Connecticut

Masaya Funakoshi Oral Physiology Department, School of Dentristy, Asahi University, Motosu, Japan

Kimberley S. Gannon Department of Psychology, Florida State University, Tallahassee, Florida

Thomas V. Getchell College of Medicine, Department of Physiology and Biophysics, Chandler Medical Center, University of Kentucky, Lexington, Kentucky

Thomas J. Glover Department of Biology, Hobart and William Smith Colleges, Geneva, New York

Charles A. Greer Sections of Neurosurgery and Neuroanatomy, School of Medicine, Yale University, New Haven, Connecticut

David B. Harder Department of Psychology, Florida State University, Tallahassee, Florida

Ary A. Hoffmann Department of Genetics and Human Variation, La Trobe University, Bundoora, Victoria, Australia

Morley R. Kare[†] Monell Chemical Senses Center, Philadelphia, Pennsylvania

Hideo Katsukawa Oral Physiology Department, School of Dentistry, Asahi University, Motosu, Japan

Ian E. Lush Department of Genetics and Biometry, University College London, London, England

Foteos Macrides Worcester Foundation for Experimental Biology, Shrewsbury, Massachusetts

John C. Maggio Department of Psychology, Florida State University, Tallahassee, Florida

Richard W. Mankin Insect Attractants, Behavior, and Basic Biology Laboratory, Agricultural Research Service, U.S. Department of Agriculture, Gainesville, Florida

Inglis J. Miller, Jr. Department of Neurobiology and Anatomy, Bowman Gray School of Medicine, Wake Forest University, Winston-Salem, North Carolina

---

[†]Deceased

CONTRIBUTORS

Paula Monte Biology Department, Yale University, New Haven, Connecticut

Yuzo Ninomiya Department of Oral Physiology, School of Dentistry, Asahi University, Motosu, Japan

John S. Parkinson Department of Biology, University of Utah, Salt Lake City, Utah

Wendell Roelofs Department of Entomology, Cornell University, Geneva, New York

Bruce Roser Cambridge Research Laboratories, Quadrant Research Foundation, Cambridge, England

Noritaka Sako Oral Physiology Department, School of Dentistry, Asahi University, Motosu, Japan

Ichiro Shimada Department of Biological Science, Tohoku University, Sendai, Japan

Obaid Siddiqi Molecular Biology Unit, Tata Institute of Fundamental Research, Bombay, India

Alan G. Singer Monell Chemical Senses Center, Philadelphia, Pennsylvania

Prim B. Singh Institute of Animal Physiology, Cambridge, England

Shelton Steinle Olfacto-Labs, El Cerrito, California

Teiichi Tanimura Biological Laboratory, Kyushu University, Fukuoka, Japan

Benjamin A. Taylor The Jackson Laboratory, Bar Harbor, Maine

Lewis Thomas Cornell Medical Center, New York Hospital, New York, New York

Judith Van Houten Department of Zoology, University of Vermont, Burlington, Vermont

Glayde Whitney Department of Psychology, Florida State University, Tallahassee, Florida

H. N. Wright Department of Otolaryngology and Communication Sciences, SUNY Health Science Center at Syracuse, Syracuse, New York

Charles J. Wysocki Monell Chemical Senses Center, Philadelphia, Pennsylvania

Kunio Yamazaki Monell Chemical Senses Center, Philadelphia, Pennsylvania

Part I
Genetic and Molecular Genetic Approaches

1
Current Methods in Mouse Genetics

Benjamin A. Taylor

The Jackson Laboratory, Bar Harbor, Maine

During the past 60 years steady progress has been made in murine genetics. Many inbred strains with unique characteristics have been developed. Numerous mutations and genetic variants have been described. An extensive linkage map exists, which is rapidly expanding (Davisson and Roderick, 1987). Special-purpose stocks, such as congenic and recombinant inbred strains and stocks bearing mutations, translocations, and inversions, are available for many kinds of genetic analysis. A rather good taxonomy now exists for the multiple subspecies of wild mice, and representatives of these genetically divergent populations have been propagated and inbred in the laboratory. These resources and accumulated data provide a strong base for what will undoubtedly be an information explosion in the next few years.

The technology for gene cloning and sequencing has had a profound influence on all aspects of genetics, and the mouse has been a major beneficiary of this revolution. Thus the first mammalian gene to be cloned was the mouse hemoglobin gene. The mouse was the organism of choice for the analysis of immunoglobulin gene structure and function. Many murine genomic and cDNA libraries have been constructed, and from these large numbers of specific clones have been isolated and analyzed. Numerous restriction fragment-length polymorphisms (RFLPs) have been identified and mapped. A rapidly growing number of cloned genes have been implicated as the site of specific spontaneous or induced mutations (Table 1). Examples of such mutations

TABLE 1 Mutations at Cloned Genetic Loci

Mutation	Phenotype	Affected gene
W	Dominant spotting, recessive anemia, and sterility	c-kit Proto-oncogene
Hba^{th-1}	Mild anemia	Hemoglobin β-major
spf	Runted, hair loss	Ornithine transcarbamylase
c	Albino	Tyrosinase
Hc^o	Complement deficient	Complement component-5
sph	Anemia	β-Spectrin
hpg	Underdeveloped reproductive tract	Gonadotropin releasing hormone
cog	Hypothyroid with goiter	Thyroglobulin
mdx	Muscle weakness	Dystrophin
shi	Tremors, myelin deficient	Myelin basic protein
jp	Tremors	Myelin proteolipid protein

are the shiverer mutation (shi), which has a deletion in the myelin basic protein gene (Roach et al., 1985), and mutations at the dominant spotting locus (W), some of which have a detectably rearranged c-kit proto-oncogene (Geissler et al., 1988). Because of its complex pleiotropic effects on hematopoiesis, pigmentation, and germ cells, the W locus has fascinated geneticists and developmental biologists for decades. Other cloned genes that are known, or are presumed to be, mutated in specific mouse mutations are hemoglobin β-chain (Skow et al., 1983), ornithine transcarbamylase (Veres et al., 1987), tyrosinase (Kwon et al., 1987), complement component 5 (D'Eustachio et al., 1986), α-spectrin (Bodine et al., 1984), gonadotropin releasing hormone (Mason et al., 1986), thyroglobulin (Taylor and Rowe, 1987), dystrophin (Chamberlain et al., 1987), and myelin proteolipid protein (Dautigny et al., 1986). For the first time it is now possible to understand the causes of many genetic disorders, from the nature of the lesion at the DNA level to the physiologic basis for its phenotypic manifestation. Many of these mutations provide useful animal models for human disease.

Transgenic mice constitute an exciting new tool for molecular genetic analysis (Palmiter and Brinster, 1986). These mice are produced by microinjecting DNA into the pronucleus of the fertilized egg, which is then transferred to a pseudopregnant female. In a percentage of the live-born mice resulting from microinjected zygotes some of the exogenous DNA has integrated into the chromosomes at one or more sites (Gordon et al., 1980). Usually the introduced gene integrates in a head-to-tail array of multiple copies. Although success varies,, usually the introduced gene (native or engineered) is expressed in the transgenic mouse. One of the initially surprising findings was that many human genes can be expressed in the transgenic animal, often in an appropriate tissue-specific manner. For example, mice bearing a human fetal hemoglobin transgene exhibited an embryonic pattern of expression (Chada et al., 1986). This indicates that not only is there conservation of structural genes in evolution, but the regulatory molecules that control the expression of structural genes are also strongly conserved over mammalian divergence. Another implication is that most of the cis-regulatory sequences required for tissue-specific expression are located within a few kilobases of the structural gene. Often the protein-encoding sequences of one gene are fused with the promoter-enhancer region of another gene. By placing a strong inducible promoter, such as the metallothioneine promoter, in front of the coding region of a second gene, the experimenter gains control over the timing and level of expression of the transgene. Thus one of the major uses of transgenic mice has been to identify and study the regulatory elements adjacent to and/or within genes. Among the more dramatic results obtained with this technology was the demonstration that "giant" mice could be created by the introduction of the human growth hormone gene into mice (Palmiter et al., 1982). This at least caught the attention of breeders of domestic livestock! Another dramatic experiment was the construction of a mouse stock highly susceptible to mammary tumors by the introduction of the c-*myc* oncogene under the control of the mammary tumor virus promoter (Leder et al., 1986). Comparable results were obtained by introducing a *myc* oncogene coupled with a strong enhancer from the switch region of the immunoglobulin heavy-chain locus (Harris et al., 1988). These transgenic mice are highly prone to development of B cell lymphomas. Transgenic mice bearing an oncogene in their germline are useful for determining the additional somatic events required for transformation. By targeting oncogene expression to specific tissues it has been possible to create stocks of mice prone to particular neoplasms that are rarely, if ever, observed in ordinary mouse strains (Hanahan, 1985). Construction of transgenic stocks often leads to mice with unpredicted pathologies. Thus, overexpression of the proto-oncogene *fos* resulted in mice with abnormalities of bone and thymus (Ruther et al., 1987, 1988). Transgenic mice have also been made that have specific cell lineages ablated as a result of the tissue-specific expression of a transgene encoding diphtheria toxin A polypeptide. By fusing the toxin gene to the enhancer-promoter regions of other genes it is possible to target specific cell lineages for ablation (Palmiter et al., 1987; Breitman et al., 1987). Another potential use of transgenic mice is for assaying the function of DNA whose function, if any, is unknown. Thus candidate DNA sequences postulated to contain a dominant gene could be evaluated in transgenic mice. Genes responsible for specific chemosensory capacities might be identified by this means.

When foreign DNA is inserted into a chromosome, there is the possibility that the inserted DNA will interrupt a gene, causing an insertional mutation. Many transgenes have proven to be associated with recessive mutations, indicating that the insertion of the foreign DNA has probably disrupted the function of a gene at the site of the insertion. It is estimated that about 7% of transgenic insertions are associated with mutations (Palmiter and Brinster, 1986). If the mutation appears interesting, it is possible to clone the DNA sequences flanking the insertion. An interesting example of such an insertional mutation is one affecting limb development (Woychik et al., 1985). This insertional mutation proved to be allelic to a previously described spontaneous mutation called limb deformity (*ld*). Another transgene could only be transmitted through females (Wilkie and Palmiter, 1987). Males carrying the transgene were fertile but did not transmit the transgene. However, the identification and cloning of the interrupted gene can be complicated because there may be many copies of the transgene and insertion of a transgene may be accompanied by complex genetic rearrangements that delete large stretches of DNA. In such cases it may be difficult to identify the gene or genes associated with the recessive phenotype.

Another way of creating insertional mutations that avoids such complications is retroviral insertion. The mouse genome contains many copies of retrovirus genomes belonging to several different families. Some of these are still active in the sense that the viruses have the potential to infect the germline. One such event led to the dilute coat color mutation (Jenkins et al., 1981). Recently, the mutation hairless (*hr*) was shown to have been caused by retrovirus integration (Stoye et al., 1988). In both these cases wild-type revertants were found in which most of the viral DNA sequences had been deleted, leaving only a single copy of the viral long terminal repeat. Exogenous infection of the early embryo with a retrovirus has been used successfully to create insertional mutations. The most notable example is the insertion of the Moloney murine leukemia virus into the α_1-collagen gene, leading to embryonic lethality (Jaenisch et al., 1985). Another way of generating germline insertions is by inoculating newborn females with infectious murine leukemia virus. Progeny of such females frequently transmit novel germline insertions (Panthier et al., 1988). In all examples of retrovirus-induced mutations studied to date, the virus has been shown to have inserted within, or very near, the affected gene. Of course the experimenter has no control over where the virus inserts, and the probability of insertion into any specific gene is very small. At present it is not practical to set up a screening program to detect insertional mutants at specific loci.

An exciting development has been the successful culturing of pluripotent embryonic stem cells (ES cells). These cells can then be transferred into a developing blastocyst to create an allophenic mouse, part of which is derived from the ES cells. If the mosaicism extends to the germline then the genetic alteration of the ES cells can be propagated. The ES cells can be infected in vitro with a retrovirus, leading to the incorporation of multiple (up to 50) copies of the provirus (Robertson et al., 1987). When it is possible to select ES cells in culture for a particular mutation, this approach provides a method for producing germline mutations at selected genes. Two groups have employed this approach to create a mouse deficient for the enzyme hypoxanthine phosphoribosyltransferase (HPRT) (Hooper et al., 1987; Kuehn et al., 1987). This was feasible because cells deficient for this housekeeping gene can be selected in culture. It also helped that this gene is X-linked and, therefore, hemizygous in male ES cells. One group selected for spontaneous mutations; the other group used retroviral insertion to obtain HPRT-deficient clones. The most surprising thing about these studies, apart from the fact that they worked, was that the HPRT-deficient mice recovered did not manifest any of the clinical symptoms of Lesch-Nyhan syndrome. Although the exact reason that mice are apparently immune to this genetic lesion is unknown, it seems likely the answer will be that the mouse's purine metabolism differs in some important way from that of the human.

In parallel with the development of ES cell technology has been the development of methods for gene replacement by homologous recombination. This technology allows the replacement of the native gene with cloned homologous sequences differing with respect to one or more bases, providing a means of creating specific mutations. Homologous recombination has been successful in ES cells, and a sib-selection scheme, taking advantage of the polymerase chain reaction to amplify the sequences of interest, allows screening of clones with the desired gene (Doetschman et al., 1987; Thomas and Capecchi, 1987). The first mice with replaced genes should be appearing in journals very soon. In principle it should now be possible to create specific germline mutations for any cloned gene. Thus the experimental possibilities are endless. Obviously this technology is not easy, so the number of mutations created in this way is likely to be small at first.

Recombinant inbred (RI) strains, derived by inbreeding from the F_2 generation of the cross of two progenitor strains, provide a powerful tool for mapping polymorphic loci (Bailey, 1971; Taylor, 1978). Such RI strains have now been typed for marker loci that cover most of the autosomal genome. Thus, if one can identify a variant that discriminates between a pair of progenitor strains, there is an excellent chance that a map position can be assigned once the appropriate RI set has been classified. Although there are still some gaps in the "RI map," these are filling in steadily.

The use of interspecific backcrosses provides an excellent way to map genes for which a DNA probe exists (Robert et al., 1985). Crosses between inbred mice and a distantly related mouse relative, *Mus spretus*, yield fertile F_1 females that can be backcrossed to either parental stock. Because of the great divergence between *M. spretus* and laboratory mice, one is almost assured of finding a useful restriction fragment length polymorphism by screening only one or two enzymes. It is feasible to type the backcross DNAs for many RFLPs. Therefore multipoint backcrosses are the rule once a particular set of DNAs has been used extensively. This approach is being used to develop detailed maps of specific parts of the genome, sometimes as part of an effort to clone a specific mutation by chromosome walking. The major limitations of this system (relative to RI strains, for instance) are that one is effectively limited to mapping DNA RFLPs and that the supply of DNA from each backcross mouse is finite. Thus this approach does not provide a general method for mapping traits that must be observed in the living animal or cells.

To overcome some of the limitations inherent to available mapping methods, I am developing a new linkage testing stock based on DNA markers. This stock, designated MEV (multiple ecotropic virus), contains 11 copies of the ecotropic murine leukemia virus provirus. Since most mouse stocks possess few if any of this particular class of provirus, a cross between MEV and another stock can usually be scored for most of the proviruses with a single Southern blot. (Most of the proviruses can be identified uniquely in a Souther blot of *Pvu*II-digested DNAs.) We have introduced three dominant visible markers into this strain background to provide additional markers. Thus by typing 50 backcross progeny for the 11 proviral and 3 visible markers, it is possible to screen about one-half of the autosomal genome. Since novel proviruses occasionally become integrated into the germline of this stock, we can expand the number of viral markers without outcrossing to another strain. Soon it may be possible to sweep most of the autosomal genome with this system. The MEV stock is expected to be particularly useful for mapping new recessive visible mutations and for the detection of major genes affecting complex traits. In collaboration with others, we have already been successful in mapping three recessive visible mutations using this stock.

Somatic cell hybrids continue to provide a useful means of assigning genes to chromosomes. This method is generally limited to mapping cloned genes and other genes that are expressed in tissue culture. Recently the method of chromosome reduction has been used to develop more detailed mapping of human genes (Graw et al., 1988). The procedure requires that there be a selectable marker on a specific human chromosome retained in a particular human-rodent hybrid cell line. Cells are irradiated to create chromosomal rearrangements, including deletions. Clones are screened for markers on the chromosome of interest to detect deletions of specific regions. From the pattern of markers retained or deleted it is possible to work out the gene order. The level or resolution can be greater than that achieved by RFLP analysis of human pedigrees and does not depend on polymorphism. By starting with a hybrid cell line that carries a single human chromosome, one can take advantage of human-specific DNA repeats to provide an almost unlimited number of markers. There is no reason this approach could not be applied to mapping the mouse genome as well.

Another method of gene mapping that has been used occasionally in the mouse is in situ hybridization to metaphase chromosomes (Harper and Saunders, 1981). Although this has been very successful in mapping the well-differentiated human chromosomes, mouse chromosomes are difficult to distinguish by banding after having been through the hybridization protocol. To overcome this difficulty, mouse stocks carrying a single Robertsonian fusion chromosome are used (Caccia et al., 1984; Adolph et al., 1987). These Robertsonian chromosomes, which are common in certain wild mouse populations, consist of two nonhomologous chromosomes joined at the centromere. These contrast with the normally acrocentric mouse chromosomes and are easily identified. By choosing a Robertsonian chromosome that combines large and small acrocentrics, it is easy to tell which arm of the metacentric bears the target sequence. The problem, of course, is that one needs previous knowledge of which chromosome contains the sequence of interest. Thus this approach is useful for physical localization of a gene once it has been assigned to a particular chromosome by another method, such as somatic cell hybridization. Even so, the resolution obtained by this method is not very great. Because of these limitations, relatively few mouse genes have been mapped by this method. However, if there is a reciprocal translocation with a breakpoint near the locus of interest and these translocation chromosomes can be reliably identified after in situ hybridization, then this approach can be very powerful. An example of this approach is provided by Mary Lyon and her collaborators, who used it to map sequences within the t complex on chromosome 17 (Lyon et al., 1986).

Ethylnitrosourea (ENU) has been demonstrated to be a potent inducer of point mutations when administered to male premeiotic germ cells. This has been studied both in the specific (visible) locus test (Russell et al., 1979) and in the induction of null or electrophoretic variants at allozyme loci (Johnson and Lewis, 1981). Both these approaches allow mutation detection in the first generation. ENU has also been used in three-generation screening tests to detect embryonic lethals and to screen for specific metabolic abnormalities. Thus it is feasible to attempt to create recessive mutations at any locus, provided that one has an efficient screening method to identify mutants. Screening can be made more efficient if one knows the location of the target locus and suitable linked markers can be incorporated into the breeding plan. ENU mutagenesis has been used in an attempt to saturate a specific region of mouse chromosome 17 (Shedlovsky et al., 1986).

Extensive linkage homology is now recognized between mouse and human (Nadeau and Taylor, 1984; Nadeau and Reiner, 1989). More than 250 homologous loci have been assigned to chromosomes in both species, and 163 of these have been mapped within the mouse linkage map. These 250 loci define 60 regions of homology. It is estimated that the total number of conserved segments may be as few as 140, with the mean length of these segments about 8 centimorgans. The ability to predict the linkage of homologous loci, once linkage has been established in one species, is helpful in planning mapping strategies and in recognizing mouse models of human genetic diseases.

Several computer data bases containing information on inbred strains, recombinant inbred strains, linkage homologies, and DNA probes are maintained at the Jackson Laboratory. *Mouse News Letter*, published three times yearly by Oxford University Press, contains updated listings of mutant genes, polymorphic loci, chromosomal rearrangements, inbred strains, DNA probes, clone libraries, and linkage maps, as well as research news from many laboratories. The second edition of the monograph *Genetic Variants and Strains of the Laboratory Mouse* (Lyon and Searle, 1989) is a treasure trove of useful information for anyone engaged in mouse genetics.

ACKNOWLEDGMENTS

This work was supported in part by NIH Research Grants GM18684 and CA33093.

REFERENCES

Adolph, S., Bartram, C. R., and Hameister, H. (1987). Mapping of the oncogenes *Myc, Sis* and *int*-1 to the distal part of mouse chromosome 15. *Cytogenet. Cell Genet.* **44**:65-68.

Bailey, D. W. (1971). Recombinant inbred strains. An aid to finding identity, linkage, and function of histocompatibility and other genes. *Transplantation* **11**:325-327.

Bodine, D. M., IV, Birkenmeier, C. S., and Barker, J. E. (1984). Spectrin deficient inherited hemolytic anemias in the mouse: Characterization by spectrin synthesis and mRNA activity in reticulocytes. *Cell* **37**:721-729.

Breitman, M. L., Clapoff, S., Rossant, J., Tsui, L. C., Glode, L. M., Maxwell, I. H., and Bernstein, A. (1987). Genetic ablation: Targeted expression of a toxin gene causes micropthalmia in transgenic mice. *Science* **238**:1563-1656.

Caccia, N., Kronenberg, M., Saxe, D., Hears, R., Bruns, G. A. P., Goverman, J., Mallissen, M., Willard, H., Yoshikai, Y., Simon, M., Hood, L., and Mak, T. W. (1984). The T cell receptor β-chain genes are located on chromosome 6 in mice and chromosome 7 in humans. *Cell* **37**:1091-1099.

Chada, K., Magram, J., and Costantini, F. (1986). An embryonic pattern of expression of human fetal globin gene in transgenic mice. *Nature* **31**:685-688.

Chamberlain J. S., Grant, S. G., Reeves, A. A., Mullins, L. J., Stephenson, D. A., Hoffman, E. P., Monaco, A. P., Kunkel, L. M., Caskey, C. T., and Chapman, V. M. (1987). Regional localization of the murine Duchenne muscular dystrophy gene on the mouse X chromosome. *Somat. Cell Mol. Genet.* **13**:671-678.

Dautigny, A., Mattei, M. G., Morello, D., Alliel, P. M., Pham-Dinh, D., Amar, L., Arnaud, D., Simon, D., Mattei, J. F., Guenet, J.-L., and Jolles, P. (1986). The structural gene coding for myelin-associated proteolipid protein is mutated in jimpy mice. *Nature* **321**:867-869.

Davisson, M. T., and Roderick, T. H. (1987). Locus map of the mouse (*Mus musculus*). In *Genetic Maps 1987: A Compilation of Linkage and Restriction Maps of Genetically Studied Organisms*, Vol. 4, S. J. O'Brien (Ed.). Cold Spring Harbor Laboratory, Cold Spring Harbor, New York, pp. 430-452.

D'Eustachio, P., Kristensen, T., Wetsel, R. A., Riblet, R., Taylor, B. A., and Tack, B. F. (1986). Chromosomal location of the genes encoding complement components C5 and factor H in the mouse. *J. Immunol.* **137**:3990-3995.

Doetschman, T., Gregg, R. G., Maeda, N., Hooper, M. L., Melton, D. W., Thompson, S., and Smithies, O. (1987). Targeted correction of a mutant Hprt gene in mouse embryonic stem cells. *Nature* **330**:586-578.

Geissler, E. N., Ryan, M. A., and Housman, D. E. (1988). The dominant-white spotting (*W*) locus of the mouse encodes the c-*kit* proto-oncogene. *Cell* **55**:185-192.

Gordon, J. W., Scangos, G. A., Plotkin, D. J., Barbosa, J. A., and Ruddle, F. H. (1980). Genetic transformation of mouse embryos by micro-injection of purified DNA. *Proc. Natl. Acad. Sci. USA* **77**:7380-7384.

Graw, S., Davidson, J., Gusella, J., Watkins, P., Tanzi, R., Neve, R., and Patterson, D. (1988). Irradiation-reduced human chromosome 21 hybrids. *Somat. Cell Mol. Genet.* **14**:233-242.

Hanahan, D. (1985). Heritable formation of pancreatic β-cell tumors in transgenic mice expressing recombinant insulin-simian virus 40 oncogenes. *Nature* **315**:115-122.

Harper, M. E., and Saunders, G. N. (1981). Localization of single-copy DNA sequences on G-banded chromosomes by in situ hybridization. *Chromosoma* **83**:431-439.

Harris, A. W., Pinkert, C. A., Crawford, M., Langdon, W. Y., Brinster, R. L., and Adams, J. M. (1988). The Eμ-*myc* transgenic mouse: A model for high incidence spontaneous lymphoma and leukemia of early B cells. *J. Exp. Med.* **167**:353-371.

Hooper, M., Hardy, K., Handyside, A., Hunter, S., and Monk, M. (1987). HPRT-deficient (Lesch-Nyhan) mouse embryos derived by germline colonization by culture cells. *Nature* **326**:292-295.

Jaenisch, R., Breindl, M., Harbers, K., Jahner, D., and Lohler, J. (1985). Retroviruses and insertional mutagenesis. *Cold Spring Harbor Symp. Quant. Biol.* **50**:439-445.

Jenkins, N. A., Copeland, N. G., Taylor, B. A., and Lee, B. K. (1981). Dilute (d) coat colour mutation of DBA/2J mice is associated with the site of an ecotropic MuLV genome. *Nature* **293**:370-374.

Johnson, F. M., and Lewis, S. E. (1981). Electrophoretically detected germinal mutations induced in the mouse by ethylnitrosourea. *Proc. Natl. Acad. Sci. USA* **78**:3138-3141.

Kuehn, M. R., Bradley, A., Robertson, E. J., and Evans, M. J. (1987). A potential animal model for Lesch-Nyhan syndrome through introduction of HPRT mutations into mice. *Nature* **326**:295-298.

Kwon, B. S., Haq, A. K., Pomerantz, S. H., and Halaban, R. (1987). Isolation and sequence of a cDNA clone for human tyrosinase that maps to the mouse c-albino locus. *Proc. Natl. Acad. Sci. USA* **84**:7473-7477.

Leder, A., Pattengale, P. K., Kuo, A., Stewart, T. A., and Leder, P. (1986). Consequences of widespread deregulation of the c-*myc* gene in transgenic mice: Multiple neoplasms and normal development. *Cell* **45**:485-495.

Lyon, M. F., and Searle, A. G. (1989). *Genetic Variants and Strains of the Laboratory Mouse*. Oxford University Press, New York.

Lyon, M. F., Zenthon, J., Evans, E. P., Burtenshaw, M. D., Dudley, K., and Williamson, K. R. (1986). Locations of the *t*-complex on mouse chromosome 17 by in situ hybridization with *Tcp*-1. *Immunogenetics* **24**:125-127.

Mason, A. J., Hayflick, J. S., Zoeller, R. T., Young, W. S., III, Phillips, H. S., Nikolics, K., and Seeburg, P. H. (1986). A deletion truncating the gonadotropin-releasing hormone gene is responsible for hypogonadism in the *hpg* mouse. *Science* **234**:1366-1371.

Nadeau, J. H., and Reiner, A. H. (1989). Linkage and synteny homologies between mouse and man. In *Genetic Variants and Strains of the Laboratory Mouse*, M. F. Lyon and A. G. Searle (Eds.). Oxford University Press, New York, pp. 506-536.

Nadeau, J. H., and Taylor, B. A. (1984). Lengths of chromosomal segments conserved since divergence of man and mouse. *Proc. Natl. Acad. Sci. USA* **81**:814-818.

Palmiter, R. D., and Brinster, R. L. (1986). Germ-line transformation of mice. *Annu. Rev. Genet.* **20**:465-499.

Palmiter, R. D., Brinster, R. L., Hammer, R. E., Trumbauer, M. E., Rosenfeld, M. G., Birnberg, N. C., and Evans, R. M. (1982). Dramatic growth of mice that develops from eggs micro-injected with metallothionein-growth hormone fusion genes. *Nature* **300**:611-615.

Palmiter, R. D., Behringer, R. R., Quaife, C. J., Maxwell, F., Maxwell, I. H., and Brinster, R. L. (1987). Cell lineage ablation in transgenic mice by cell-specific expression of a toxin gene. *Cell* **50**:435-443.

Panthier, J.-J., Condamine, H., and Jacob, F. (1988). Inoculation of newborn SWR/J females with an ecotropic murine leukemia virus can produce transgenic mice. *Proc. Natl. Acad. Sci. USA* **85**:1156-1160.

Roach, J., Takahashi, N., Pravtcheva, Ruddle, R. H., and Hood, L. (1985). Chromosomal mapping of mouse myelin basic protein gene and structure and transcription of the partially deleted gene in shiverer mutant mice. *Cell* **42**:149-155.

Robert, B., Barton, P., Minty, A., Daubas, P., Weydert, A., Bonhomme, F., Catalan, J., Chazottes, D., Guenet, J.-L., and Buckingham, M. (1985). Investigation of genetic linkage between myosin and actin genes using an interspecies mouse backcross. *Nature* **314**:181-183.

Robertson, E., Bradley, A., Kuehn, M., and Evans, M. (1986). Germ-line transmission of genes introduced into cultured pluripotent cells by retroviral vector. *Nature* **323**:445-448.

Russell, W. L., Kelly, E. M., Hunsicker, P. R., Bangham, J. W., Maddux, C., and Phipps, E. L. (1979). Specific-locus test shows ethylnitrosourea to be the most potent mutagen in the mouse. *Proc. Natl. Acad. Sci. USA* **76**:5818-5819.

Ruther, V., Garber, C., Komitowski, D., Muller, R., and Wagner, E. F. (1987). Deregulated c-*fos* expression interferes with normal bone development in transgenic mice. *Nature* **325**:412-416.

Ruther, V., Muller, W., Sumida, T., Tokuhisa, T., Rajewsky, K., and Wagner, E. F. (1988). c-*fos* Expression interferes with thymus development in transgenic mice. *Cell* **53**:847-856.

Shedlovsky, A., Guenet, J.-L., Johnson, L. L., and Dove, W. F. (1986). Induction of recessive lethal mutations in the *T/t-H-2* region of the mouse genome by a point mutagen. *Genet. Res.* **47**:135-142.

Skow, L. C., Burkhart, B. A., Johnson, F. M., Popp, R. A., Popp, D. M., Goldberg, S. Z., Anderson, W. F., Barnett, L. B., and Lewis, S. E. (1983). A mouse model for β-thallasemia. *Cell* **34**:1043-1052.

Stoye, J. P., Fenner, S., Greenoak, G. E., Moran, C., and Coffin, J. M. (1988). Role of endogenous retroviruses as mutagens: The hairless mutation of mice. *Cell* **54**:383-391.

Taylor, B. A. (1978). Recombinant inbred strains: Use in gene mapping. In *Origins of Inbred Mice*, H. C. Morse, III (Ed.). Academic Press, New York, pp. 423-438.

Taylor, B. A., and Rowe, L. (1987). The congenital goiter mutation is linked to the thyroglobulin gene in the mouse. *Proc. Natl. Acad. Sci. USA* **84**:1986-1990.

Thomas, K. R., and Capecchi, M. R. (1987). Site directed mutagenesis by gene targeting in mouse embryo-derived stem cells. *Cell* **51**:503-512.

Veres, G., Gibbs, R. A., Scherer, S. E., and Caskey, C. T. (1987). The molecular basis of the sparse fur mouse mutation. *Science* **237**:415-417.

Wilkie, T. M., and Palmiter, R. D. (1987). Analysis of the integrant in the Myk-103 transgenic mice in which males fail to transmit the integrant. *Mol. Cell. Biol.* **7**: 1646-1655.

Woychik, R. P., Stewart, T. A., David, L. G., D'Eustachio, P., and Leder, P. (1985). An inherited limb deformity created by insertional mutagenesis in a transgenic mouse. *Nature* **318**:36-40.

2
Genetic and Immunologic Probes of Signal Transduction in Olfaction

Richard C. Bruch

Northwestern University, Evanston, Illinois

I. INTRODUCTION

It is generally accepted that activation of peripheral olfactory neurons in vertebrates is initiated by stimulus interaction with the apical dendritic cilia and microvilli of the receptor cells (Rhein and Cagan, 1981). The molecular nature of this interaction is generally assumed to involve the reversible binding of stimuli to macromolecular receptors localized predominantly in the apical cilia and microvilli (Bruch et al., 1988; Getchell, 1986; Getchell et al., 1985; Lancet, 1986, 1988; Rhein and Cagan, 1981). Additional mechanisms, independent of specific stimulus-receptor interaction, may also be involved in olfactory stimulus-response coupling (Kurihara et al., 1986; Lerner et al., 1988; Nomura and Kurihara, 1987a,b). Although the putative receptors have not yet been identified at the molecular level, it is well established from neurophysiologic experiments that stimulus interaction with the apical neuronal membrane elicits depolarization of the chemosensory membrane and subsequent action potential generation and synaptic transmission (Getchell, 1986). Recent biochemical and neurophysiologic evidence indicates that signal-transducing GTP binding proteins (G proteins) in olfactory neurons link the initial stimulus-receptor interaction to regulation of the ion channels underlying membrane depolarization. This chapter summarizes currently available evidence regarding the role of G protein-linked receptor-effector systems in olfactory signal transduction. Since detailed reviews of this subject have been published recently (Bruch, 1990a,b; Bruch and Teeter, 1989; Bruch et al., 1988; Lancet, 1986, 1988), this chapter briefly describes the role of G proteins in olfactory stimulus-response coupling. In addition, the applicability of recently developed genetic and immunologic probes of G protein-linked receptor-effector systems to study the molecular components involved in olfactory reception and transduction is also considered.

II. G PROTEIN-LINKED RECEPTOR-EFFECTOR SYSTEMS IN OLFACTION

A. Receptor-G Protein Coupling

Cell surface receptors can be broadly grouped into two types: those that are functionally coupled to members of the family of signal-transducing G proteins, and those that function by mechanisms independent of G proteins. Many hormone and neurotransmitter receptors are coupled to G proteins; growth factor receptors and the nicotinic acetylcholine receptor function by alternative mechanisms. Recent biochemical and neurophysiologic evidence indicates that olfactory receptors are G protein linked. In a variety of cell types, G protein-linked receptors share many structural and functional properties (Lefkowitz and Caron, 1988). Thus, olfactory receptors may be structurally and functionally homologous to many well-characterized hormone and neurotransmitter receptors. Although this conclusion must be rigorously confirmed by identification and isolation of the receptors, it is consistent with biochemical data obtained from radioligand binding studies and the use of G protein-selective probes.

The stimulus recognition characteristics of olfactory receptors have been most extensively studied by radioligand binding techniques in aquatic species. The feasibility and applicability of ligand binding studies in olfaction was first demonstrated by the pioneering work of Cagan and coworkers (Rhein and Cagan, 1981, 1983). These initial experiments showed that the interactions of stimulus amino acids with the binding sites in isolated olfactory cilia from trout were specific, saturable, and reversible. These minimal criteria expected for a membrane-associated receptor were subsequently described for stimulus amino acid binding in catfish (Bruch and Teeter, 1989), salmon (Rehnberg and Schreck, 1986), and skate (Novoselov et al., 1988). Stimulus binding in isolated cilia preparations from catfish was also shown to exhibit the appropriate selectivity, potency, and stereoselectivity expected for the receptors from neurophysiologic data (Caprio, 1978; Caprio and Byrd, 1984). Stimulus binding in these isolated preparations was also selectively inhibited by lectins, consistent with the likely glycoprotein nature of the receptors (Kalinoski et al., 1987). These combined results, together with the excellent correlation between ligand binding studies in vitro and neurophysiologic recordings in vivo, support the conclusion that the binding of stimulus amino acids represents the interaction of stimuli with physiologically relevant receptors.

Functional coupling of the olfactory amino acid receptors to G proteins was also demonstrated in isolated cilia from catfish (Bruch and Kalinoski, 1987). In the presence of GTP or a hydrolysis-resistant analog, the affinity of the receptors for ligand was decreased, reflecting uncoupling of the receptors from G proteins (Figure 1). This observation further supported the physiologic relevance of the binding measurements, since stimulus binding in the presence of guanine nucleotides satisfied a well-established pharmacologic criterion for a G protein-linked receptor (Gilman, 1987). G proteins were also identified in isolated olfactory cilia from several vertebrate species by bacterial toxin-catalyzed ADP ribosylation and immunoblotting analysis with subunit-specific antisera (Bruch, 1990b). G_s, the G protein associated with receptor-mediated stimulation of adenylate cyclase, was identified in all species examined. At least three pertussis toxin-sensitive G proteins were identified in different species, including G_0, a G protein of unknown function, and two forms of G_i, the G protein associated with inhibition of adenylate

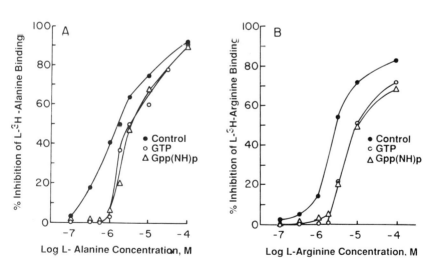

FIGURE 1 Effect of exogenous guanine nucleotides on olfactory receptor binding. [L-^{3}H]Alanine and [L-^{3}H]arginine binding was determined in the absence (control) and presence of 100 μM GTP or Gpp(NH)p with 1 μM radioligand in the presence of the indicated concentrations of unlabeled amino acids. The rightward shift of the binding curves (i.e., lower affinity) is indicative of uncoupling of the receptors from G proteins. (From Bruch and Kalinoski, 1987, with permission.)

cyclase. All these G proteins, identified by established biochemical methods, and an additional G protein were also identified in cDNA clones derived from rat olfactory epithelium (Jones and Reed, 1987).

B. G Protein-Effector Coupling

G proteins function to couple cell surface receptors with intracellular effectors. These intracellular effectors may be enzymes that catalyze the release of second messengers, ion channels, or other cellular components (Gilman, 1987). In olfaction, the identification of G proteins in olfactory cilia and the demonstration of functional coupling of the receptors to G proteins (Section II.A) have encouraged speculation that G protein-linked effectors, particularly second messengers, may mediate olfactory signal transduction. Although stimuli may directly gate ion channel activity (Labarca et al., 1988), two stimulus-regulated second messenger effector systems have been described in isolated olfactory cilia (Bruch, 1990a,b; Bruch et al., 1988; Lancet, 1986, 1988). Biochemical and neurophysiologic evidence has indicated the likely involvement of cyclic nucleotide- and phosphoinositide-derived second messengers in olfactory signal transduction. However, both these effector systems have been shown to be G protein linked since both systems are guanine nucleotide dependent.

The involvement of cyclic nucleotides, particularly cyclic AMP, in olfactory signal transduction has been anticipated for many years. A role for cyclic AMP in olfactory stimulus-response coupling was suggested by initial neurophysiologic experiments indicating that membrane-permeable cyclic AMP analogs and cyclic nucleotide phosphodiesterase inhibitors attenuated stimulus-evoked responses (Menevse et al., 1977). Recently, adenylate cyclase in isolated olfactory cilia was shown to exhibit many of the characteristics associated with the classic hormone-regulated enzyme (Gilman, 1984). These characteristics included sensitivity to forskolin, a direct activator of the catalyst, and to G protein effectors, such as cholera toxin, fluoride ion, and guanine nucleotides (reviewed in Bruch, 1990a,b; Bruch and Teeter, 1989; Lancet, 1986, 1988). Stimulus activation of the enzyme required guanine nucleotides, indicating that stimulus-receptor interaction was coupled to adenylate cyclase by a G protein, presumably G_s. Olfactory stimuli associated with inhibition of adenylate cyclase have not yet been reported. Cyclic nucleotide protein kinase activity and endogenous protein substrates for the enzyme(s) were also identified in isolated olfactory cilia (Heldman and Lancet, 1986; Kropf et al., 1987). Neurophysiologic studies also showed that cyclic nucleotides modulated cation channel activity in olfactory cilia, either by direct gating of the channels (Bruch and Teeter, 1990b; Nakamura and Gold, 1987) or by activation of protein kinase (Vodyanoy and Vodyanoy, 1987). Thus, the cellular components required for a candidate chemosensory

GENETIC AND IMMUNOLOGIC PROBES OF SIGNAL TRANSDUCTION

signal transduction pathway, linking stimulus-receptor interaction to ion channels mediating membrane depolarization by cyclic nucleotides, were identified (Figure 2). However, since stimulus activation of adenylate cyclase requires high stimulus concentrations and prolonged exposure to stimulus, it is likely that cyclic AMP mediates tonic or adaptive responses (Bruch and Teeter, 1989; Pace et al., 1987).

Although many olfactory stimuli activate adenylate cyclase, some stimuli do not affect cyclic AMP levels in isolated cilia (Sklar et al., 1986), implying that at least one additional transduction pathway must be available to account for responses evoked by these stimuli. An additional G protein-linked transduction pathway involving phosphoinositide-derived second messengers was also identified in isolated cilia from catfish (reviewed in Bruch, 1990a,b; Bruch and Teeter, 1989; Bruch et al., 1988). Like stimulus regulation of adenylate cyclase, stimulus activation of phospholipase C also required

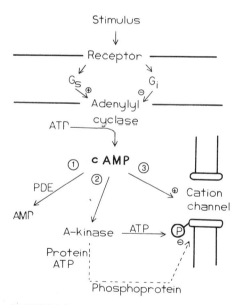

FIGURE 2 Second messenger role of cyclic AMP in olfactory signal transduction. Stimulation of adenylate cyclase following stimulus-receptor interaction, presumably mediated by G_s, leads to elevation of intracellular cyclic AMP levels. Three metabolic fates of cyclic AMP are indicated: (1) degradation catalyzed by cyclic nucleotide phosphodiesterase (PDE); (2) stimulation of cyclic AMP-dependent protein kinase (A-kinase); and (3) direct activation of cation channels in olfactory cilia. Stimulation of A kinase may deactivate cyclic AMP-gated channels by phosphorylation of an intermediary modulatory protein that subsequently regulates channel activity (dashed line). (From Bruch, 1990a, with permission.)

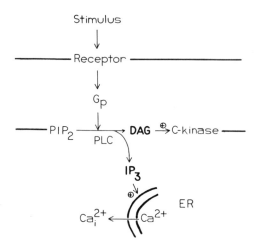

FIGURE 3 Phosphoinositide-derived second messengers in olfactory signal transduction. Stimulus-receptor interaction stimulates phospholipase C (PLC) by a G protein (G_p)-dependent mechanism. Hydrolysis of phosphatidylinositol-4,5-bisphosphate (PIP_2) generates the second messengers diacylglycerol (DAG) and inositol trisphosphate (IP_3). DAG is expected to activate protein kinase C (C kinase); IP_3 has been shown to release calcium from isolated olfactory microsomes (ER). (From Bruch, 1990a, with permission.)

guanine nucleotides, indicating that stimulus-receptor interaction was coupled to the enzyme by a G protein (Figure 3). As in other cells (Fain et al., 1988), the identity of the G protein involved in receptor-mediated stimulation of phospholipase C in olfactory cilia has not yet been determined. Stimuli rapidly (10-15 s) enhanced inositol trisphosphate (IP_3) formation in isolated cilia. The well-established second messenger role of IP_3 in mediating calcium release from intracellular stores (Putney, 1987) was also confirmed in isolated microsomes from the olfactory epithelium (Bruch, 1989b). Protein kinase C activity was also identified in isolated olfactory cilia by phorbol ester binding and immunoreactivity (Anholt et al., 1987). Stimulus regulation of protein kinase C and endogenous protein substrates for the enzyme have not yet been reported in the olfactory system.

III. GENETIC AND IMMUNOLOGIC PROBES IN OLFACTORY SIGNAL TRANSDUCTION

As described in Section II, it is apparent that olfactory receptors are coupled to G protein-linked effector systems. However, perhaps the most important unresolved problem in olfaction is the identification of the receptors at the molecular level and their isolation to study their properties. Although several candidate olfactory receptor proteins have been isolated (Bruch et al., 1988; Fesenko et al., 1987, 1988; Novoselov et al., 1988; Price and Willey, 1988), confirmatory evidence that the isolated proteins retain the appropriate biologic activity (i.e., ligand binding and/or regulation of intracellular transduction events) has not yet been reported. In addition, it has been suggested that traditional biochemical methods that depend on ligand binding to identify the receptors may be inappropriate as a result of low-affinity binding of stimuli (Lancet, 1988). These considerations suggest that novel approaches may be required to unambiguously identify olfactory receptors. Since olfactory receptors are G protein linked, they may be structurally and functionally homologous to other members of the family of G protein-linked receptors. Although somewhat speculative, this conclusion is supported by genetic evidence indicating that G protein-linked receptors share many structural features. Adrenergic receptors and rhodopsin, the visual receptor, share common structural and functional features. These common features include coupling to G proteins, membrane topology, sequence homology, and regulatory mechanisms (Lefkowitz and Caron, 1988). Molecular cloning and site-directed mutagenesis experiments have demonstrated that these receptors, despite different tissue localization and biologic function, share similar sequence and functional domain structure (Kobilka et al., 1988). It may therefore be anticipated that olfactory receptors, by virtue of their coupling to G proteins, may be identified by application of cDNA probes originally designed for other G protein-linked receptors.

Molecular biology techniques have already been applied to identification of G proteins in the olfactory system (Jones and Reed, 1987). In the rat, cDNA clones encoding five G proteins were identified and sequenced. As described in Section II.A, only four of these proteins were detected by traditional biochemical methods in a variety of species. Although subunit-specific antisera to mammalian G protein subunits have been applied to the vertebrate olfactory system (Section II.A), these antisera may be limited in their usefulness in different vertebrate species. It should also be noted that these antisera have recently been shown to discriminate multiple G protein subunits that were not detected by traditional purification and characterization methods (Gilman, 1987). It may therefore be anticipated that the uncertainties associated with traditional biochemical methods may be abrogated by the application of currently available molecular biology technology to determine the identity, developmental expression, and function of G proteins in the olfactory system.

Similar considerations are also applicable to other cellular components involved in signal transduction in olfactory neurons. Little doubt remains that ion channels ultimately account for olfactory neuron activation, but the identity of the channels, their neuronal distribution, and their expression and regulation remain to be established. In this regard it should be noted that site-directed antibodies to ion channels have been described (e.g., Gordon et al., 1988). The ambiguities associated with ion channels in olfactory neurons also extend to other components of the transduction sequence. For example, as in many other tissues, multiple forms of phospholipase C were detected in the olfactory system (Boyle et al., 1987). The role of these isoenzymes in olfactory signal transduction, their expression, and regulation remain to be established. The recently described cloning of phospholipase isoenzymes in other tissues should also be applicable to olfaction (Bennett et al., 1988; Katan et al., 1988; Suh et al., 1988). Immunologic and genetic methods, together with biochemical and electrophysiologic techniques, may therefore be anticipated to eliminate many of the long-standing ambiguities associated with the molecular basis of olfaction.

ACKNOWLEDGMENTS

Work in the author's laboratory was supported by grants from the National Institutes of Health, the National Science Foundation, and the Veterans Administration.

REFERENCES

Anholt, R. R. H., Mumby, S. M., Stoffers, D. A., Girard, P. R., Kuo, J. F., and Snyder, S. H. (1987). Transduction proteins of olfactory receptor cells: Identification of guanine nucleotide binding proteins and protein kinase C. *Biochemistry* **26**:788-795.

Bennett, C. F., Balcarek, J. M., Varrichio, A., and Crooke, S. T. (1988). Molecular cloning and complete amino acid sequence of form-I phosphoinositide-specific phospholipase C. *Nature* **334**:268-270.

Boyle, A. G., Park, Y. S., Huque, T., and Bruch, R. C. (1987). Properties of phospholipase C in isolated olfactory cilia from the channel catfish (*Ictalurus punctatus*). *Comp. Biochem. Physiol.* **88B**:767-775.

Bruch, R. C. (1990a). Signal transduction in taste and olfaction. In *G-Proteins*, L. Birnbaumer and R. Iyengar (Eds.). Academic Press, New York, pp. 411-428.

Bruch, R. C. (1990b). G-proteins in olfactory neurons. In *G-Proteins and Calcium Signaling*, P. H. Naccache (Ed.). CRC Press, Boca Raton, Florida, pp. 123-134.

Bruch, R. C., and Kalinoski, D. L. (1987). Interaction of GTP-binding regulatory proteins with chemosensory receptors. *J. Biol. Chem.* **262**:2401-2404.

Bruch, R. C., and Teeter, J. H. (1989). Second messenger signalling mechanisms in olfaction. In *Receptor Events and Transduction in Taste and Olfaction*, J. G. Brand, J. H. Teeter, M. R. Kare, and R. H. Cagan (Eds.). Marcel Dekker, New York, pp. 283-298.

Bruch, R. C., and Teeter, J. H. (1990). Cyclic AMP links amino acid chemoreceptors to ion channels in olfactory cilia. *Chem. Senses*, in press.

Bruch, R. C., Kalinoski, D. L., and Kare, M. R. (1988). Biochemistry of vertebrate olfaction and taste. *Annu. Rev. Nutr.* **8**:21-42.

Caprio, J. (1978). Olfaction and taste in the channel catfish: An electrophysiological study of the responses to amino acids and derivatives. *J. Comp. Physiol.* **123**:357-371.

Caprio, J., and Byrd, R. P., Jr. (1984). Electrophysiological evidence for acidic, basic, and neutral amino acid olfactory receptor sites in the catfish. *J. Gen. Physiol.* **84**:403-422.

Fain, J. N., Wallace, M. A., and Wojcikiewicz, R. J. H. (1988). Evidence for involvement of guanine nucleotide-binding regulatory proteins in the activation of phospholipases by hormones. *FASEB J.* **2**:2569-2574.

Fesenko, E. E., Novoselov, V. I., and Bystrova, M. F. (1987). The subunits of specific odor-binding glycoproteins from rat olfactory epithelium. *FEBS Lett.* **219**:224-226.

Fesenko, E. E., Novoselov, V. I., and Bystrova, M. F. (1988). Properties of odour-binding glycoproteins from rat olfactory epithelium. *Biochim. Biophys. Acta* **937**:369-378.

Getchell, T. V. (1986). Functional properties of olfactory receptor neurons. *Physiol. Rev.* **66**:772-818.

Getchell, T. V., Margolis, F. L., and Getchell, M. L. (1985). Perireceptor and receptor events in vertebrate olfaction. *Prog. Neurobiol.* **23**:317-345.

Gilman, A. G. (1984). G-Proteins and dual control of adenylate cyclase. *Cell* **36**:577-579.

Gilman, A. G. (1987). G-proteins: Transducers of receptor-generated signals. *Annu. Rev. Biochem.* **56**:615-649.

Gordon, D., Merrick, D., Wollner, D. A., and Catterall, W. A. (1988). Biochemical properties of sodium channels in a wide range of excitable tissues studied with site-directed antibodies. *Biochemistry* **27**:7032-7038.

Heldman, J., and Lancet, D. (1986). Cyclic AMP-dependent protein phosphorylation in chemosensory neurons: Identification of cyclic nucleotide-regulated phosphoproteins in olfactory cilia. *J. Neurochem.* **47**:1527-1533.

Jones, D. T., and Reed, R. R. (1987). Molecular cloning of five GTP-binding protein cDNA species from rat olfactory neuroepithelium. *J. Biol. Chem.* **262**:14241-14249.

Kalinoski, D. L., Bruch, R. C., and Brand, J. G. (1987). Differential interaction of lectins with chemosensory receptors. *Brain Res.* **418**:34-40.

Katan, M., Kriz, R. W., Totty, N., Philp, R., Meldrum, E., Aldape, R. A., Knopf, J. L., and Parker, P. J. (1988). Determination of the primary structure of PLC-154 demonstrates diversity of phosphoinositide-specific phospholipase C activities. *Cell* **54**:171-177.

Kobilka, B. K., Kobilka, T. S., Daniel, K., Regan, J. W., Caron, M. G., and Lefkowitz, R. J. (1988). Chimeric α_2-, β_2-adrenergic receptors: Delineation of domains involved in effector coupling and ligand binding specificity. *Science* **240**:1310-1316.

Kropf, R., Lancet, D., and Lazard, D. (1987). A bovine olfactory cilia preparation: Specific transmembrane glycoproteins and phosphoproteins. *Soc. Neurosci. Abstr.* **13**(Part 2):1410.

Kurihara, K., Yoshii, K., and Kashiwayanagi, M. (1986). Transduction mechanisms in chemoreception. *Comp. Biochem. Physiol.* **85A**:1-22.

Labarca, P., Simon, S. A., and Anholt, R. R. H. (1988). Activation by odorants of a multistate cation channel from olfactory cilia. *Proc. Natl Acad. Sci. USA* **85**: 944-947.

Lancet, D. (1986). Vertebrate olfactory reception. *Annu. Rev. Neurosci.* **9**:329-355.

Lancet, D. (1988). Molecular components of olfactory reception and transduction. In *Molecular Neurobiology of the Olfactory System*, F. L. Margolis and T. V. Getchell (Eds.). Plenum Press, New York, pp. 25-50.

Lefkowitz, R. J., and Caron, M. G. (1988). Adrenergic receptors: Models for the study of receptors coupled to guanine nucleotide regulatory proteins. *J. Biol. Chem.* **263**:4993-4996.

Lerner, M. R., Reagan, J., Gyorgyi, T., and Roby, A. (1988). Olfaction by melanophores: What does it mean? *Proc. Natl. Acad. Sci. USA* **85**:261-264.

Menevse, A., Dodd, G. H., and Poynder, T. M. (1977). Evidence for the specific involvement of cyclic AMP in the olfactory transduction mechanism. *Biochem. Biophys. Res. Commun.* **77**:671-677.

Nakamura, T., and Gold, G. H. (1987). A cyclic nucleotide-gated conductance in olfactory receptor cilia. *Nature* **325**:442-444.

Nomura, T., and Kurihara, K. (1987a). Liposomes as a model for olfactory cells: Changes in membrane potential in response to various odorants. *Biochemistry* **26**: 6135-6140.

Nomura, T., and Kurihara, K. (1987b). Effects of changed lipid composition on response of liposomes to various odorants: Possible mechanism of odor discrimination. *Biochemistry* **26**:6141-6145.

Novoselov, V. I., Krapivinskaya, L. D., and Fesenko, E. E. (1988). Amino acid binding glycoproteins from the olfactory epithelium of skate (*Dasyatis pastinaca*). *Chem. Senses* **13**:267-278.

Pace, U., Heldman, J., Shafir, I., Rimon, G., and Lancet, D. (1987). Molecular correlates of olfactory adaptation: Adenylate cyclase and protein phosphorylation. *Soc. Neurosci. Abstr.* **13**(Part 1):362.

Price, S., and Willey, A. (1988). Effects of antibodies against odorant binding proteins on electrophysiological responses to odorants. *Biochim. Biophys. Acta* **965**: 127-129.

Putney, J. W., Jr. (1987). Formation and actions of calcium-mobilizing messenger, inositol 1,4,5-trisphosphate. *Am. J. Physiol.* **252**:G149-G157.

Rehnberg, B. G., and Schreck, C. B. (1986). The olfactory L-serine receptor in coho salmon: Biochemical specificity and behavioral response. *J. Comp. Physiol.* **159**: 61-67.

Rhein, L. D., and Cagan, R. H. (1981). Role of cilia in olfactory recognition. In *Biochemistry of Taste and Olfaction*, R. H. Cagan and M. R. Kare (Eds.). Academic Press, New York, pp. 47-68.

Rhein, L. D., and Cagan, R. H. (1983). Biochemical studies of olfaction: Binding specificity of odorants to a cilia preparation from rainbow trout olfactory rosettes. *J. Neurochem.* **41**:569-577.

Sklar, P. B., Anholt, R. R. H., and Snyder, S. H. (1986). The odorant-sensitive adenylate cyclase of olfactory receptor cells: Differential stimulation by distinct classes of odorants. *J. Biol. Chem.* **261**:15538-15543.

Suh, P. G., Ryu, S. H., Moon, K. H., Suh, H. W., and Rhee, S. G. (1988). Cloning and sequence of multiple forms of phospholipase C. *Cell* **54**:161-169.

Vodyanoy, V., and Vodyanoy, I. (1987). Ion channel modulation by cAMP and protein kinase inhibitor. *Soc. Neurosci. Abstr.* **13**(Part 2):1410.

3
Induction of Odorant-Evoked Current Transients In Ovo by RNA Isolated from the Olfactory Mucosa

Thomas V. Getchell

University of Kentucky, Lexington, Kentucky

Substantial progress is being achieved in electrophysiologic studies of vertebrate olfactory transduction in which the results of earlier biochemical studies (Lancet, 1986; Anholt, 1987; Snyder et al., 1988) are being extended to the cellular level in physiologically intact preparations (Firestein and Werblin, 1987; Persaud et al., 1987, 1988). Also, initial studies indicate that the *Xenopus* oocyte is an advantageous translation system in which to examine the functional expression of olfactory-specific gene products associated with transduction (Lancet, 1987; Getchell, 1988; Margolis, 1988). In this report, I summarize the basic results from two series of ongoing experiments in which we have identified an amiloride-sensitive odorant-activated Na^+ conductance in the olfactory epithelium and examined the induction of odorant-evoked current transients in oocytes.

We have adapted the Ussing method to identify and characterize pharmacologically the standing (sIsc) and odorant-evoked (oIsc) short-circuit currents in the isolated bullfrog olfactory mucosa. When bathed symmetrically in Ringer's solution, the tissue developed an open-circuit potential of -3.6 ± 0.3 mV (N = 18), with the ciliated side electronegative. Under voltage clamp conditions (± 10 mV), the sIsc was 53.0 ± 14.5 $\mu A/cm^2$ and the tissue resistance was 67 ± 26.5 $\Omega \cdot cm^2$. The sIsc decayed to zero when the tissue was oxygen deprived and was reduced substantially when the serosal side was incubated with 10^{-4} M ouabain or the ciliated side with 10^{-4} M furosemide.

These properties provide evidence for electrogenic ion transport, the presence of a Na^+,K^+-ATPase pump, and a Na^+,Cl^- cotransport site in the olfactory mucosa. An aliquot of the odorant 1,8-cineole evoked a concentration-dependent influx of positive current (oIsc) that ranged in magnitude from 0.1 to 1.0 μA. Incubation of the ciliary side with BrcAMP (10^{-4} M), forskolin (10^{-6} M), GTPγS (10^{-5} M), and IBMX (10^{-4} M) enhanced oIsc, whereas GDP βS (10^{-5} M) and BrcGMP (10^{-6} M) reduced it. These results confirm the biochemical studies and extend them to the cellular level. Voltage clamp analysis of sIsc indicated that the direction of ion translocation was a direct function of the electrochemical gradient across the mucosa, suggesting activation of a passive ion channel. Simultaneous delivery of the odorant (10^{-4} M) and amiloride (10^{-4} M) reversibly blocked the oIsc but had no effect on the sIsc. Symmetrical replacement of the Na^+ ion with N-methyl-D-glucammonium reduced the oIsc by 80%, indicating that amiloride blocked a passive Na^+ conductance associated with odorant transduction.

We have also utilized the *Xenopus* mRNA translation system to examine the functional expression of odorant-specific gene products. Total RNA was isolated from the olfactory mucosae of grass frogs and the brains of rat pups. We estimated that 30 μg RNA per 130 mg wwt olfactory tissue and 8.2 μg per eight brains was obtained. Stage V and VI oocytes were isolated from mature *Xenopus* and then microinjected with RNA. The control oocytes were injected with 50 nl sterile H_2O; the experimental oocytes were injected with 50 ng brain total RNA per 50 nl sterile H_2O, or about 150 ng olfactory mucosal total mRNA per 50 nl sterile H_2O. The responsiveness of the voltage-clamped oocytes was then examined by introducing representative neurotransmitters and an odorant cocktail into the perfusing medium. They were nonresponsive on days 1 and 2. On day 3, the oocytes injected with brain RNA were responsive to serotonin (1 μM) but not to an odorant cocktail (1-10 mM). The response consisted of a spikelike current transient with oscillations similar to those recorded in other studies (Snutch, 1988). The oocytes injected with olfactory mucosal RNA were responsive to the odorant cocktail but not neurotransmitters. The odorant-evoked responses consisted of a long latency (40-50 s) and a slowly rising inward current transient that peaked to maximum levels (10-20 nA) in 2-3 minutes. The transient typically recovered to near baseline values within 3 minutes. Following the initial response, three oocytes were incubated in situ with 1 mM 8-bromo-cAMP. The magnitude of subsequent odorant-evoked responses increased by approximately 30-50% to a maximum of about 50 nA. The current transients also recovered slowly to baseline with about the same time course. The oocytes were nonresponsive to alanine, arginine, and an amino acid cocktail. The results of this initial study confirm the usefulness of the *Xenopus* oocyte as an in ovo mRNA translation system and extends its use to the olfactory system. They

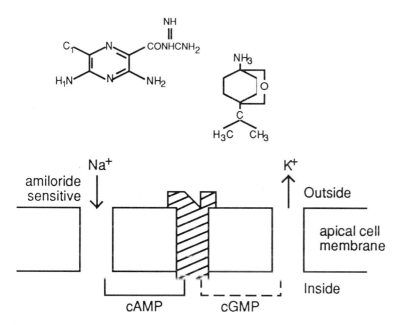

FIGURE 1

also suggest that functional odorant receptor molecule-gated ion channels were inserted into the oocyte membrane.

Current experiments utilizing the oocyte preparation are examining the effectiveness of amiloride in blocking the odorant-evoked inward current. Specifically, does Ca^{2+} mediate an initial influx of amiloride sensitive Na^+ current, and are the adenylate cyclase or the $InsP_3$ second messenger systems utilized as intermediates to initiate the conductance changes? The results of these experiments will determine the usefulness of the oocyte as a functional expression system to investigate the role of olfactory-specific gene products in transduction.

ACKNOWLEDGMENT

Supported by NIH Grants RO 1-NS-16340 and F33-NS-08486.

REFERENCES

Anholt, R. R. H. (1987). Primary events in olfactory reception. *Trends Biochem. Sci.* **12**:58-62.

Firestein, S., and Werblin, F. S. (1987). Gated currents in isolated olfactory receptor neurons of the larval tiger salamander. *Proc. Natl. Acad. Sci. USA* **84**:6292-6296.

Getchell, T. V. (1988). Induction of odorant-evoked current transients in *Xenopus* oocytes injected with mRNA isolated from the olfactory mucosa of *Rana pipiens*. *Neurosci. Lett.* **91**:217-221.

Lancet, D. (1986). Vertebrate olfactory reception. *Annu. Rev. Neurosci.* **9**:329-355.

Lancet, D. (1987). Proteins of olfactory cilia and cAMP-mediated transduction. *Disc. Neurosci.* **IV**(3):68-74.

Margolis, F. L. (1988). Molecular cloning of olfactory-specific gene products. In *Molecular Neurobiology of the Olfactory System*, F. L. Margolis and T. V. Getchell (Eds.). Plenum Press, New York, pp. 237-265.

Persaud, K. C., De Simone, J. A., Getchell, M. L., Heck, G. L., and Getchell, T. V. (1987). Ion transport across the frog olfactory mucosa: The basal and odorant-stimulated states. *Biochim. Biophys. Acta* **902**:65-79.

Persaud, K. C., Heck, G. L., De Simone, S. K., Getchell, T. V., and De Simone, J. A. (1988). Ion transport across the frog olfactory mucosa: The action of cyclic nucleotides on the basal and odorant-stimulated states. *Biochim. Biophys. Acta* **944**:49-62.

Snutch, T. P. (1988). The use of *Xenopus* oocytes to probe synaptic communication. *TINS* **11**:250-256.

Snyder, S. H., et al. (1988). Odorant-binding protein and adenylate cyclase. In *Molecular Neurobiology of the Olfactory System*, F. L. Margolis and T. V. Getchell (Eds.). Plenum Press, New York, pp. 3-24.

Part II
Invertebrate Inquiries

4
Molecular Genetics of Sensory Signaling in Bacterial Chemotaxis

John S. Parkinson

University of Utah, Salt Lake City, Utah

I. INTRODUCTION

Motile bacteria migrate toward favorable chemical environments and away from unfavorable ones. Chemotactic movements in bacteria, as in other organisms, require a locomotor system, receptors for detecting chemical stimuli, and a means of transducing stimulus information into signals that modulate locomotor behavior. Thus, the chemotactic behavior of bacteria represents a simple chemosensory system that is amenable to sophisticated genetic and biochemical analysis. Modern work on bacterial chemotaxis, initiated about 20 years ago, has yielded a detailed molecular view of how bacteria detect and process sensory information. Although there remain significant gaps in our understanding of these events, it is now clear that the basic information-handling strategies in prokaryotes are not very different from those of higher organisms. In particular, the transmembrane chemoreceptor proteins of bacteria are both structurally and functionally similar to their eukaryotic counterparts, such as hormone receptors. This chapter summarizes current views of chemoreception, sensory transduction, and intracellular signaling in *Escherichia coli* and *Salmonella typhimurium*, with emphasis on the use of genetic approaches to investigate the molecular basis of these events. Interested readers should consult recent reviews (Macnab, 1987b; Stewart and Dahlquist, 1987) for more comprehensive treatment of this subject and for references to the background information mentioned here.

II. OVERVIEW OF THE BACTERIAL SENSORY SYSTEM

A. Locomotor Response

E. coli cells swim by rotating their flagellar fragments: counterclockwise (CCW) rotation produces episodes of smooth swimming, whereas clockwise (CW) reversals cause abrupt directional changes or tumbles. In the absence of chemotactic stimuli, the flagellar motors reverse about once per second, causing the cells to move about in a three-dimensional random walk. In chemical gradients the cells execute chemotactic movements by modulating flagellar rotational bias (reversal probability) in response to changes in attractant or repellent concentration. Increasing attractant levels and decreasing repellent levels enhance the probability of CCW rotation, whereas the converse stimuli enhance the probability of CW rotation. Thus, the cells migrate toward favorable environments in biased random walk fashion by prolonging runs that carry them toward more favorable conditions, such as an attractant source.

B. Stimulus Detection

E. coli cells are attracted to potentially beneficial chemicals, such as amino acids, sugars, and oxygen, and are repelled by a variety of potentially harmful compounds, including pH extremes, fatty acids, hydrophobic amino acids, alcohols, glycerol, and several divalent cations. The amino acid and sugar attractants are sensed by means of specific chemoreceptors whose ligand binding sites are arrayed on the periplasmic side of the cytoplasmic membrane. These receptor proteins have binding affinities in the micromolar range and enable the cells to monitor attractant concentrations as they swim. Changes in receptor occupancy, rather than the ensuing physiologic consequences of chemoeffectors, are responsible for triggering chemotactic responses. In contrast, oxygen and other electron acceptors are probably not sensed by specific chemoreceptors but, rather, through their effects on the electron transport system and the cellular energy level. It is not yet clear whether repellent chemicals are detected by specific receptor proteins or through their effects on the physiology of the cell.

C. Flagellar Signaling

The rotational behavior of the flagellar motors appears to be controlled by an internal signal that is modulated in response to perceived temporal changes in chemoeffector levels as the cell moves about in spatial gradients. The ability to make temporal distinctions is conferred by a sensory adaptation system that in effect enables the cell to compare present chemoeffector concentration with that averaged over the past few seconds. The adaptation machinery

SENSORY SIGNALING IN BACTERIAL CHEMOTAXIS

cancels excitatory signals elicited by changes in receptor occupancy state but operates at a slow rate, thus enabling the cell to "remember" its recent past until adaptation is complete. Thus, flagellar signals reflect the net difference between an excitatory component representing current chemoeffector levels and a slower adaptative component representing previous conditions. Since multiple sensory inputs have roughly additive effects on flagellar behavior, it seems that all sensory transducers may modulate the same flagellar signal.

D. Transduction Pathways and Components

Three different transduction systems are known to generate or regulate flagellar signals in *E. coli* (Figure 1), but two, the PMF and PTS pathways, are still only poorly understood. The PMF pathway mediates responses to oxygen and alternative electron acceptors. These stimuli appear to be sensed through their effects on the level of proton-motive force (PMF), the energy source for flagellar rotation, but little is known about the "protometer" responsible for monitoring cellular PMF and how it communicates with the locomotor apparatus (reviewed by Taylor, 1983). The PTS pathway handles responses to glucose, mannitol, and related sugars that are transported into the cell by a phosphoenolpyruvate-dependent uptake system. The PTS transport system consists of two general protein components, HPr and enzyme I, and various membrane-associated proteins, each of which serves as a specific receptor for sugar transport and chemotaxis. PTS stimuli may be sensed as

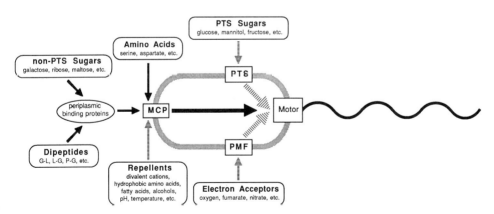

FIGURE 1 Transduction pathways for chemotaxis in *E. coli*. Three types of sensory transducers, methyl-accepting chemotaxis proteins (MCP), phosphotransferase system components (PTS), and proton-motive force detectors (PMF) sense changes in the organism's chemical environment and generate signals that control the rotary motors of the flagella.

changes in the relative amounts of the phosphorylated and unphosphorylated forms of these receptor proteins or in the rate of phosphate cycling through the general PTS components (reviewed by Postma and Lengeler, 1985).

Most of the chemotactic responses exhibited by *E. coli* are mediated by a family of inner membrane proteins known as methyl-accepting chemotaxis proteins (MCPs) (see Figure 1) (reviewed by Springer et al., 1979). MCP molecules contain an extracellular receptor domain that monitors the chemical environment and an intracellular signaling domain that controls the rotational behavior of the flagellar motors (Figure 2). Four MCP structural genes have been identified in *E. coli*, and *S. typhimurium* probably has a similar number. Each MCP handles a subset of attractant inputs, so mutants lacking any one species of MCP still exhibit normal responses to stimuli processed by other MCP or non-MCP pathways. Tar handles aspartate and maltose, Trg handles ribose and galactose, Tap handles dipeptides, and Tsr handles serine. Stimulus detection can occur in two different ways. Amino acid attractants (e.g., serine and aspartate) bind directly to the receptor domains of specific MCP molecules. Non-PTS sugar attractants (e.g., galactose, maltose, and ribose) first complex with specific soluble binding proteins located in the periplasmic space; then the binding protein-sugar complex interacts with a recognition site on the appropriate class of MCP molecules. Both detection mechanisms enable the MCPs to monitor chemoeffector concentrations through the proportion of occupied MCP binding sites. Changes in MCP occupancy state elicit excitatory signals that trigger changes in flagellar

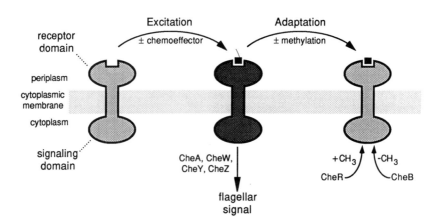

FIGURE 2 Signaling properties of MCP transducers.

rotational bias, and subsequent changes in MCP methylation state cancel these signals, leading to sensory adaptation (Figure 2).

Although some 30 gene products are involved in the structure and function of the flagellar motors (reviewed by Macnab, 1987a), only a few soluble proteins, in concert with the MCP transducers, are needed to generate or regulate the signals that control flagellar rotation (reviewed by Parkinson, 1981). The CheA, CheW, CheY, and CheZ proteins are involved in excitatory signaling, whereas the CheR and CheB proteins are involved in sensory adaptation (see Figure 2). Defects in these intracellular signaling components cause aberrant swimming patterns and a general loss of chemotactic ability.

III. STRUCTURE-FUNCTION STUDIES OF MCP TRANSDUCERS

A. Domain Organization and Membrane Topology

The primary structures of the Tar, Tap, Tsr, and Trg proteins of *E. coli* and the Tar protein of *S. typhimurium*, deduced from their respective gene sequences, exhibit a common structural theme (Figure 3) (Krikos et al., 1983; Russo and Koshland, 1983; Bollinger et al., 1984). MCP transducers are approximately 550 amino acid residues in length and organized into two discrete structural domains. The N-terminal half of the molecule comprises the periplasmic receptor domain, which is flanked at each end by membrane-spanning segments (TM1 and TM2). The C-terminal half of the molecule comprises the cytoplasmic signaling domain, which is flanked at each end by segments (K1 and R1) containing the methylation sites. This portion of MCPs is highly conserved in primary structure, consistent with physiologic evidence that all MCP transducers modulate the same internal signal (Berg and Tedesco, 1975; Spudich and Koshland, 1975). In contrast, the chemoreceptor domains, as well as the "linker" segment that joins the periplasmic and cytoplasmic portions of the molecule, vary considerably in primary structure.

B. Covalent Modifications

MCP molecules undergo reversible methylation reactions catalyzed by two cytoplasmic enzymes. The CheR protein (methyltransferase) transfers methyl groups from S-adenosylmethionine to MCP glutamic acid residues, forming a glutamyl methyl ester (Springer and Koshland, 1977). The CheB protein (methylesterase) promotes hydrolysis of the methylated sites, liberating methanol and reforming a glutamyl residue (Stock and Koshland, 1978). The MCP molecules in *cheR* mutants are devoid of methyl groups, whereas those in *cheB* mutants are methylated, in some cases to a greater extent than

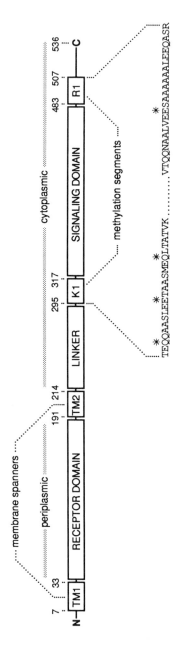

FIGURE 3 Functional architecture of MCP molecules. The amino acid coordinates given are those for the Tsr transducer (Boyd et al., 1983). Other MCPs are similar in both size and domain organization. The sites of reversible carboxymethylation involved in sensory adaptation are indicated by asterisks (Kehry and Dahlquist, 1982). Note that two of the sites are glutamine residues, which must be deamidated to glutamic acids before they are competent to accept methyl groups (Kehry et al., 1983).

in wild-type cells. Both types of mutants can respond to chemotactic stimuli, but cannot carry out sensory adaptation, and continue to respond until the stimulus is removed (Parkinson and Revello, 1978; Yonekawa et al., 1983). Thus, methylation and demethylation of MCP molecules are not involved in stimulus excitation, but changes in MCP methylation state are required for sensory adaptation.

Several of the MCP methylation sites are encoded in the DNA as glutamine residues, which are converted to glutamic acids by posttranslational deamidation (see Figure 3) (Kehry et al., 1983). This reaction is dependent on, and presumably catalyzed by, the CheB enzyme. The signaling properties of undeamidated MCP molecules are similar to those of methylated molecules, indicating that glutamine residues at the methylation sites are probably chemically and functionally homologous to glutamyl methyl esters. The purpose behind MCP deamidation is not known, but it could enable the organism to synthesize transducers in a "neutral" (rather than CW- or CCW-biased) signaling state. This would ensure that newly made MCP molecules do not perturb the optimal pattern of flagellar reversals under a variety of growth rates.

MCP methylation levels are determined by the relative activities of the CheR and CheB enzymes, and changes in methylation state during the adaptation phase of a chemotactic response are brought about in two ways. First, occupied and unoccupied receptors appear to have different substrate properties toward these enzymes: ligand-bound MCP molecules are better substrates for methylation, whereas unliganded molecules are more readily demethylated (Sanders and Koshland, 1988). Second, transducer signals regulate CheB activity through a cytoplasmic feedback circuit: CW-enhancing signals stimulate CheB, resulting in a net increase in methylation state; CCW-enhancing signals inhibit CheB, causing a net decrease in MCP methylation levels (Kehry et al., 1985).

C. Mutant Studies

Although little is yet known about the higher order structures of MCP molecules, it might be possible to deduce some basic principles of their operation through studies of transducer mutants. However, with few biochemical approaches available, the incisiveness of such analyses is limited by the ability to ascertain the functional defects of the mutants. My research group has studied transducer mutants with specific defects in chemoreception or in the control of output signals. These types of mutants have enabled us to identify the various MCP structural features involved in ligand binding, flagellar signaling, and sensory adaptation and in turn to devise testable working models of how these transducers might function in transmembrane signaling.

Transducers with Chemoreception Defects

The Tar chemoreceptor of *E. coli* mediates chemotactic responses to aspartate and maltose. Aspartate is detected by direct binding, whereas maltose is sensed through interactions with the periplasmic maltose binding protein (MPB). We have isolated *tar* mutants with a greatly reduced ability to detect aspartate and its attractive analogs but with relatively minor deficits in maltose sensing (Wolff and Parkinson, 1988). These mutants failed to make chemotactic rings on aspartate swarm plates, but some still exhibited detectable responses to aspartate family compounds (aspartate, α-methylaspartate, glutamate, methionine, and succinate) in more sensitive behavioral assays. The detection thresholds for these residual responses were shifted to higher concentrations, indicating that the mutations affect the affinity of the aspartate receptor for ligand but do not eliminate its ability to generate flagellar signals. Maltose responses in the mutants ranged from 10 to 80% of wild type, but unlike the aspartate results, the response thresholds were not demonstrably shifted to higher concentrations, indicating that the affinity of the mutant transducers for maltose-MBP is not substantially altered. These properties suggest that the mutants have specific defects in the aspartate binding site of Tar.

The mutational changes in these mutants affected three closely spaced codons in the *tar* gene, which specify arginine residues at positions 64, 69, and 73 in the receptor domain of Tar. Each mutation results in replacement of one of the arginines with an uncharged amino acid. The proximity and periplasmic location of these arginine residues imply that they may play critical roles in aspartate detection. On the one hand, they could be directly involved in ligand binding, for example through electrostatic interactions with the negatively charged carboxyl groups of aspartate. On the other hand, they could be responsible for maintaining proper conformation of the aspartate binding site, which might be located elsewhere in the receptor domain.

The Tsr transducer, which serves as the chemoreceptor for serine, alanine, and glycine, contains an arginine segment very similar to that of Tar. Arginine residues are found at the same positions in both transducers, and many of the flanking residues are identical or chemically similar as well. Moreover, several serine-blind mutants of the Tsr transducer have amino acid replacements at arginine residue 64 (Lee et al., 1988), suggesting that the arginine segment motif may be specifically involved in amino acid sensing. It seems probable that one or more of the homologous arginine residues in the Tar and Tsr receptor domains may be responsible for detecting determinants common to the aspartate and serine attractant families, for example through charge interactions with the α-carboxyl group.

Tar molecules undergo substantial conformational changes upon ligand binding (Falke and Koshland, 1987). It seems likely that changes in receptor site occupancy first trigger conformational shifts in the periplasmic domain, which in turn are propagated to the cytoplasmic domain to modulate its signaling activity. The arginine segment of the Tar transducer may represent a major conformational control point in the receptor domain that serves to amplify or propagate local structural changes triggered by ligand binding. For example, ligand-free receptor might be held in a "strained" conformation by electrostatic repulsion between the positively charged arginine residues. If aspartate detection occurred primarily through compensating charge interactions with these arginines, it could serve to alleviate some of those forces. Thus, aspartate binding might enable the receptor site to assume a less strained conformation, which in turn could precipitate conformational changes throughout the molecule.

Transducers with Locked Output Signals

Mutations that inactivate the Tsr transducer result in loss of serine taxis, but such mutants are still capable of responding to stimuli processed through other transducer pathways. However, some *tsr* mutations cause a dominant, generally nonchemotactic phenotype, and several lines of evidence indicate that these mutant transducers are "locked" in active signaling states that interfere with other sensory input pathways (Parkinson, 1980; Callahan and Parkinson, 1985; Ames and Parkinson, 1988). First, they cause predominantly CW or CCW rotation of the flagellar motors, consistent with a constant production of reversal-suppressing signals like those generated transiently during chemotactic responses. Second, the mutant transducers have abnormal methylation states, consistent with permanent activation of the sensory adaptation machinery through the feedback control circuitry. These locked transducers are evidently able to generate flagellar signals in the absence of overt stimuli and to resist the signal-damping action of the sensory adaptation machinery. In contrast to null mutants, many of which are likely to be defective in signal generation, transducers with locked outputs should be defective in signal control and should identify regions of the molecule that are important in modulating its signaling behavior.

Locked transducers can clamp the flagellar motors in either a CW or CCW rotational mode, indicating the MCP molecules are capable of generating signals that augment rotation in either direction. The dominant phenotypes of these mutants imply that both CW- and CCW-biased behaviors are due to aberrant signaling activities rather than a loss of function. Thus, we propose that MCP molecules normally alternate between two active conformations that correspond to CCW and CW signaling states. This in turn implies that net signal output from a population of transducers should reflect the time-averaged proportion of molecules in each signaling state. Any transducer alteration that shifts this equilibrium should lead to a locked output condition. In principle, this could happen through changes in the signal control mechanism used in the excitatory or adaptive responses, through structural alterations that prevent conformational transitions, and possibly through changes in the signal-generating sites themselves. The transducer regions susceptible to locked output mutations indicate that most of these defects may be represented among our mutants (Figure 4, top).

The most striking feature of the mutation distribution is the clustering of lesions in the middle of the signaling domain. Alterations in this portion of the transducer can lead to both CW- and CCW-biased mutants and to slow switchers in which different individuals are stuck in one or the other rotational mode. The variety of signaling defects and the interspersion of the CW and CCW types suggest that this segment plays an important role in mediating transitions between the CW and CCW conformations. Since the slow switchers seem to be specifically defective in carrying out these transitions, this region may serve as a "conformational hinge" that enables the molecule to switch signaling modes. This "switch" region is flanked by segments that yielded relatively few locked mutants (Figure 4), suggesting that those portions of the transducer may be essential for generating CCW and CW output signals. Thus, the majority of mutational lesions in these output segments would be expected to reduce or eliminate signaling activity, whereas much more specific alterations would be required to shift the equilibrium proportions of the CCW and CW signaling modes sufficiently to cause a locked output condition.

The K1 and R1 methylation segments modulate signal output as part of the sensory adaptation mechanism. The few locked signal mutations found in these regions may represent alterations that mimic or interfere with the control exerted by changes in methylation state. The remainder of the locked output mutations, which lie in the N-terminal half of the transducer, probably act through the communication mechanism that modulates signal output in response to receptor input. We surmise that both the linker region and the first membrane spanner play an important role in coupling the receptor domain to the signaling domain.

Transducers with Biased Output Signals

Since transducers with locked output signals no longer undergo sensory adaptation, their defects must somehow override the effects of receptor methylation on signaling. We have also studied more subtle transducer mutants that generate biased output signals in the absence of the adaptation machinery but function normally in its presence (Ames et al., 1988). These mutants were obtained as phenotypic suppressors of *cheR* or *cheB* mutants. For example, mutants defective in *cheR* function cannot methylate their MCPs and consequently have a CCW flagellar rotational bias that prevents chemotactic responses. Modest chemotactic ability can be restored by mutations

SENSORY SIGNALING IN BACTERIAL CHEMOTAXIS

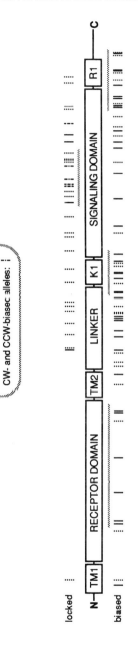

FIGURE 4 Distribution of locked and biased output mutations in the Tsr transducer.

that impart some CW rotational behavior to the mutant cells, and MCP alterations that generate a CW signal in the absence of methylation represent a common source of such suppressors. Conversely, selecting for suppression of *cheB* defects yields MCP mutants with CCW-biased signaling properties. Both the CW- and CCW-biased transducers obtained in this fashion function normally in wild-type genetic backgrounds, indicating that their signal output can be modulated by the sensory adaptation machinery.

The distribution of biased output mutations is summarized in Figure 4 (bottom). In striking contrast to those in locked transducers, a number of these mutations were located in the chemoreceptor domain, including three within the Tsr arginine segment, which may be the serine binding site. These receptor alterations presumably mimic the action of chemoeffectors, generating a persistent transmembrane signal when the adaptation system is nonfunctional. It should be interesting to examine the ligand binding properties of these mutants, which could prove useful in identifying major conformational control points in the receptor domain. Many of the other biased signal mutations occur within or near the K1 or R1 methylation segments. Relatively few are located in the putative switch region that was identified with locked signal mutants. These findings support the idea that the methylation regions are involved in fine-tuning transducer signal output. Amino acid replacements within these segments probably elicit shifts in the distribution of CW and CCW transducer conformations in much the same manner as do changes in methylation state.

IV. TRANSMEMBRANE SIGNALING

Our analysis of transducer mutants with biased or locked output signals indicates that the cytoplasmic signaling domain contains regions essential for generating both CW and CCW output and a conformational hinge that seems to permit transitions between the CW and CCW signaling states. Both transducer conformations probably interact with the cytoplasmic signaling machinery to transmit sensory information to the flagella. We propose that transducer output is controlled by modulating the equilibrium proportions of the CW and CCW forms and that the methylation segments flanking the signaling domain are primarily responsible for this control. Since the methylation sites are not utilized in strict sequential order (Kehry and Dahlquist, 1982) and their effects on signaling behavior are roughly additive (Springer et al., 1979), signal output may be regulated by interactions between the methylation regions. For example, the K1 and R1 regions may associate through electrostatic or hydrophobic interactions along complementary faces of α-helices. Changes in methylation state would alter the mutual affinity or relative alignment of these segments, leading to a shift in the CW-CCW equilibrium of the signaling domain. Alterations in the K1 and R1 segments that cause biased output signals presumably affect these interactions, perhaps by mimicking the structural rearrangements elicited by changes in methylation state. Accordingly, other parts of the transducer molecule that are capable of perturbing the interaction of the methylation segments should also affect transducer output. This could account for the signaling defects caused by mutations in the linker and membrane-spanning segments and suggests a simple mechanism for transmembrane signaling.

Communication between the receptor and signaling domains may occur by direct propagation of conformational changes through the linker segment (Figure 5). The conformation of the receptor domain may be maintained by interactions between the membrane-spanning segments (Oosawa and Simon, 1986), just as interactions between the methylation regions may regulate the conformation of the signaling domain. Conformational changes in the receptor domain, triggered by changes in ligand occupancy, could affect the alignment of TM1 and TM2, leading to a push or pull on the linker segment, which in turn influences the signaling domain. The linker may modulate signal output by influencing the relative alignment of the methylation segments, which in turn control conformation of the signaling domain. Subsequent changes in transducer methylation state during the adaptation phase of the response probably induce compensating changes in the methylation segments

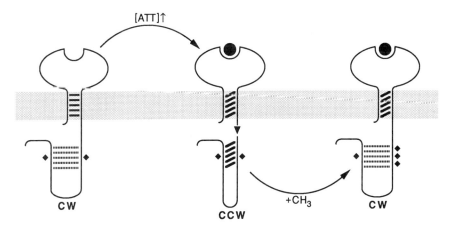

FIGURE 5 Model of transmembrane signaling by MCP molecules. Methylation sites are indicated by solid diamonds. Interactions between the methylation segments may directly control transducer signal output, whereas interactions between the membrane-spanning segments may propagate conformational changes through the linker to elicit stimulus responses.

that counteract the conformational effects of the linker, canceling the output signal.

V. INTRACELLULAR SIGNALING

A. Protein Phosphorylation

MCP transducers probably mediate their behavioral effects through a phosphorylation cascade involving the cytoplasmic CheA, CheB, and CheY proteins. The signal relay begins with CheA, which is capable of autophosphorylation (Hess et al., 1987). CheA phosphoryl groups are subsequently transferred to CheB and CheY and probably serve to regulate their functional activities (Hess et al., 1988b). Phosphorylated CheB may be the catalytically active form of the MCP methylesterase, accounting for feedback control of the adaptation system. Phosphorylated CheY may produce CW rotation of the flagellar motors, providing the mechanism for eliciting changes in rotational behavior. Since the phosphoryl groups on CheB and CheY are labile, the steady-state levels of the phosphorylated forms should reflect the rate at which phosphate can be cycled through CheA. The CheA autophosphorylation reaction is therefore the most likely point of transducer control (Oosawa et al., 1988). The autophosphorylation rate of CheA may somehow be increased by transducer molecules in the CW signaling mode and reduced by those in the CCW conformation.

B. Communication Modules

Although the mechanism that couples MCP transducers to the cytoplasmic phosphorylation cascade is not understood, analyses of the primary structures of chemotaxis proteins have provided some intriguing leads. Bacterial proteins with known or suspected signaling roles often contain sequence motifs that seem to act as transmitter or receiver modules in mediating protein-protein communication (Ronson et al., 1987; Kofoid and Parkinson, 1988). These communication modules appear to retain their functional identities in many protein hosts, implying that they are structurally independent elements. It seems likely that transmitter modules correspond to kinase domains and receiver modules to phosphate acceptor sites. However, since *E. coli* contains dozens of transmitter and receiver proteins, receivers must somehow be "tuned" to their cognate transmitters to avoid cross talk between signaling pathways. Thus, matched transmitters and receivers should be capable of specifically recognizing and interacting with one another while avoiding or reducing interactions with inappropriate modules.

Transmitter proteins typically contain an input domain responsible for modulating transmitter activity in response to sensory stimuli. Activation presumably enables the transmitter to recognize and modify its cognate receiver in another protein, leading to signal propagation. Receiver proteins exhibit a variety of output domains, whose functional activity is somehow controlled by modification of the adjoining receiver. These output regions are often DNA binding domains targeted to specific promoters and are involved in regulating patterns of gene expression in response to environmental changes. However, receivers can also modulate output domains with enzymatic activities, as is the case for the CheB methylesterase (see later). In a few cases, the output domain is yet another transmitter, indicating that communication modules may be used to construct signaling networks with sophisticated information-processing capabilities. For example, proteins that contain coupled receivers and transmitters should be versatile signaling elements and may be a characteristic feature of signaling systems with multiple inputs or feedback controls.

Nearly all the chemotaxis proteins discussed in this chapter contain transmitter or receiver modules (Figure 6). The signaling domain of MCP transducers corresponds to a transmitter, but it appears that the flanking methylation regions are not part of the transmitter proper, but instead an embellishment of the basic transmitter motif. This finding implies that most transmitter proteins might function as tonic receptors that are incapable of undergoing sensory adaptation. The CheA protein contains not only a transmitter, but

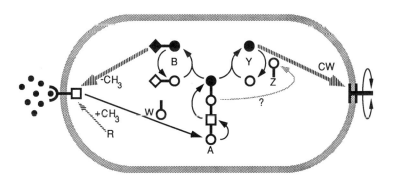

FIGURE 6 Intracellular signaling via a protein phosphorylation cascade. Signaling transactions between the membrane receptors and the soluble proteins of the chemotaxis machinery appear to be mediated by communication modules: transmitters (squares), which may be protein kinase domains, and receivers (circles), which may be phosphorylation sites. Receivers in CheA, CheB, and CheY (indicated by filled circles) are known to be phosphorylated; however, the other communication links may not involve phosphorylation events and are still largely hypothetical. Phosphorylation of CheY probably enables it to elicit CW rotation at the flagellar motor, whereas phosphorylation of CheB probably activates its enzymatic domain (indicated by a diamond), resulting in a net removal of MCP groups and culminating in sensory adaptation. See text for additional details.

three potential receivers (Figure 6). The modular complexity of CheA is consistent with a central role in processing chemotactic signals, and it may serve to integrate sensory inputs, not only from MCP molecules, but from other sensory pathways as well. The autophosphorylation site in CheA is a histidine residue within the N-terminal receiver (Hess et al., 1988a). This high-energy phosphoryl group is transferred to CheY and CheB, both of which contain receiver modules, evidently through receiver-receiver interactions. The CheZ protein, which also contains a receiver module, probably serves as a functional antagonist of CheY by accelerating its rate of dephosphorylation. The CheZ receiver may play a direct role in this process. Alternatively, it might be involved in modulating CheZ function in response to sensory stimuli, possibly through interactions with one of the CheA receiver modules. Finally, the CheW protein, which is needed to generate MCP signals, also contains a receiverlike module, which might enable CheW to interact with one of the communication modules in CheA or the MCP transducers. It seems unlikely that all these module interactions involve concomitant phosphate transfer reactions, but this remains to be shown. However, the key to working out the intracellular signaling pathway in *E. coli* clearly requires an understanding of the molecular determinants underlying the specificity of transmitter-receiver interactions.

VI. CONCLUSIONS

Studies of chemotaxis and other behaviors in bacteria have demonstrated that even relatively "simple" organisms possess sophisticated biochemical strategies for detecting and processing sensory information. These systems are useful for exploring the molecular basis of chemoreception and intracellular signal processing and promise to teach us lessons that should be applicable to sensory transduction processes in higher organisms.

ACKNOWLEDGMENTS

I wish to thank the members of my research group who performed the experiments and contributed many of the ideas discussed in this article: Peter Ames, Emma Lou Bardall, Jin Chen, Eric Kofoid, Jingdong Liu, Hamid Sanatinia, and Claudia Wolff. Our research was supported by Grants GM19559 and GM28706 from the National Institutes of Health.

REFERENCES

Ames, P., and Parkinson, J. S. (1988). Transmembrane signaling by bacterial chemoreceptors: *Escherichia coli* transducers with locked signal output. *Cell* **55**:817-826.

Ames, P., Chen, J., Wolff, C., and Parkinson, J. S. (1988). Structure-function studies of bacterial chemosensors. *Cold Spring Harbor Symp. Quant. Biol.* **53**:59-65.

Berg, H. C., and Tedesco, P. M. (1975). Transient response to chemotactic stimuli in *Escherichia coli*. *Proc. Natl. Acad. Sci. USA* **72**:3235-3239.

Bollinger, J., Park, C., Harayama, S., and Hazelbauer, G. L. (1984). Structure of the Trg protein: Homologies with and differences from other sensory transducers of *Escherichia coli*. *Proc. Natl. Acad. Sci. USA* **81**:3287-3291.

Boyd, A., Kendall, K., and Simon, M. I. (1983). Structure of the serine chemoreceptor in *Escherichia coli*. *Nature* **301**:623-626.

Callahan, A. M., and Parkinson, J. S. (1985). Genetics of methyl-accepting chemotaxis proteins in *Escherichia coli: cheD* Mutations affect the structure and function of the Tsr transducer. *J. Bacteriol.* **161**:96-104.

Falke, J. J., and Koshland, D. E., Jr. (1986). Global flexibility in a sensory transducer: A site-directed cross-linking approach. *Science* **237**:1596-1600.

Hess, J. F., Oosawa, K., Matsumura, P., and Simon, M. I. (1987). Protein phosphorylation is involved in bacterial chemotaxis. *Proc. Natl. Acad. Sci. USA* **84**: 7609-7613.

Hess, J. F., Bourret, R. B., and Simon, M. I. (1988a). Histidine phosphorylation and phosphoryl group transfer in bacterial chemotaxis. *Nature* **336**:139-143.

Hess, J. F., Oosawa, K., Kaplan, N., and Simon, M. I. (1988b). Phosphorylation of three proteins in the signaling pathway of bacterial chemotaxis. *Cell* **53**:79-87.

Kehry, M. R., and Dahlquist, F. W. (1982). The methyl-accepting chemotaxis proteins of *Escherichia coli*. Identification of the multiple methylation sites on methyl-accepting protein. I. *J. Biol. Chem.* **257**:10378-10386.

Kehry, M. R., Bond, M. W., Hunkapiller, M. W., and Dahlquist, F. W. (1983). Enzymatic deamidation of methyl-accepting chemotaxis proteins in *Escherichia coli* catalyzed by the *cheB* gene product. *Proc. Natl. Acad. Sci. USA* **80**:3599-3603.

Kehry, M. R., Doak, T. G., and Dahlquist, F. W. (1985). Sensory adaptation in bacterial chemotaxis: Regulation of demethylation. *J. Bacteriol.* **163**:983-990.

Kofoid, E. C., and Parkinson, J. S. (1988). Transmitter and receiver modules in bacterial signaling proteins. *Proc. Natl. Acad. Sci. USA* **85**:4981-4985.

Krikos, A., Mutoh, N., Boyd, A., and Simon, M. I. (1983). Sensory transducers of *E. coli* are composed of discrete structural and functional domains. *Cell* **33**:615-622.

Lee, L., Mizuno, T., and Imae, Y. (1988). Thermosensing properties of *Escherichia coli tsr* mutants defective in serine chemoreception. *J. Bacteriol.* **170**:4769-4774.

Macnab, R. M. (1987a). Flagella. *In Escherichia coli and Salmonella typhimurium*, F. C. Neidhardt (Ed.). ASM, Washington, D. C., pp. 70-83.

Macnab, R. M. (1987b). Motility and chemotaxis. *In* Escherichia coli and Salmonella typhimurium, F. C. Neidhardt (Ed.). ASM, Washington, D. C., pp. 732-759.

Oosawa, K., and Simon, M. I. (1986). Analysis of mutations in the transmembrane region of the aspartate chemoreceptor in *Escherichia coli*. *Proc. Natl. Acad. Sci. USA* **83**:6930-6934.

Oosawa, K., Hess, J. F., and Simon, M. I. (1988). Mutants defective in bacterial chemotaxis show modified protein phosphorylation. *Cell* **53**:89-96.

Parkinson, J. S. (1980). Novel mutations affecting a signaling component for chemotaxis of *Escherichia coli*. *J. Bacteriol.* **142**:953-961.

Parkinson, J. S. (1981). Genetics of bacterial chemotaxis. *Soc. Gen. Microbiol.* **31**: 265-290.

Parkinson, J. S., and Revello, P. T. (1978). Sensory adaptation mutants of *E. coli*. *Cell* **15**:106-113.

Postma, P. W., and Lengeler, J. (1985). The bacterial phosphoenolpyruvate:carbohydrate phosphotransferase system. *Biochim. Biophys. Acta* **457**:213-257.

Ronson, C. W., Nixon, B. T., and Ausubel, F. M. (1987). Conserved domains in bacterial regulatory proteins that respond to environmental stimuli. *Cell* **49**:579-581.

Russo, A. F., and Koshland, D. E., Jr. (1983). Separation of signal transduction and adaptation functions of the aspartate receptor in bacterial sensing. *Science* **220**: 1016-1020.

Sanders, D. A., and Koshland, D. E., Jr. (1988). Receptor interactions through phosphorylation and methylation pathways in bacterial chemotaxis. *Proc. Natl. Acad. Sci. USA* **85**:8425-8429.

Springer, M. S., Goy, M. F., and Adler, J. (1979). Protein methylation in behavioral control mechanisms and in signal transduction. *Nature* **280**:279-284.

Springer, W. R., and Koshland, D. E., Jr. (1977). Identification of a protein methyltransferase as the *cheR* gene product in the bacterial sensing system. *Proc. Natl. Acad. Sci. USA* **74**:533-537.

Spudich, J. L., and Koshland, D. E., Jr. (1975). Quantitation of the sensory response in bacterial chemotaxis. *Proc. Natl. Acad. Sci. USA* **72**:710-713.

Stewart, R. C., and Dahlquist, F. W. (1987). Molecular components of bacterial chemotaxis. *Chem. Rev.* **87**:997-1025.

Stock, J. B., and Koshland, D. E., Jr. (1978). A protein methylesterase involved in bacterial sensing. *Proc. Natl. Acad. Sci. USA* **75**:3659-3663.

Taylor, B. L. (1983). Role of proton motive force in sensory transduction in bacteria. *Annu. Rev. Microbiol.* **37**:551-573.

Wolff, C., and Parkinson, J. S. (1988). Aspartate taxis mutants of the *E. coli* Tar chemoreceptor. *J. Bacteriol.* **170**:4509-4515.

Yonekawa, H., Hayashi, H., and Parkinson, J. S. (1983). Requirement of the *cheB* function for sensory adaptation in *Escherichia coli*. *J. Bacteriol.* **156**:1228-1235.

5
Chemoreception in *Paramecium*: A Genetic Approach

Judith Van Houten

University of Vermont, Burlington, Vermont

I. INTRODUCTION

Paramecia respond to chemical stimuli, particularly chemicals that signal bacteria, their foodstuff, are in evidence. The cells change their swimming patterns and eventually accumulate in or disperse from attractants or repellents. The cell response behavior is specific and saturable, suggesting that receptors mediate the information flow from external stimulus to internal second messages that produce the altered ciliary beating and altered swimming patterns. Since this pathway promises to be complex, it would be helpful to have specific pharmacologic or genetic blocks to dissect the pathway. I have chosen a genetic approach, and this chapter describes the role genetics has played in this research.

II. CHEMORESPONSE BEHAVIOR

The chemoresponse of *Paramecium tetraurelia* is measured easily yet reliably in a T maze, a modified three-way stopcock (Van Houten et al., 1975, 1982) The cells are allowed to disperse for 30 minutes between two arms containing control buffer or buffer with test stimulus. Ideally, only one ion pair differs between the two conditions; for example, potassium acetate (KOAc) in a buffer is tested relative to KCl in the same buffer and pH. An index of chemokinesis is determined by counting the cells in the two arms and dividing the total cell count into the number in the arm containing the test stimulus. In this way, 0.5 indicates neutrality with numbers > 0.5 or < 0.5 indicating attraction or repulsion.

Small organisms accumulate or disperse by oriented movements (taxis) or movements with no orientation relative to the stimulus source (kinesis) (Fraenkel and Gunn, 1961). Paramecia show no obvious orientation and, therefore, respond to chemical stimuli by indirect means, kineses. There are two general kinesis mechanisms: (1) a biased random walk with a long mean free path while the organism moves up the positive stimulus or down a negative stimulus gradient (klinokinesis) and (2) speed modulation with fast and slow swimming causing dispersal and accumulation, respectively (orthokinesis).

Paramecia have two components to their normal swimming behavior: frequency of turning and speed. Probably because a turn in a smooth path is so noticeable relative to a subtle change in swimming speed, for many years it was assumed that paramecia modulate only their frequency of turning to respond to all manner of stimuli, including chemicals (Jennings, 1906). However, in examining a mutant that was repelled by a stimulus (acetate) that normally attracts *P. tetraurelia*, I found the first evidence for a second mechanism of chemoresponse at work in paramecia (Van Houten, 1977). Normal cells swim smoothly in acetate and make many more turns in the relative control solutions with chloride substituted for the acetate. The expectation for this mutant with reversed sign of behavior was a reversal of the normal swimming response: frequent turning in acetate and smooth swimming in chloride. Interestingly, this mutant swam very smoothly and rapidly in chloride and even more smoothly and rapidly in acetate. In other words, the mutant gave the same kind of responses as normal cell, but to an extreme.

Repulsion can be achieved by modulation of speed (Fraenkel and Gunn, (1961). Therefore, it appeared that smooth, fast swimming could cause dispersal, with the cells swimming so smoothly and fast that they move through and out of the area of acetate. The acetate mutant implied that paramecia could indeed modulate speed to be attracted and repelled and led to the description of chemoresponse mechanisms that integrated both turning and speed and to the recognition of a second mechanism (orthokinesis) that had not been considered before for *Paramecium* (Van Houten, 1978).

Other Mendelian mutants were used to confirm that turning frequency provided only one of at least two conditions for attraction and repulsion (Van Houten, 1978). Pawn mutants that, as the chess piece, could only swim forward with no abrupt turns (Saimi and Kung, 1987), could not be attracted or repelled by some stimuli, yet were normally responsive to others. Fast-2, then considered a "sodium pawn," showed smooth swimming (Kung et al., 1975), that is, pawnlike behavior, in sodium buffers and was not attracted or repelled by the same compounds that also did not elicit a response from the pawns. However, the turning behavior of the fast-2 and also attraction and repulsion were normal in potassium solutions. This mutant in particular provided a very satisfying controlled test for the role of turning in chemoresponse.

III. MEMBRANE POTENTIAL HYPOTHESIS

Once it was apparent that cells could be attracted and repelled by modulating the frequency of turning or the speed of swimming, it was possible to produce a unifying hypothesis for chemokinesis based on principles of *Paramecium* physiology. Paramecia swim by beating cilia. The frequency of ciliary beating and angle to the body determine swimming speed, and the frequency and angle are a function of the membrane potential (Machemer, 1976, 1989). A turn in the swimming path is produced by the transient reversal of cilia brought about by a calcium action potential that causes intraciliary calcium to rise above 10^{-6} M for a very short time (Eckert, 1972). Therefore, a new level of the membrane potential determines both the frequency and angle of ciliary beating and the frequency of action potentials: hyperpolarization increases ciliary beat frequency and moves the membrane potential away from the threshold for action potentials, causing a suppression of turns; depolarization decreases the frequency of ciliary beating and increases the frequency of action potentials. Hence, the membrane potential determines both the speed of swimming and the frequency of turns.

Observations of normal and mutant paramecia made it clear that there were attractants and repellents that required the turning response of the cells to be intact (Van Houten, 1978). Mutants, such as pawns and fast-2, could not accumulate or disperse from some attractants and repellents, yet could respond to others. These mutants could not make the abrupt action potential-based turn, but they could modulate speed; therefore stimuli fell into two categories: attractants and repellents requiring the ability to generate action potentials and hence turns, and attractants and repellents requiring only intact control of speed.

Upon graphing this information (Figure 1) the normal behavior of *Paramecium* fell into only two quadrants, and it became clear that the membrane potential was the unifying theme. Attractants and repellents requiring turns produce behavior characteristic of cells experiencing small hyperpolarizations or small depolarizations, respectively. Attractants and repellents requiring speed modulation, but not the ability to turn, produce behavior characteristic of large hyperpolarizations and depolarizations, respectively. For example, acetate elicits a small hyperpolarization relative to chloride and the hyperpolarization suppresses action potentials and increases speed to produce the characteristic smooth swimming in acetate. However, alkaline pH (OH$^-$

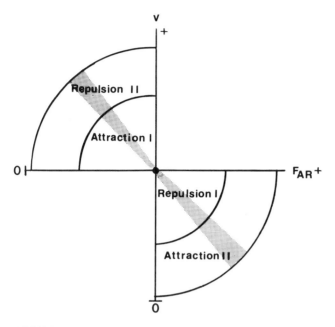

FIGURE 1 A graphic description of *Paramecium* behavior. Behavior of cells in control solution is at the origin. An increase or decrease in velocity v from control is plotted on the y axis. An increase or decrease in the frequency of turns (also called avoiding reactions, hence F_{AR}) is plotted on the x axis. The behavior of normal animals falls in the upper left or lower right quadrants. Behavior is restricted to an area represented by the shaded sector, which is determined by the membrane potential. (From Van Houten, 1978, with permission.)

relative to chloride) elicits a larger hyperpolarization and causes the extremely smooth and fast swimming that causes the cells to disperse from areas of high pH.

Direct tests by electrophysiologic measurement of normal cells indicated that the hypothesis of membrane potential control was essentially correct (Figure 2) (Van Houten, 1979). Mutants, such as pawns, were used to confirm that the changes in potential and speed were as expected in these cells, but since they lack the ability to generate action potentials and turns, they were not attracted to the so-called type I attractants and repellents (Figure 1) (Van Houten, 1980).

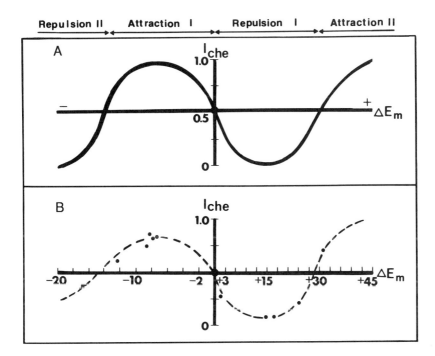

FIGURE 2 A graphic description of membrane potential control of chemokinesis. (A) Change in membrane potential E_m from control (at origin) is plotted against the index of chemokinesis; $I_{che} > 0.5$ indicates attraction; $I_{che} < 0.5$ indicates repulsion. As chemical stimuli change E_m relative to control, animals are attracted or repelled, depending on the magnitude and direction of the change. (B) Data from measurements of E_m in stimulus and control solutions plotted as E_m produced by attractants or repellents versus I_{che}. Scale of E_m is different for depolarizing and hyperpolarizing stimuli. (From Van Houten, 1979, with permission.)

IV. IONIC MECHANISMS

In the description of the ionic mechanisms of the hyperpolarization induced by attractant stimuli, mutants once again played a part. I anticipated a classic electrophysiologic approach to a description of a conductance that was induced by attractants. However, the hyperpolarizations in response to folate and acetate were not affected by membrane voltage, external K or Na, presence of surface charge inhibitors, or agents that block voltage-dependent conductances (Preston and Van Houten, 1987). Mutants with specific defects in characteristic conductances (voltage-dependent Ca, voltage-dependent K,

Ca-dependent K, and Ca-dependent Na) and in surface charge (courtesy of Kung and Richard) showed normal stimulus-induced hyperpolarizations and helped to rapidly eliminate possible explanations of the 0.2 nA current that appeared to be the source of the hyperpolarization (Preston and Van Houten, 1987). Of the "borrowed" mutants, only restless (Richard et al., 1986) showed some alteration in the size of the hyperpolarizations, suggesting the involvement of calcium (Preston and Van Houten, 1987). A folate chemoresponse mutant with a single Mendelian mutation (DiNallo et al., 1982) was both unable to respond to folate and to hyperpolarize in folate more than 1-2 mV compared to the 12 mV hyperpolarization of normal cells (Schulz et al., 1984), but this mutant divulged no clues to the ion carrying the hyperpolarizing current.

The many conductance and other mutants at our disposal helped to narrow our hypothesis for the stimulus-induced hyperpolarization to the stimulus activation of a calcium or proton extrusion pump that then generates the hyperpolarizing current. The calcium pump was favored because of the wide external pH range over which the hyperpolarization remained constant (Preston and Van Houten, 1987). Likewise, other evidence pointed to calcium, such as the observation that lithium inhibited both chemoresponse and calcium efflux from whole cells (Wright and Van Houten, 1989). Additionally, the lack of Mendelian chemoresponse mutants with a generally defective stimulus-induced hyperpolarization hinted that the alteration of the mechanism of hyperpolarization could be a lethal event for the cell, much as the alteration of the calcium homeostasis pump would be.

V. EVIDENCE FOR RECEPTORS

Figure 3 shows the chemosensory transduction pathway in highly simplified form. Detection of the stimulus by the interaction of receptor with ligand is transduced into a change in membrane potential. Although internal second messengers, such as cAMP, are implicated in addition to membrane potential in ciliary beat control (Bonini et al., 1986; Bonini and Nelson, 1988; Schultz et al., 1984), internal cyclic nucleotide levels do not seem to play a direct role in chemoresponse information flowing from the outside to the cilia (Van Houten and Preston, 1987). Therefore, the chemosensory transduction pathway for *Paramecium*, envisioned as in Figure 3, ends in the generation of membrane potential changes because the cilia appear to be directly affected

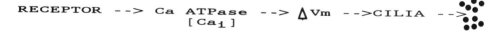

FIGURE 3 *Paramecium* chemosensory tranduction pathway.

by potential. Although not explicit in Figure 3, sometime after the change in potential and before the cilia there must be an adaptation step(s) because behavior of the cells in an unchanging chemical environment returns to basal levels, even though the stimulus is still present (Van Houten et al., 1982) and the cells remain hyperpolarized (Van Houten, 1979). Adaptation is an important factor in the chemoresponse behavior as demonstrated by computer simulations (Van Houten and Van Houten, 1982) and is a time to accommodate to the new chemical environment by resetting the threshold for action potentials (Machemer and DePeyer, 1977). However, Mendelian mutants that were initially able to respond to ammonium but unable to sustain the response demonstrate much more dramatically the requirement for adaptation in the normal chemoresponse (Van Houten et al., 1982).

If the pathway ends with membrane potential, presumably it begins with receptors (Figure 3). Structure-activity studies of the behavior and the electrophysiologic responses suggested a specificity that receptors would afford. Similarly, behavioral responses and potential changes showed saturation, again implying receptors as part of the pathway. Other indirect evidence for receptors has come from photoaffinity labeling of whole cells with 8-N_3-cAMP, which covalently binds to proteins upon ultraviolet irradiation, and from the covalent attachment of folate to whole cells (Baez et al., 1988; Sasner and Van Houten, 1989; Van Houten et al., 1987). Both these treatments render the cells unable to respond to cAMP or folate, respectively, but leaves all other responses assayed intact. Presumably, the receptors are among the surface-exposed sites on whole cells and the cells are rendered insensitive to gradients of attractant stimuli (cAMP or folate) by the permanent occupation of receptors by covalently bound ligand.

Perhaps the most cogent arguments for receptors are found among the many lines of Mendelian mutants that have been isolated and genetically characterized (Van Houten, 1978; Van Houten, 1977; DiNallo et al., 1982; Schulz et al., 1984; Smith et al., 1987). For example, the folate chemoresponse mutant used in the electrophysiologic studies (DiNallo et al., 1982) was found to have no specific, saturable binding of [^{3}H]folate to whole cells (Schulz et al., 1984). Likewise, a loss of folate surface binding was dramatically demonstrated by the loss of fluorescein folate labeling of whole cells (Van Houten et al., 1985). Another mutant unable to detect cAMP as a result of a single Mendelian mutation showed absolutely no specific binding (Smith et al., 1987). The loss of chemoresponse to a single ligand and the concomitant losses of chemoresponse and specific ligand binding due to a single-site mutation are perhaps the strongest lines of evidence for receptors in the chemosensory transduction pathway.

The specific mutants were indeed a welcome assistance in the binding studies, which were technically difficult as a result of the low affinity of ligand for whole cells. The mutants provided confidence that the binding we assayed was meaningful and specific and was a measure of the number and affinity of the receptors. However, the mutants were inevitably conditional, as were more recently isolated folate chemoresponse mutants that were selected not for abnormal behavior but for abnormal binding of fluorescein folate (Sasner, 1989; Sasner et al., 1988). As a consequence, the mutants have been of limited help in the identification of protein receptors.

VI. THE RECEPTOR PROTEINS

External, eukaryotic chemoreceptors, such as those of olfactory or taste receptor cells, are difficult molecules for the biochemist to purify because of the receptors' relatively low affinity for ligand, lack of tightly or covalently binding pharmacologic agents for labeling (e.g., bungarotoxin for the acetylcholine receptor), and lack of large quantities of tissues that are highly enriched in the receptors. Nonetheless, there has been some success in the identification, isolation, and cloning of genes for the receptors of five unicellular eukaryotes, in part because large quantities of homogeneous cell cultures can be harvested for biochemical analysis. The *Dictyostelium* cAMP receptor, *Arbacia* spermatozoan resact receptor, and two yeast mating-type pheromone receptors have been identified and cloned (Klein et al., 1988; Singh et al., 1988; Hagen et al., 1986; Burkholder and Hartwell, 1985). The fifth receptor now identified is the *P. tetraurelia* cAMP receptor.

Null mutants of *P. tetraurelia* that were totally lacking binding sites would provide important evidence in the identification of membrane receptor proteins. However, the conditional nature of the *Paramecium* binding and chemoresponse mutants did not make it likely that these mutants could be used in a simple comparison of normal and mutant membrane proteins to identify the receptor. Nonetheless, genetics continues to be an essential part of the approach to this receptor, except now the genetics are molecular instead of Mendelian. As with other high-affinity and more tractable receptors, molecular genetics is the route of choice to the control and study of receptor proteins (Marx, 1987; Birdsall, 1989).

Affinity chromatography was used to identify a prominent doublet of proteins of 48,000 kD from the cell body membranes of *P. tetraurelia*. This doublet was specifically eluted with cAMP, not with high salt or cGMP. The partial agonist for behavior and electrophysiologic response, 5'-AMP, eluted the protein, as did acetate, a partial inhibitor of the cAMP chemoresponse behavior. The specificity of elution matched that expected from the specificity of behavior and hyperpolarization by cAMP (Smith et al., 1987).

Both proteins of the doublet are most likely to be surface exposed because they are glycosylated, as shown by wheat germ aglutinin-biotin overlays of blots of the proteins and development of the blots with avidin-alkaline phosphatase. Other evidence of surface exposure from $^{125}I_2$ labeling of whole cells was negative but was taken to mean that tyrosines were not exposed for surface labeling.

Antibodies were produced in rabbits by excising gel slices containing the doublet of proteins and immunizing rabbits with the proteins in acrylamide and Freund's adjuvant. The resulting serum was diluted 1:40,000 and used to treat whole cells before testing them in behavior assays for their response to cAMP, folate, and lactate. Cells with no pretreatment and with exposure to a 1:40,000 dilution of preimmune serum were used as two different controls. There was some small loss of motility due to serum treatment as indicated by the slight decrease in the index of chemokinesis between untreated and preimmune serum-treated cells (Table 1). However, the loss of chemoresponse to cAMP after treatment of immune serum was specific and complete (An index of 0.5 indicates a neutral response to cAMP.) This same result was found upon each replication of the experiment, and the responses to the control stimuli were always found to be within normal ranges. These data were taken as good evidence that the doublet of proteins identified by affinity chromatography comprised the receptor.

The current approach to the study of this doublet is to clone the gene for the protein. (Total amino acid analyses indicate that the two proteins are closely related, perhaps by covalent modification as in the *Dictyostelium* receptor doublet. Therefore, I refer to the doublet as a protein.) The determination of the N-terminus sequence is in progress and will provide the information for oligonucleotides for screening libraries of *P. tetraurelia* genomic DNA. The antiserum could be used for screening expression libraries, but

TABLE 1 Effect of Antireceptor Antiserum on Behavior[a]

Pretreatment	Index of chemokinesis attraction to:		n
	Na-cAMP versus NaCl	Na-lactate versus NaCl	
None	0.74 ± 0.06	0.79 ± 0.06	15,8
Incubation with preimmune serum diluted 1:40,000	0.67 ± 0.06	0.69 ± 0.07	12,8
Incubation with immune serum diluted 1:40,000	0.49 ± 0.03	0.65 ± 0.06	13,8

[a]Index of chemokinesis: 0.5 is neutral response; >0.5 indicates attraction. Data are averages of n experiments ± 1 SD. Other attractants, such as NH_4Cl and acetate, give similar results. Sera were diluted with standard buffer used for chemoresponse analysis (Van Houten, 1979).

expression screening may be difficult because *Paramecium* utilizes two stop codons to code for glutamine (Preer et al., 1985).

VII. SUMMARY

In the initial studies of *P. tetraurelia* chemoreception, Mendelian mutants were isolated for their defects in chemoresponse swimming behavior (Van Houten, 1977; DiNallo et al., 1982). The process of autogamy in *P. tetraurelia* made it possible to render the genome homozygous merely by starving the cells after about 20 fissions and thereby made it rather simple to generate F_2 cells and to express recessive mutants (Kung et al., 1975). Additionally, the link between behavior and membrane electrical events made it possible to select for very clear-cut alterations in behavior in hope of isolating mutants with important defects or at least alterations in the excitable membrane functions (see Kung et al., 1975, for a description of this rationale for isolating interesting mutants.)

Mendelian mutants have been useful in the study of the *P. tetraurelia* chemosensory transduction pathway, particularly in the early stages, in dissecting the behavioral mechanisms that lead to accumulation or dispersal of populations of cells and the role of adaptation in the population response. In particular, a chemoreception mutant opened up the possibility for a new, second mechanism of chemoresponse (by orthokinesis) and was useful, along with pawns and other behavioral mutants with defects in turning, in the development of a membrane potential hypothesis that unified chemoresponse behavior. Additionally, mutants with specific losses of responses to single ligands were utilized in the binding studies, which were technically difficult and benefited from the controls that these mutants could provide.

Although a wide variety of mutants with specific conductance or other membrane property alterations were available, it was not possible to more than eliminate a large number of possible mechanisms for the membrane hyperpolarization induced by attractant stimuli. Nonetheless, through this process of elimination, it was possible to generate a hypothesis in which such attractants as folate stimulate a calcium extrusion pump, whose current is the hyperpolarizing conductance that can be measured. This hypothesis continues to be tested.

The usefulness of Mendelian mutants was limited because they inevitably were conditional mutants. For identification of gene products, it would have been useful to compare normal and mutant cells in search of missing proteins. However, the null mutants required for this were not available, despite changing the screening protocol away from behavioral screens to those for binding mutants (Sasner et al., 1989; Sasner, 1989).

The cAMP receptor protein has been identified by affinity chromatography and verified by the use of polyclonal antibodies raised against the receptor doublet. At this point, molecular genetics is being employed to clone the gene for the receptor and thereby circumvent the problems of producing this protein in quantities necessary for biochemistry. Additionally, the cloned gene will afford the possibilities of site-directed mutagensis, comparison of normal and mutant sequences, antisense constructs to produce phenocopies, and more. As with the study of internal neurotransmitter receptors, molecular genetics will be the next important genetic approach for the study of *P. tetraurelia* external chemoreceptors and other components of the chemosensory transduction pathway.

ACKNOWLEDGMENTS

I would like to acknowledge the help of Brian Cote and other lab members whose names appear in the citations throughout the text. This work was supported by NSF and the VRCC.

REFERENCES

Baez, J., Zhang, J., Cote, B., and Van Houten, J. (1988). The putative cAMP chemoreceptor of *Paramecium*. *Chem. Senses* **13**:672.

Birdsall, N. J. (1989). Receptor structure: The accelerating impact of molecular biology. *Trends Pharmacol. Sci.* **10**:50-52.

Bonini, N. M., and Nelson, D. L. (1988). Differential regulation of *Paramecium* ciliary motility by cAMP and cGMP. *J. Cell. Biol.* **106**:1615-1623.

Bonini, N., Gustin, M. C., and Nelson, D. L. (1986). Regulation of ciliary motility in *Paramecium*: A role for cyclic AMP. *Cell Motil. Cytoskel.* **6**:256-272.

Burkholder, A. C., and Hartwell, L. H. (1985). The yeast alpha factor receptor. *Nucleic Acids Res.* **13**:8463-8473.

DiNallo, M., Wohlford, M., and Van Houten, J. (1982). Mutants of *Paramecium* defective in chemokinesis to folate. *Genetics* **102**:149-158.

Eckert, R. (1972). Bioelectric control of ciliary activity. *Science* **176**:473-481.

Fraenkel, G. S., and Gunn, D. L. (1961). *The Orientation of Animals*. Dover, New York.

Hagen, D. C., McCaffrey, G., and Sprague, G. F. (1986). Evidence the yeast STE3 gene encodes a receptor for the peptide pheromone a factor: Gene sequence and implications for the structure of the presumed receptor. *Proc. Natl. Acad. Sci. USA* **83**:1418-1422.

Jennings, H. S. (1906). *Behavior of the Lower Organisms*. Indiana University Press, Bloomington, Indiana.

Klein, P., Sun, T. J., Saxe, C. L., Kimmel, A. R., Johnson, R. L., and Devreotes, P. N. (1988). A chemoattractant receptor controls development in *Dictyostelium discoideum*. *Science* **241**:1467-1472.

Kung, C., Chang, S. Y., Satow, Y., Van Houten, J., and Hansma, H. (1975). Genetic dissection of behavior in *Paramecium*. *Science* **188**:898-904.

Machemer, H. (1976). Interactions of membrane potential and cations in regulation of ciliary activity in *Paramecium. J. Exp. Biol.* **65**:427-448.

Machemer, H. (1989). Cellular behavior modulated by ions: Electrophysiological implications. *J. Protozool.* **36**:463-487.

Machemer, H., and De Peyer, J. E. (1977). Swimming sensory cells: Electrical membrane parameters, receptor properties and motor control in ciliated protozoa. *Verh. Dtsch. Zool. Ges.* **1977**:86-110.

Marx, J. L. (1987). Receptor gene family is growing. *Science* **238**:615-616.

Preer, J., Preer, L., Rudman, B., and Barnett, A. (1985). Deviation from the universal code shown by the gene for surface protein 51 A in *Paramecium. Nature* **314**:188-190.

Preston, R. R., and Van Houten, J. L. (1987). Chemoreception in *Paramecium tetraurelia*: Acetate and folate-induced membrane hyperpolarization. *J. Comp. Physiol.* **160**:525-535.

Richard, E. A., Saimi, Y., and Kung, C. (1986). A mutation that increases a novel calcium-activated potassium conductance of *Paramecium tetraurelia. J. Membr. Biol.* **91**:173-181.

Saimi, Y., and Kung, C. (1987). Behavioral genetics of *Paramecium. Annu. Rev. Genet.* **21**:47-65.

Sasner, J. (1989). Ph.D. dissertation, University of Vermont.

Sasner, J., and Van Houten, J. (1989). Evidence for a *Paramecium* folate chemoreceptor. *Chem. Senses* **14**:587-595.

Sasner, J. M., Isaksen, J., and Van Houten, J. (1988). Immunological and genetic approach to the folate chemoreceptor of *Paramecium. Chem. Senses* **13**:732.

Schultz, J., Grunemund, R., Von Hirschhausen, R., and Schonefeld, U. (1984). Ionic regulation of cyclic AMP levels in *Paramecium tetraurelia* in vivo. *FEBS Lett.* **167**:113-116.

Schulz, S., Denaro, M., and Van Houten, J. (1984). Relationship of folate binding and uptake to chemoreception in *Paramecium. J. Comp. Physiol.* [A] **155**:113-119.

Singh, S., Lowe, D. G., Thorpe, D. S., Rodriguez, H., Kuang, W.-J., Daggott, L., Chinkers, M., Goeddel, D. V., and Garbers, D. L. (1988). Membrane guanylate cyclase is a cell-surface receptor with homology to protein kinases. *Nature* **334**:708-712.

Smith, R., Preston, R., Schulz, S., and Van Houten, J. (1987). Correlation of cyclic adenosine monophosphate binding and chemoresponse in *Paramecium. Biochim. Biophys. Acta* **928**:171-178.

Van Houten, J. (1977). A mutant of *Paramecium* defective in chemotaxis. *Science* **198**:746-749.

Van Houten, J. (1978). Two mechanisms of chemotaxis in *Paramecium. J. Comp. Physiol.* [A] **152**:232-238.

Van Houten, J. (1979). Membrane potential changes during chemokinesis in *Paramecium. Science* **240**:1100-1103.

Van Houten, J. (1980). Chemosensory transduction in *Paramecium*: Role of membrane potential. In *Olfaction and Taste VII*, H. van der Starre (Ed.). IRL Press, London, pp. 53-56.

Van Houten, J., and Preston, R. R. (1987). Chemoreception: *Paramecium* as a receptor cell. In *Molecular Mechanisms of Neuronal Responsiveness*, Y. Ehrlich, R. Lenox, E. Kornecki, and W. Berry (Eds.). Plenum Press, New York, pp. 375-384.

Van Houten, J., and Van Houten, J. C. (1982). Computer analysis of *Paramecium* chemokinesis behavior. *J. Theor. Biol.* **98**:453-468.

Van Houten, J., Hansma, H., and Kung, C. (1975). Two quantitative assays for chemotaxis in *Paramecium. J. Comp. Physiol.* [*A*] **104**:211-223.

Van Houten, J., Martel, E., and Kasch, T. (1982). Kinetic analysis of chemokinesis of *Paramecium. J. Protozool.* **29**:226-230.

Van Houten, J., Smith, R., Wymer, J., Palmer, B., and Denaro, M. (1985). Fluorescein-conjugated folate as an indicator of specific folate binding to *Paramecium. J. Protozool.* **32**:613-616.

Van Houten, J., Zhang, J., Cote, B., Baez, J., and Sasner, J. M. (1987). Cyclic AMP chemoreceptor of *Paramecium. Soc. Neurosci. Abstr.* **13**:1404.

Wright, M. V., and Van Houten, J. (1989). Calcium ATPase as part of the transduction pathway in *Paramecium* chemoreception (abstract). *Chem. Senses* **14**:762-763.

6
Evolution of Pheromonal Specificity in Insect Chemoreceptors

Richard W. Mankin

U.S. Department of Agriculture, Gainesville, Florida

I. INTRODUCTION

Chemosensation, the most ancient of the senses, plays an important role in the orientation of insects to food and potential mates. The importance of this role is reflected in the diversity of numbers and types of chemoreceptors that have evolved on the antennae, mouthparts, legs, ovipositor, and epipharynx (Chapman, 1982). Notable examples are (1) the pheromone-sensitive olfactory receptor, specialized to detect sexual odors emitted by conspecifics, and (2) the contact (taste) chemoreceptor of a phytophagous insect, specialized to detect specific host and nonhost plant compounds. In general, these two chemoreceptors lie at opposite ends on a generalist-specialist spectrum of response specificity. Pheromone receptor neurons usually respond to only a few compounds that resemble their key pheromone components (e.g., Mayer and Mankin, 1985). Contact chemoreceptors tend to be generalists (Frazier, 1986), responding to a large number of different sugars, salts, or alkaloids. Exceptions to this rule do occur, however. Some insects have relatively unspecific pheromone receptor neurons (e.g., Hansson, 1988), and some contact chemoreceptor neurons in caterpillars are specialists for detecting certain secondary plant compounds (Schoonhoven and Dethier, 1966; Schoonhoven, 1967; Dethier and Crnjar, 1982). This chapter addresses some questions about the evolution of chemoreceptor neurons that are specialists in the detection of key stimulus chemicals. A number of such questions have already been raised with respect to the coevolution of insect herbivores and host plants (Dethier, 1980). Here the problem is extended to consider the coevolution between senders and receivers in pheromone communication systems.

In many moth species, the antennae of male (and in some cases female) adults contain sensilla with olfactory receptor neurons that respond selectively to separate components of the pheromone blend emitted by conspecific females. Other neurons in these sensilla respond to components emitted by females of other species. The blend of pheromone components is different for each species, but blends emitted by closely related species tend to be more similar than those emitted by distantly related species (Roelofs and Brown, 1982; Renou et al., 1988; Horak et al., 1988). Several moth species in the subfamily Plusiinae (Figure 1), share (Z)-7-dodecen-1-ol acetate (Z7-12:Ac) as a major sex pheromone component (Steck et al., 1982) and share a number of minor components as attractant or inhibitory chemicals (Table 1, Figure 2). Three of these species, the cabbage looper [*Trichoplusia ni* (Hübner)], the soybean looper [*Pseudoplusia includens* (Walker)], and the celery looper [*Anagrapha falcifera* (Kirby)] are of particular interest because their geographic distributions overlap (Figure 3) and they are somewhat cross-attractive

TABLE 1 Pheromone Blends of *A. falcifera*, *T. ni*, and *P. includens*[a]

Chemical	*T. ni* [b]	*P. includens* [c]	*A. falcifera* [d]
Z7-12:Ac	+a	+a	a
Z5-12:Ac	+a	−i	
Z7-14:Ac	+a	−	i
Z9-14:Ac	+a	−i	
11-12:Ac	+a	+a	
12:Ac	+a	+a	
Z7-12:Pro	−	+a	
Z7-12:But	−	+a	
Z7-12:OH	−i	−	a

[a] A 12 or 14 indicates the number of carbon atoms in the backbone of the molecule, Z indicates a cis double-bond at the specified carbon, and Ac, OH, Pro, and But are acetate, alcohol, propionate, and butyrate moieties, respectively, attached to the first carbon. A + or − indicates the presence or absence in the blend; a or i indicates behavioral attraction or inhibition.
[b] From Bjostad et al. (1984); Linn et al. (1984).
[c] From Linn et al. (1988).
[d] From Steck et al. (1979b), based on captures by pheromone traps.

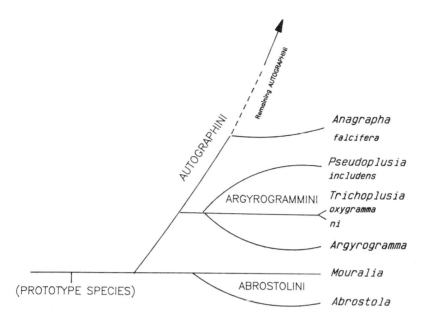

FIGURE 1 Phylogenetic tree (Eichlin and Cunningham, 1978) showing the ancestry of the three Plusiinae moth species in the study: *Anagrapha falcifera*, *Pseudoplusia includens*, and *Trichoplusia ni*. The tropical ancestors of these three species diverged into three tribes, of which Abrostolini is the most primitive and Autographini the least primitive. *P. includens* and *T. ni* are in the same tribe, Agryrogrammini, so they are more closely related to each other than to *A. falcifera*.

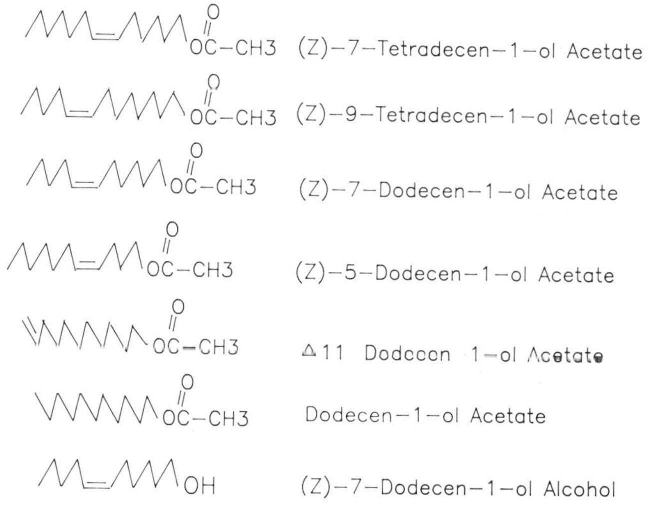

FIGURE 2 Chemical structure of pheromone components identified for *T. ni*, *P. includens*, and *A. falcifera*.

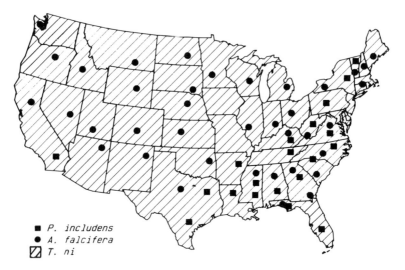

FIGURE 3 Distribution of *A. falcifera, P. includens,* and *T. ni* in the United States. *T. ni* occurs throughout the United States and has the largest population. *A. falcifera* occurs throughout the United States except in desert areas and south Florida. *P. includens* occurs in the Southwest and the Eastern Seaboard. Extensive areas of overlapping populations occur.

to each other (Kaae et al., 1973; Leppla, 1983). Consequently, these species are probably under selection to maintain reproductive isolation by changes in their pheromone communication systems. This makes them excellent candidates for a study of potential mechanisms by which pheromone senders and receivers coevolve in closely related species. Surveys of responses by pheromone receptor neurons on the antennae of *T. ni* (Mayer and Mankin, 1987) and *P. includens* (Grant et al., 1988) have been published recently. To enable further evolutionary comparisons, responses of pheromone-sensitive chemoreceptors on *A. falcifera* antennae were recorded to the same pheromone components as in the two previous studies.

II. MATERIALS AND METHODS

Adult male *A. falcifera*, aged 2-3 days after eclosion, were obtained from a colony started by Dr. Peter Landolt of this laboratory with local wild stock in 1987. The small size of the colony limited the number of males that could be tested, so this should be considered only a preliminary survey of receptor neuron types rather than a comprehensive, quantitative study. The purities and sources of the chemicals were (Z)-7-dodecen-1-ol acetate, no detectable impurities, from Dr. B. Leonhardt, Insect Chemical Ecology Laboratory, Beltsville, MD; Δ-11-dodecen-1-ol acetate (11-12:Ac), 99.4%, from Dr. S. Voerman, Institute for Pesticide Research, Wageningen, The Netherlands; dodecen-1-ol acetate (12:Ac), 98.8%, from Pfalz and Bauer Corp.; (Z)-7-dodecen-1-ol (Z7-12:OH), no detectable impurities, from Dr. J. Tumlinson of this laboratory; and (Z)-5-dodecen-1-ol acetate (Z5-12:Ac), 98.3%, (Z)-7-tetradecen-1-ol acetate (Z7-14:Ac), 99.2%, and (Z)-9-tetradecen-1-ol acetate (Z9-14:Ac), 99.2%, from Farchan Corp. The analyses of purity were performed by Dr. R. Doolittle and R. Heath of this laboratory using capillary gas chromatography.

Electrophysiological recordings were obtained by inserting a tungsten electrode into the base of a sensillum and inserting a second electrode into the lumen of the distal portion of the antenna. The electrical potentials from the electrodes were amplified by a Grass P15 high-impedance preamplifier and transmitted to the 10 Hz, 12 bit Analog-Digital register of a PDP-11/23 microcomputer. User-written software stored the potentials on disk and classified the action potentials (spikes) elicited from the two neurons in each sensillum (Mankin et al., 1987). Examples of spike records generated by this procedure are found in Mankin et al. (1987) and Grant et al. (1989).

The test stimuli were delivered from glass tube dispenser assemblies (Mayer et al., 1987) that led into a stimulus delivery system described in Grant et al. (1989). Until stimulus onset, the antenna was bathed in a stream of clean carrier air at 1200 ml/minute. A separate airstream flowed concurrently through the glass tube dispenser assembly at 200 ml/minute into a vacuum pulling 250 ml/minute. When the valve was closed at stimulus onset, the stimulus stream proceeded into a glass chamber, mixed with the carrier air, and passed to the antenna. The dispensers were dosed by diluting a specified quantity (usually between 0.001 and 1.0 μg) of a pheromone component or blend of components in hexane and coating the mixture over the inside of the dispenser assembly while the hexane evaporated. The concentrations that passed over the antenna were calculated by dividing the emission rates, listed in Mayer et al. (1987), by the 1400 ml/minute mixture volume. The stimulus period was 3 s, and the spike frequencies listed in Table 1, and the figures were calculated as means over the 3 s intervals. The spontaneous activity was recorded before the first stimulus and also for 3 and 4 s periods immediately preceding and following stimulation, respectively. Stimuli were spaced about 5 minutes apart to avoid effects of adaptation.

The order of presentation was standardized on the basis of an initial survey (about 10 sensilla), which established that most of the sensilla had one neuron that responded to Z7-12:Ac and a second that responded either to Z7-12:OH or to Z5-12:Ac at dispenser doses of 0.01 μg or higher. (A dispenser dose of 1 μg emits Z7-12:Ac at approximately the same rate as a *T. ni* sex pheromone gland.) No responses were observed to 12:Ac, 11-12:Ac, or Z7-14:Ac.

Because the preparations often were viable for only 15-30 minutes, only four to six stimuli could be presented before the responses began to degrade and the presentations were standardized to quickly determine the specificities of the neurons. A blend of 0.1 μg each of 12:Ac, 11-12:Ac, and Z7-14:Ac was presented as the first stimulus. The second stimulus was a 0.032 μg dose of Z7-12:Ac. If there was no response, the next stimulus was 0.1 μg Z9-14:Ac; otherwise it was 0.1 μg Z5-12:Ac and then 0.1 μg Z7-12:OH. Once the specificity was determined, responses were recorded to a series of increasing doses of the active component. A total of 21 sensilla were tested by this procedure. About 50 others were surveyed only to determine whether they responded to Z9-14:Ac.

III. RESULTS AND DISCUSSION

A. Response Specificities of *A. falcifera* Receptor Neurons

Three different sensillum types were found on the *A. falcifera* antenna (Figure 4). One contained two neurons that responded separately to Z7-12:Ac and

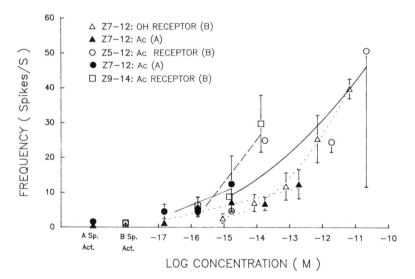

FIGURE 4 Responses of *A. falcifera* receptor neurons to the pheromone components listed in Figure 2. Responses from the A (larger amplitude) neurons are indicated by dark symbols and the B (smaller-amplitude) neurons by open symbols. The type of sensillum containing the neuron is indicated by symbol type and line type (dotted lines and triangles for sensillum found in all three species, solid lines and circles for sensillum with Z5-12:Ac receptor neuron, and dashed line for sensillum with Z9-14:Ac receptor neuron). Spontaneous activities are shown at the left of the figure.

EVOLUTION OF PHEROMONAL SPECIFICITY 67

Z7-12:OH (10 of 21 sensilla surveyed, Figure 5). The spike amplitudes and spontaneous activities (1.5 ± 0.8 spikes per s for the Z7-12:Ac-sensitive neuron and 0.6 ± 0.4 spikes per s for the Z7-12:OH-sensitive neuron) resembled those for neurons of similar specificities in *T. ni* (Mankin et al., 1987) and *P. includens* (Grant et al., 1988). The two other sensillum types, however, contained neurons whose responses were different from those in any sensillum found previously in either *T. ni* or *P. includens*. The first sensillum type contained two neurons that responded separately to Z7-12:Ac (0.7 ± 0.4 spikes per s spontaneous activity) and Z5-12:Ac (0.9 ± 0.6 spikes per s spontaneous activity, 7 of 21 sensilla). The second sensillum type also contained two neurons, one that responded to Z9-14:Ac (1.3 ± 0.7 spikes per s spontaneous activity, 4 of 21 sensilla). The other neuron in this sensillum was unresponsive to any of the chemicals tested (0.3 ± 0.2 spikes per s spontaneous activity). Because of the small sample size, the actual distribution of different types is probably different from the 10:7:4 ratio in this study.

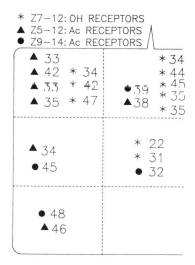

FIGURE 5 Template of a male *A. falcifera* antennal subsegment showing the distribution of recording sites for the three types of sensillum. The antennal flagellum has about 70 subsegments in total; for technical reasons most of the recordings were obtained between subsegments 30 and 45, counting from the base. Each subsegment was divided into 12 sections. The midline of the subsegment is indicated by the spike at the top of the template. The numbers indicate the subsegment number at which the recording was obtained, and the symbol indicates the sensillum type. Most of the sensilla with neurons sensitive to Z5-12:Ac were located along the edge of the subsegment. The other two types of sensillum appeared to be more evenly distributed.

B. Neurophysiologic and Behavioral Comparisons with Other Plusiinids

There are a number of similarities between the different sensillum types on these three Plusiinid moths. Each species has one sensillum type with a neuron that responds to the main pheromone component, Z7-12:Ac, and a neuron that responds to Z7-12:OH. In *T. ni* and the *P. includens*, a second type of sensillum contains two neurons of similar spontaneous activities and spike amplitudes, one that does not respond to any known pheromone component and a second that responds to either Z7-14:Ac or Z5-12:Ac. the *A. falcifera* antenna has two other types of sensillum, one that has neurons sensitive to Z7-12:Ac and Z5-12:Ac, found primarily along the outer margin of the subsegment (Figure 5). The second type is a rarely encountered sensillum with a neuron that responds to Z9-14:Ac. Rare types like the Z9-14:Ac chemoreceptor have been reported elsewhere as well (e.g., Van der Pers and Löfstedt, 1986).

Analysis of the behavioral responses of other Plusiinid moths to Z7-14:Ac, Z5-12:Ac, and Z9-14:Ac suggests that many such similarities occur in their pheromone receptor systems. *T. ni* is attracted to combinations of Z7-12:Ac with Z5-12:Ac and Z9-14:Ac, both of which inhibit the sexual response of *P. includens* to Z7-12:Ac (Table 1) (Figure 1) (Landolt and Heath, 1987; Linn et al., 1988). The closest relative of *T. ni*, *Trichoplusia oxygramma* (Geyer), is attracted to combinations of Z7-12:Ac with Z9-14:Ac (Landolt and Heath, 1986). Z9-14:Ac inhibits the response to Z7-12:Ac in *Autographa californica* (Speyer) and *A. flagellum* (Steck et al., 1979a). Z7-14:Ac inhibits the sexual response of *A. falcifera* (Steck et al., 1979b). It is not likely that these attractive or inhibitory effects occur unless receptor neurons that are sensitive to these chemicals at low stimulus intensities occur in at least small numbers on the male antennae of the respective species. One hesitates to predict from such a small sample, but it seems likely that at least a small number of receptor neurons for Z7-12:Ac, Z7-12:OH, Z5-12:Ac, Z9-14:Ac, and Z7-14:Ac will be found on the antennae of most other Plusiinid species. If so, the differences among Plusiinid pheromone chemoreceptor systems may lie primarily in the proportions of different sensillum type and how the neural responses are integrated in the central nervous system (CNS).

C. Inferences About the Evolution of Pheromone Receptor Neurons

Insect sex pheromone communication is a chain of three complex activities: (1) the production, storage, and release of a species-specific blend of pheromone components, (2) the detection and discrimination of the pheromone blend, and (3) the tracking of the pheromone plume to the potential mate.

Such a system is expected to be evolutionarily stable (Haynes et al., 1984; Mitter and Klun, 1987) because a mutation affecting any of these activities would survive in the population only if it maintained an unbroken chain of communication. Changes are expected primarily in an environment in which a mutation restores communication between conspecifics after it has been disrupted by a sudden introduction of conflicting or masking signals from other animals (Alexander, 1962; Lundberg and Löfstedt, 1987). Comparative studies of the pheromone components emitted by females of closely related Noctuid and Tortricid species are in agreement with such an expectation (Roelofs and Brown, 1982; Doré et al., 1986; Renou et al., 1988; Horak et al., 1988). Of 467 pheromones cited for female Noctuid moths, 68% include one of the four components: Z7-12:Ac, Z5-12:Ac, Z9-14:Ac, or Z11-16:Ac (Renou et al, 1988). The similarities among *T. ni, P. includens,* and *A. falcifera* pheromone chemoreceptors are also supportive of a hypothesis that evolution of the pheromone chemoreception system has been a slow process.

The evolutionary processes that may have occurred to create the differences that now exist among pheromone chemoreceptor systems are yet to be determined. A possible mechanism, similar to that proposed by Dethier (1980) for contact chemoreceptors specialized to detect particular secondary plant compounds, is that independent mutations first led to the presence of additional pheromone components and chemoreceptor types that had no initial role in the communication system. After the communication system was perturbed, a serendipitous mutation produced a change in the way input from different chemoreceptor types was integrated in the CNS. This permitted new behaviors to become associated with the previously disregarded stimuli.

Dethier proposed his original hypothesis on the basis of similarities and differences in the contact chemoreceptors of different lepidopterous larvae. The sensilla styloconica usually have three or four receptor neurons (e.g., Ma, 1972; Schoonhoven, 1972; Den Otter and Kahoro, 1983). Three of these neurons have similar specificities, for salts, water, and sucrose, respectively. In larvae of many moth species (Schoonhoven, 1973; Stadler, 1984; Frazier, 1986), the fourth styloconic neuron responds to secondary host plant substances. The specificity of this fourth neuron is considerably different in different species, even when the species exist on similar host plants (Den Otter and Kahoro, 1983). Dethier proposed that the prototype fourth neuron was a generalist. In those insects in which the fourth neuron became a deterrent receptor, an adaptive decrease occurred in the sensitivity of this neuron to compounds present in the host plant (Bernays and Chapman, 1987). Support for such a hypothesis comes from comparative studies of *Yponomeuta* chemoreceptors by Van Drongelen (1979). Similarly, in those insects in which the fourth neuron became a stimulant receptor, an adaptive decrease occurred in the sensitivity of this neuron to compounds not present in the host plant. In both cases, the evolution would be driven by changes in oviposition behavior (Van Drongelen, 1979).

To consider how the hypothesis might apply at the chemoreceptor side of the pheromone communication system, first suppose that there are multiple alleles coding for the chemoreceptor sensitive to the major sex pheromone component of the ancestral Plusiinid prototype. One example of this condition is the gene coding for green-sensing pigment in human color vision (Nathans et al., 1986). It may also occur with the gene coding for a pheromone receptor in the red-banded leafroller moth (Chapman et al., 1978). Occasionally a mutation occurs at one of these alleles, and a small group of neurons appears on the antenna with specificity for a slightly different compound. The presence of this new receptor type is only slightly deleterious because of the high degree of convergence in the deutocerebrum (e.g., Christensen and Hildebrand, 1987). The sensitivity to a particular pheromone component in the CNS increases in proportion to the square root of the number of receptor neurons sensitive to it (Mankin and Mayer, 1983), so a small reduction in the number of neurons sensitive to the major components has very little effect on the overall CNS sensitivity. Over time, random mutations lead to a proliferation of receptor neuron types in the population. All the males have receptor neurons sensitive to the major pheromone component, and different individuals have small numbers of receptor neurons sensitive to compounds that differ slightly from the original major pheromone component. Occasionally, one of these mutations enhances the sensitivity to a minor component emitted solely by conspecific females or one emitted solely by females of other related species.

Once a pool of different individuals with receptor neurons sensitive to slightly different pheromone components develops in the population, selection may occur whenever a change in the environment affects the communication channel. Because the progeny from interspecific matings are usually less viable than from conspecific matings, strong selection pressure favors males whose central nervous systems discriminate in favor of blends emitted solely by conspecific females. Likewise, selection favors males that discriminate against blends emitted solely by females of other species. Subsequent mutations in regulatory genes controlling the quantity of receptor protein (as in Jan et al., 1987) may effect an increase in the percentage of receptor neurons sensitive to this component. Conversely, other mutations may reduce the number of neurons sensitive to a given component when the original selection pressures for discrimination are eliminated (e.g., Nagy and George, 1981; Löfstedt et al., 1986). Finally, all this occurs while the blends of pheromone components emitted by conspecific females and those of other females in the environment undergo their own evolutionary changes. This results in an assemblage of several species that share one or more pheromone components in common and share a number of different types of pheromone chemoreceptor in common but discriminate for or against other components (or particular ratios of components) depending on the patterns of mutation of pheromone production of the local conspecific and interspecific females.

In applying such a hypothesis to the three insects in this report, it appears that pheromone production mutations in the ancestors of *T. ni* females led to the emission of Z7-14:Ac, Z9-14:Ac, and Z5-12:Ac in addition to the Z7-12:Ac emitted by the common ancestor of the three insects. Mutations ending the production of Z7-12:OH may have occurred in an ancestor of *T. ni* and *P. includens*, but not in *A. falcifera*. Chemoreceptors detecting these compounds either were present on antennae of ancestral males or evolved independently in all three species. The behavioral responses to the new components thereafter evolved differently for each species, depending on the pheromone blends emitted by conspecific females.

A critical concern in the hypothesis is how input from a new type of chemoreceptor becomes integrated into CNS decision processes. Are additional mutations required in the CNS before a new chemoreceptor can elicit a change in sexual behavior? How does the new input reach a "command center" where pheromone blends are discriminated? Morphologic and developmental studies of insect antennae suggest a possible answer. The sensilla may proliferate in a pattern that permits neurons of any new or ancestral chemoreceptor type to converge upon separate loci in the CNS. On some insect antennae, sensilla of different types lie approximately in rows along the longitudinal axis of the subsegment (Steinbrecht, 1970; Sanes and Hildebrand, 1976), which may be the case for the Z5-12:Ac sensillum of *A. falcifera* (Figure 5). In others, the sensilla lie in rows along the circumference of the subsegment (Van Der Pers et al., 1980) or in small bundles or patches (Esslen and Kaissling, 1976; Ramaswamy and Gupta, 1981; Hansson et al., 1986). In developmental studies, Schneiderman et al. (1986) prepared gynandromorphic female *Manduca sexta* by transplanting antennal imaginal disks from male to female fifth instar larvae. The male pheromone receptor neurons converged normally in the female deutocerebrum, and behavioral responses were observed both to sex pheromones and to ovipositional attractants. These patterns of neural distribution conform to a developmental scheme that was proposed from genetic studies of *Drosophila* homeotic mutants (Campos-Ortega and Hartenstein, 1985). In this scheme, the position of the sensillum maps to a distinct territory in the CNS. For pheromone receptor neurons, this territory is probably a locus in the macroglomerulus of the deutocerebrum (Christensen and Hildebrand, 1987). The termination sites in the CNS seem to be organized according to olfactory specificity (Stocker et al., 1983) in a segmentally repetitive fashion (Campos-Ortega and Hartenstein, 1985; Rospars, 1988). Thus, the axons from patches or rows of sensilla located at approximately the same position on each subsegment are expected to terminate at the same locus in the CNS. Because the small number of neurons on each subsegment would map together at the same locus in the CNS, this could enable individual receptor neuron responses to be integrated together for greatest effect (Light, 1986). Koontz and Schneider (1987) detected such a pattern in the macroglomerulus of *Bombyx mori*. This pattern of convergence would be particularly important when there are only a few receptors of a specific chemoreceptor type on the antenna.

Less is known about the ultimate fate of the signals from new chemosensory types that converge at a command center in the CNS. It seems plausible, however, that the command center undergoes a process of "coevolution" under separate genetic control and that selection determines whether the new input (or even older input) is excitatory, inhibitory, or neutral. In *Ostrinia nubilalis*, for example, an allele for different pheromone chemoreceptor phenotypes appears to be autosomal and independent of a sex-linked allele for different types of behavioral response (Roelofs et al., 1987).

IV. SUMMARY

Insects sense their chemical environment through receptor neurons inside sensilla on the antennae, mouthparts, and legs. Some of these neurons detect a broad spectrum of chemicals, but others are highly selective for specific chemicals used as mate or host identification cues. A relatively unexplored question about such chemoreceptors is the coevolution of receiver and sender. How do chemoreceptors evolve in correspondence with evolutionary changes in the production of their key stimuli?

Two receptor systems for which coevolution is of particular interest are the pheromone chemoreceptor system and the larval contact chemoreceptor system. Pheromone chemoreceptors are selective for sexually excitatory odorants (sex pheromones) emitted by conspecifics or for sexually inhibitory odorants emitted by individuals of closely related species. Some contact chemoreceptor (taste) neurons on larval mouthparts are selective for secondary chemicals present in (or absent from) host plants. Much of what is known now about the evolution of chemoreceptor specificity comes from studies of contact chemoreceptors, but the two systems are similar in many respects.

To consider the evolution of specificity in pheromone chemoreceptors, responses from neurons on the antennae of the celery looper moth (*Anagrapha falcifera*) were recorded to six pheromone components tested in previous surveys of two close relatives, the soybean looper (*Pseudoplusia includens*) and the cabbage looper (*Trichoplusia ni*). It was found that these three moths share in common one type of sensillum with two pheromone-sensitive neurons of the same specificity. This sensillum type may have passed unchanged from a Plusiinid ancestor, in which case it may be found on the antennae of most Plusiinid species. The antennae also contain other pheromone-sensitive receptor neurons of apparently unique specificities. The evolution of neurons with unique pheromal specificities was discussed in this chapter in reference to previous hypotheses about the coevolution of plants and insect herbivore contact chemoreceptors.

ACKNOWLEDGMENTS

Many thanks to Dr. Peter Landolt for his generous donation of the *A. falcifera* used in this study and to Drs. John Sivinski, Doug Light, and Linda Kennedy for their editorial comments. Technical assistance was provided by Jane Sharp and Pam Wilkening.

REFERENCES

Alexander, R. D. (1962). Evolutionary change in cricket acoustical communication. *Evolution* **16**:443-467.
Bernays, E. A., and Chapman, R. F. (1987). Chemical deterrence of plants. In *Molecular Entomology*, J. H. Law (Ed.). Alan R. Liss, New York, pp. 107-116.
Bjostad, L. B., Linn, C. E., Du, J. W., and Roelofs, W. T. (1984). Identification of new sex pheromone components in *Trichoplusia ni* predicted from biosynthetic precursors. *J. Chem. Ecol.* **10**:1309-1323.
Campos-Ortega, J. A., and Hartenstein, V. (1985). Development of the nervous system. In *Comprehensive Insect Physiology, Biochemistry, and Pharmacology*, Vol. V, G. A. Kerkut and L. I. Gilbert (Eds.). Pergamon Press, Oxford, pp. 49-84.
Chapman, R. F. (1982). Chemoreception: The significance of receptor numbers. In *Advances in Insect Physiology*, Vol. XVI. Academic Press, New York, pp. 247-356.
Chapman, O. L., Klun, J. A., Mattes, K. C., Sheridan, R. S., and Maini, S. (1978). Chemoreceptors in Lepidoptera: Stereochemical differentiation of dual receptors for an achiral pheromone. *Science* **201**:926-928.
Christensen, T. A., and Hildebrand, J. G. (1987). Male-specific, sex pheromone-selective projection neurons in the antennal lobes of the moth *Manduca sexta*. *J. Comp. Physiol.* [A] **160**:553-569.
Den Otter, C. J., and Kahoro, H. M. (1983). Taste cell responses of stemborer larvae, *Chilo partellus* (Swinhoe), *Eldana saccharina* Wlk. and *Maruca testulalis* (Geyer) to plant substances. *Insect Sci. Appl.* **4**:153-157.
Dethier, V. G. (1980). Evolution of receptor sensitivity to secondary plant substances with special reference to deterrents. *Am. Naturalist* **115**:45-66.
Dethier, V. G., and Crnjar, R. M. (1982). Candidate codes in the gustatory system of caterpillars. *J. Gen. Physiol.* **79**:549-569.

Doré, J. C., Michelot, D., Gordon, G., Labia, R., Zagatti, P., Renou, M., and Descoins, C. (1986). Approche factorielle des relations entre 8 triubs de Lépidoptéres Tortricidae et 41 molécules a effet attractif sur les males. *Ann. Soc. Entomol. Fr.* **22**:387-402.

Eichlin, T. D., and Cunningham, H. B. (1978). The Plusiinae (Lepidoptera:Noctuidae) of America north of Mexico: Emphasizing genitalic and larval morphology. *USDA Tech. Bull.* **1567**:1-122.

Esslen, J., and Kaissling, K.-E. (1976). Zahl und Verteilung antennaler Sensillen bei der Honigbiene (*Apis mellifera* L.). *Zoomorphologie* **83**:227-251.

Frazier, J. L. (1986). The perception of plant allelochemicals that inhibit feeding. In *Molecular Aspects of Insect-Plant Associations*, L. B. Brattsten and S. Ahmad (Eds.). Plenum, New York, pp. 1-42.

Grant, A. J., O'Connell, R. J., and Hammond, A. M. (1988). A comparative study of perception in two species of Noctuid moths. *J. Insect. Behav.* **1**:75-96.

Grant, A. J., Mankin, R. W., and Mayer, M. S. (1989). Neurophysiological responses of pheromone-sensitive receptor neurons on the antenna of *Trichoplusia ni* (Hübner) to pulsed and continuous stimulation regimens. *Chem. Senses* **14**:449-462.

Hansson, B. S. (1988). Reproductive isolation by sex pheromones in some moth species: An electrophysiological approach. Ph.D. Dissertation, Lund University, Sweden, pp. 89-98.

Hansson, B. S., Löfstedt, C., Löfqvist, J., and Hallberg, E. (1986). Spatial arrangement of different types of pheromone-sensitive sensilla in a male moth. *Naturwissenschaften* **73**:269-270.

Haynes, K. F., Gaston, L. K., Pope, M. M., and Baker, T. C. (1984). Potential for evolution of resistance to pheromones. *J. Chem. Ecol.* **10**:1551-1565.

Horak, M., Whittle, C. P., Bellas, T. E., and Rumbo, E. R. (1988). Pheromone gland components of some Australian tortricids in relation to their taxonomy. *J. Chem. Ecol.* **14**:1163-1175.

Jan, Y. N., Bodmer, R., Jan, L. Y., Ghysen, A., and Dambly-Chaudiere, C. (1987). Mutations affecting the embryonic development of the peripheral nervous system in *Drosophila*. In *Molecular Entomology*, J. H. Law (Ed.). Alan R. Liss, New York, pp. 45-56.

Kaae, R. S., Shorey, H. H., McFarland, S. U., and Gaston, L. K. (1973). Sex pheromones of Lepidoptera. XXXVII. Role of sex pheromones and other factors in reproductive isolation among ten species of noctuidae. *Ann. Entomol. Soc. Am.* **66**:444-448.

Koontz, M. A., and Schneider, D. (1987). Sexual dimorphism in neuronal projections from the antennae of silk moths (*Bombyx mori, Antheraea polyphemus*. and the gypsy moth (*Lymantria dispar*). *Cell Tissue Res.* **249**:39-59.

Landolt, P. J., and Heath, R. R. (1986). A sex attractant synergist for *Trichoplusia oxygramma* (Lepidoptera: Noctuidae). *Florida Entomol.* **69**:425-426.

Landolt, P. J., and Heath, R. R. (1987). Role of female produced sex pheromone in behavioral reproductive isolation between *Trichoplusia ni* (Hübner) and *Pseudoplusia includens* (Walker) (Lepidoptera: Noctuidae, Plusiinae). *J. Chem. Ecol.* **13**:1005-1018.

Leppla, N. C. (1983). Chemically mediated reproductive isolation between cabbage looper and soybean looper moths (Lepidoptera: Noctuidae). *Environ. Entomol.* **12**:1760-1765.

Light, D. M. (1986). Central integration of sensory signals: An exploration of processing of pheromonal and multimodal information in lepidopteran brains. In *Mechanisms in Insect Olfaction*, T. L. Payne, M. C. Birch, and C. E. J. Kennedy (Eds). Oxford University Press, Oxford, pp. 287-301.

Linn, C. E., Jr., Bjostad, L. B., Du, J. W., and Roelofs, W. L. (1984). Redundancy in a chemical signal: Behavioral responses of male *Trichoplusia ni* to a 6-component sex pheromone blend. *J. Chem. Ecol.* **10**:1635-1658.

Linn, C. E., Jr., Hammond, A., Du, J., and Roelofs, W. L. (1988). Specificity of male response to multicomponent pheromones in Noctuid moths *Trichoplusia ni* and *Pseudoplusia includens*. *J. Chem. Ecol.* **14**:47-57.

Löfstedt, C., Herrebout, W. M., and Du, J.-W. (1986). Evolution of the ermine moth pheromone tetradecyl acetate. *Nature* **323**:621-623.

Lundberg, S., and Löfstedt, C. (1987). Intra-specific competition in the sex communication channel: A selective force in the evolution of moth pheromones? *J. Theor. Biol.* **125**:15-24.

Ma, W. C. (1972). Dynamics of feeding responses in *Pieris brassicae* L. as a function of chemosensory input: A behavioral, ultrastructural and electrophysiological study. *Meded. Landbouwhogesh.* **7211**:1-162.

Mankin, R. W., and Mayer, M. S. (1983). A phenomenological model of the perceived intensity of single odorants. *J. Theor. Biol.* **100**:123-138.

Mankin, R. W., Grant, A. J., and Mayer, M. S. (1987). A microcomputer-controlled response measurement and analysis system for insect olfactory receptor neurons. *J. Neurosci. Methods* **20**:307-322.

Mayer, M. S.,, and Mankin, R. W. (1985). Neurobiology of pheromone perception. In *Comprehensive Insect Physiology, Biochemistry, and Pharmacology,* Vol. IX. G. A. Kerkut and L. I. Gilbert (Eds.) Pergamon Press, Oxford, pp. 95-144.

Mayer, M. S., and Mankin, R. W. (1987). A linkage between coding of quantify and quality of pheromone gland components by receptor cells of *Trichoplusia ni*. *Am. N.Y. Acad. Sci.* **510**:483-484.

Mayer, M. S., Mankin, R. W., and Grant, A. J. (1987). Quantitative comparison of behavioral and neurophysiological responses of insects to odorants: Inferences about central nervous system processes. *J. Chem. Ecol.* **13**:509-531.

Mitter, C., and Klun, J. A. (1987). Evidence of pheromonal constancy among sexual and asexual females in a population of fall cankerworm, *Alsophila pometaria* (Geometridae). *J. Chem. Ecol.* **13**:1823-1831.

Nagy, B. A., and George, J. A. (1981). Differences in the numbers of sensilla trichodea between reared and wild adults of the oriental fruit moth *Grapholitha molesta* (Lepidoptera: Tortricidae). *Proc. Entomol. Soc. Ont.* **112**:67-72.

Nathans, J., Thomas, D., and Hogness, D. S. (1986). Molecular genetics of human color vision: The genes encoding blue, green, and red pigments. *Science* **232**:193-202.

Ramaswamy, S. B., and Gupta, P. (1981). Sensilla of the antennae and the labial and maxillary palps of *Blatella germanica* (L.) (Dictyoptera: Blatellidae): Their classification and distribution. *J. Morphol.* **168**:269-279.

Renou, M., Lelanne-Cassou, B., Michelot, D., Gordon, G., and Doré, J.-C. (1988). Multivariate analysis of the correlation between noctuidae subfamilies and the chemical structure of their sex pheromones or male attractants. *J. Chem. Ecol.* **14**:1187-1215.

Roelofs, W. L., and Brown, R. L. (1982). Pheromones and evolutionary relationships of Tortricidae. *Annu. Rev. Ecol. Syst.* **13**:395-422.

Roelofs, W. L., Glover, T., Tang, X. H., Sreng, I., Robbins, P., Eckenrode, C., Löfstedt, C., Hansson, B. S., and Bengtsson, B. O. (1987). Sex pheromone production and perception in European corn borer moths is determined by both autosomal and sex-linked genes. *Proc. Natl. Acad. Sci. USA* **84**:7585-7589.

Rospars, J. P. (1988). Structure and development of the insect antennodeutocerebral system. *Int. J. Morphol. Embryol.* **17**:243-294.

Sanes, J. R., and Hildebrand, J. G. (1976). Structure and development of antennae in a moth, *Manduca sexta*. *Dev. Biol.* **51**:282-299.

Schneiderman, A. M., Hildebrand, J. G., Brennan, M. M., and Tumlinson, J. H. (1986). Trans-sexually grafted antennae alter pheromone-directed behavior in a moth. *Nature* **323**:801-803.

Schoonhoven, L. M. (1967). Chemoreception of mustard oil glucosides in larvae of *Pieris brassicae*. *Proc. K. Ned. Akad. Wet.* [C] *Biol. Med. Sci.* **70**:556-568.

Schoonhoven, L. M. (1972). Secondary plant substances and insects. In *Structural and Functional Aspects of Phytochemistry*, V. C. Runeckless and T. C. Tso (Eds.). *Recent Adv. Phytochem.* **4**:197-224.

Schoonhoven, L. M. (1973). Plant recognition by lepidopterous larvae. *Symp. R. Entomol. Soc. Lond.* **6**:87-99.

Schoonhoven, L. M., and Dethier, V. G. (1966). Sensory aspects of host-plant discrimination by lepidopterous larvae. *Arch. Neer. Zool.* **16**:497-530.

Stadler, E. (1984). Contact chemoreceptors. In *Chemical Ecology of Insects*, W. Bell and R. Cardé (Eds.). Sinauer Associates, Sunderland, New York, pp. 3-36.

Steck, W. F., Chisholm, M. D., Bacley, B. K., and Underhill, E. W. (1979a). Moth sex attractants found by systematic field testing of 3-component acetate-aldehyde candidate lines. *Can. Entomol.* **111**:1263-1269.

Steck, W. F., Underhill, E. W., Chisholm, M. D., and Gerber, H. S. (1979b). Sex attractant for male alfalfa looper moths, *Autographa californica*. *Environ. Entomol.* **8**:373-375.

Steck, W. F., Underhill, E. W., and Chisholm, M. D. (1982). Structure-activity relationships in sex attractants for North American Noctuid moths. *J. Chem. Ecol.* **8**:731-754.

Steinbrecht, R. A. (1970). Zur Morphometrie der Antenne des Seidenspinners, *Bombyx mori* L.: Zahl und Verteilung der Reichsensillen (Insecta: Lepidoptera). *Z. Morph. Tiere* **68**:93-126.

Stocker, R. F., Singh, R. N., Schorderet, M., and Siddiqi, O. (1983). Projection patterns of different types of antennal sensilla in the antennal glomeruli of *Drosophila melanogaster*. *Cell Tissue Res.* **232**:237-248.

Van Der Pers, J. N. C., and Löfstedt, C. (1986). Signal-response relationship in sex pheromone communication. In *Mechanisms in Insect Olfaction*, T. L. Payne, M. C. Birch, and C. E. J. Kennedy (Eds.). Clarendon Press, Oxford, pp. 235-241.

Van Der Pers, J. N. C., Cuperus, P. L., and Den Otter, C. J. (1980). Distribution of sense organs on male antennae of small ermine moths, *Yponomeuta* spp. (Lepidoptera: Yponomeutidae). *Int. J. Insect Morphol. Embryol.* **9**:15-23.

Van Drongelen, W. (1979). Contact chemoreception of host-plant specific chemicals in larvae of various *Yponomeuta* species (Lepidoptera). *J. Compl. Physiol.* [*A*] **135**:265-280.

7
Olfaction in *Drosophila*

Obaid Siddiqi

*Tata Institute of Fundamental Research
Bombay, India*

I. INTRODUCTION

Some years ago Veronica Rodrigues and I described a set of X-linked olfactory (*olf*) genes in *Drosophila melanogaster* whose mutations cause specific anosmias (Rodrigues and Siddiqi, 1978). It was known that *Drosophila* has a well-developed olfactory sense and a rich repertoire of chemosensory behavior (Barrows, 1907; Thorpe, 1939; Begg and Hogben, 1946). We wished to take advantage of the highly developed genetics and molecular biology of Drosophila to gain some insight into odor perception. In particular, it seemed to us that the study of electrophysiologic and biochemical defects in anosmic mutants was likely to throw light on the vexing problem of olfactory receptors and on the mechanism of odor discrimination.

To detect lesions in olfactory mutants it was necessary to obtain a sufficiently detailed account of the wild type. Toward this end, our group has studied the Canton special strain from which most of the mutants are derived. I first describe the salient features of olfaction in normal flies before discussing mutants. In several respects *Drosophila* combines richness of chemosensory behavior with an underlying simplicity of structure and physiology that makes it eminently suitable as a model animal for the study of olfactory perception.

II. OLFACTORY BEHAVIOR OF *DROSOPHILA*

The reactions of *Drosophila* to volatiles are not indiscriminate. At the beginning of our studies, we examined the olfactory responses of *Drosophila* to a set of about 50 odorants that were known to be either potent attractants or repellents to other insects or constituents of *Drosophila* food sources. The fly responds selectively to a small subset of these (Figure 1) (Rodrigues, 1980; Siddiqi, 1983). Chemicals of the same group often elicit very unequal responses. Ethyl acetate, isoamyl acetate, and butanol are strong attractants, but citronellyl acetate and geraniol are relatively ineffective. Benzaldehyde is a strong repellent but other aldehydes are only weakly so (Figure 2).

We then investigated the chemical specificity of the presumptive receptors by studying mutual interactions between odors. If the response to an odor is measured with increasing concentrations of a second odor in the background, the background is expected to inhibit the response when the two chemicals compete for a common receptor. If the two odors are detected by independent receptors, there should be no interference. Such odor "masking" or "jamming" experiments were carried out using larval behavior (Figure 3) or the adult electroantennogram (Figure 4). Strong interferences were encountered only when the odors belonged to the same chemical group. We inferred that *Drosophila* employs a small number of odor receptors corresponding to the functional groups of the volatiles in which the fly is most interested, namely acetates, alcohols, ketones, aldehydes, and certain fatty acids. For some groups, such as acetates and alcohols, there seem to be more than one receptor (Siddiqi, 1983; Borst, 1983) so that the functional group is not the only factor determining receptor specificity. Nevertheless, the range of specificity of the olfactory receptors is broad and accommodates considerable variation in size and shape. We have chosen a set of six chemicals representing the chief receptor classes to study the olfactory behavior of *Drosophila*. This canonical set of six odors does not exhaust the olfactory world of the fly. There clearly remain other molecules, for instance the sex pheromones, which are important to *Drosophila*. The role of olfaction in courtship has been extensively studied by Hall, Tompkins, and Jallon (Jallon, 1984; Tompkins, 1984; Gailey et al., 1986). The odors we investigated provide an overview of the fly's olfactory system and the strategy for detecting odors that it employs.

The larvae of *Drosophila* do not show any repulsion at lower concentrations of odors. They are attracted by all volatiles, including such noxious substances as pyridine. In the case of some chemicals, on reaching the source the larvae do not run into the source but form a ring around it. The diameter of the ring is directly proportional to the concentration at the source (Rodrigues, 1980). It is not known how the larva orients toward the odor. At elevated concentrations it makes a beeline toward the source, but with decreasing strength of the cue its track assumes the character of biased wandering (Figure 5).

OLFACTION IN DROSOPHILA

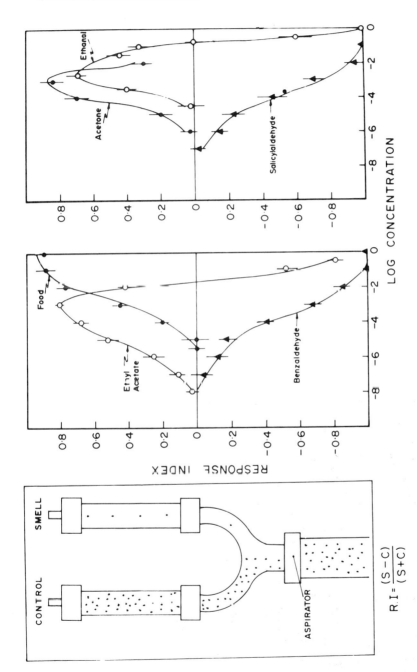

FIGURE 1 Olfactory responses of *Drosophila*. The response of adult flies is measured in a Y olfactometer (left). One arm of the olfactometer is connected to the odor source (S) and the other to control air (C). The response index RI is computed from the number of flies in the collecting arms: $RI = (S - C)/(S + C)$. The curves show the responses of normal Canton special strain to a few chemicals. Note the reversal of attraction above 1000-fold dilution. (From Rodrigues, 1980).

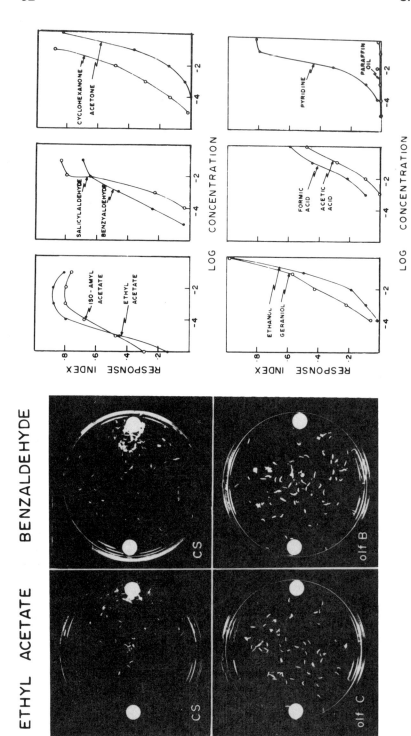

FIGURE 2 Larval response. The photograph on the left shows the response of wild-type and mutant strains to ethyl acetate and benzaldehyde. The response curves on the right show attraction to all tested odors. (From Rodrigues, 1980.)

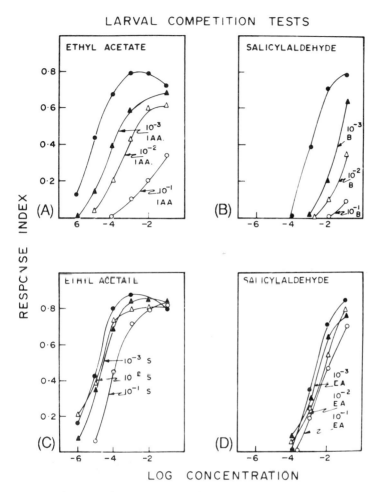

FIGURE 3 Effect of background odor on larval taxis. The response curves for ethyl acetate and salicylaldehyde against increasing concentrations of the background odors: (A) isoamyl acetate (IAA); (B), benzaldehyde (B); (C) salicylaldehyde (S); (D) ethyl acetate (EA). (From Rodrigues, 1980.)

The adult fly discriminates between attractants and repellents. Benzaldehyde, the strongest among repellents, is avoided at extremely low concentrations. Many of the attractants elicit a mixed response. At higher concentration the response changes from attractant to repulsion. Even the most potent attractants, such as ethyl acetate and isoamyl acetate, turn into repellents at concentrations higher than a 1000-fold dilution at source. In this sense it is

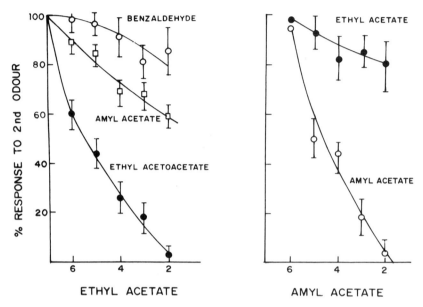

FIGURE 4 Inhibitory effect of background on the amplitude of electroantennogram (EAG). Abccissa, negative logarithm of dilution of the background stimulus. Ordinate, response to the test stimulus. (From Siddiqi, 1983.)

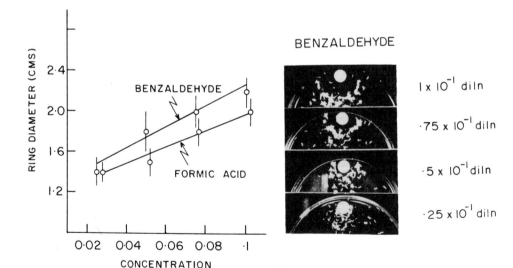

FIGURE 5 Larval avoidance reaction at close quarters. Repulsion to high source concentrations could be due to contact with diffusing chemical. (From Rodrigues, 1980.)

somewhat misleading to describe the chemicals as attractants or repellents without specifying the concentration. The volatiles used in olfactometric experiments are lipophilic, and one must bear in mind the possibility that the response at nonphysiologic concentrations may involve mechanisms different from those that underlie normal responses at lower concentrations. This is particularly important in choosing methods for screening of mutants.

The sensitivity to a particular odorant is measured by the threshold for the response and the maximal response that can be obtained. The exact measurements of these indices depend upon the design of the olfactometer and the "state" or "disposition" of the fly. Y mazes, which introduce odors in an airstream, give low detection thresholds compared to mazes without airflow. The optimal conditions for measuring attraction and repulsion are not the same. Repulsion is best measured in rapid tests when alarmed flies are forced to make a quick binary choice. On the other hand, attraction is stronger when the flies are allowed to explore the maze relatively at ease.

Borst and Heisenberg (1982) have shown that a concentration of attractant odor on one antenna 1.67 times the concentration on the other antenna evokes a positive turning response. It is doubtful whether, in nature, the two antennae encounter a concentration difference as large as this. The exact mechanism of this osmotropotaxis thus remains to be understood.

III. SENSORY PHYSIOLOGY

The third antennal segment of *Drosophila* carries the three types of sensory hairs usually found in insects. The sensilla basiconica and sensilla coeloconica respond to odors. The sensilla trichoidea are probably also olfactory, but their function remains to be determined.

The structure of the olfactory sensilla and their distribution on the funicular surface have been described by Singh and his associates, who have also reported a group of olfactory hairs on the maxillary palp (Venkatesh and Singh, 1983; Singh and Nayak, 1985). The larva possesses well-developed olfactory organs (Singh and Singh, 1984), but their physiology has not been studied.

Anil Gupta and I have examined the electrophysiologic responses of sensilla basiconica, the principle sensors of general odors in adult flies. The single-unit responses of over a hundred neurons in a sample of more than half the *basiconica* on the funiculus is summarized in Figure 7. The hairs can be classified into recognizable response types. Neglecting types V and VII, which are represented by a small number of units, at least six major types of response patterns are distinguishable. Type I hairs respond to all six test odors. Type II respond to the four attractants. Type I and type II hairs are highly sensitive, three or four orders of magnitude more sensitive than the other hairs. They also exhibit a high rate of background firing. It is possible that these

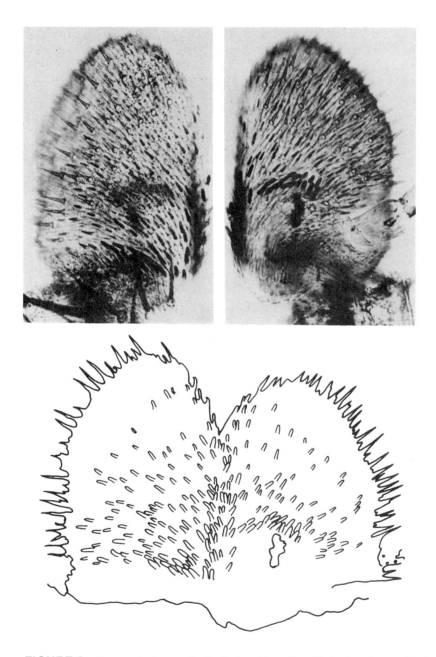

FIGURE 6 Sensory hairs on the funiculus. Top: Sensilla basiconica on the three faces of the antenna. The hairs have been impregnated with silver. Bottom: Map of the antennal surface showing the pattern of distribution in basiconica.

OLFACTION IN DROSOPHILA

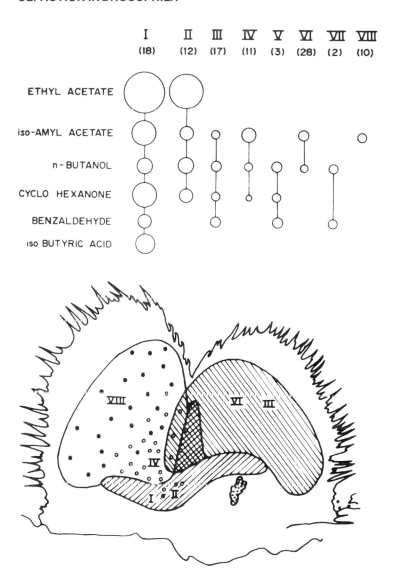

FIGURE 7 Response spectra of sensilla basiconica: Top: Responses of 101 single units. Diameter of circle denotes sensitivity on a logarithmic scale. Bottom: Map showing topographic distribution of six different response types.

hairs constitute a high-sensitivity-low-acuity system of detection and the other hairs mediate the recognition of odors at higher concentrations. The remaining hair types III, IV, VI, and VIII respond selectively to other subsets of odors: four, three, two, or one. The responses are not distributed in random combinations. There is an orderly organization of receptor specificity, and only a small subset of all possible combinations of sensitivity is actually present.

The orderliness of the receptor system is also seen in the spatial distribution of the response types (Figure 7). Each hair type occupies a continuous patch on the antennal surface. These patches overlap, giving rise to areas that contain different combinations of sensillary types. The pattern of distribution of basiconica is fairly invariant. One can recognize hairs of known response characteristics from their positions and record from a given type of hair repeatedly. This makes the *Drosophila* antenna particularly suitable for electrophysiologic investigations on mutants. Another convenient feature of the olfactory neurons of *Drosophila* is the absence of peripheral inhibition. The neurons, as a rule, respond to odors with an increased firing. The classification of response spectra is therefore comparatively easy.

IV. CENTRAL PROJECTIONS OF ODORS

The neurons of the antennal sensilla project to the antennal lobes; some 1100 axons converge to 22 glomeruli of each lobe (Pinto et al., 1989). The connections of the antennal neurons to the glomeruli are very simple. Each neuron is connected to a specific glomerulus. Some project only ipsilaterally, but others send a branch along the antennal commissure to the same glomerulus on the contralateral side (Stocker et al., 1983). The antennal glomeruli can be recognized by their position, shape, and size. The three types of olfactory sensilla are connected to different subsets of glomeruli (Figure 8).

Rodrigues, in collaboration with Buchner, used the deoxyglucose (DOG) uptake method to map the activity pattern in the antennal lobe generated by elementary odors. (Rodrigues 1988; Rodrigues and Buchner, 1984; Rodrigues and Bulthoff, 1985). They have shown that odor quality is represented by a discrete spatial pattern of excitation in the antennal lobe (Figure 9). Four of the six test odors project to different but overlapping subsets of a set of 12 glomeruli (Figure 10).

An unexpected finding of the DOG experiments is the striking difference in the pattern of activity generated by the repellent benzaldehyde compared to other odorants. Other odors cause mainly ipsilateral excitation, but unilateral stimulation with benzaldehyde excites the lobes on both sides. This observation provides a neurologic correlate of the finding by Borst and Heisenberg (1982) that benzaldehyde fails to evoke an orientation response. At present the DOG results are of a somewhat qualitative nature. The technique needs

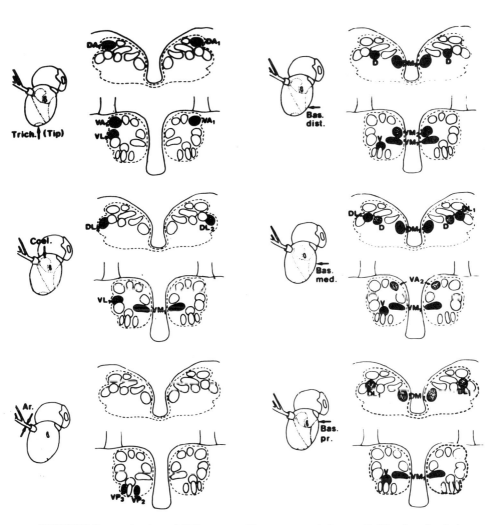

FIGURE 8 Projection of different sensilla to antennal glomeruli. The site of cobalt injection is indicated by arrows. A total of 19 glomeruli are shown in two sections. The upper section shows 8 dorsal glomeruli; the lower section shows 11 ventral glomeruli. Altogether, 22 glomeruli can be identified. (From Stocker et al., 1983; Pinto et al., 1989.)

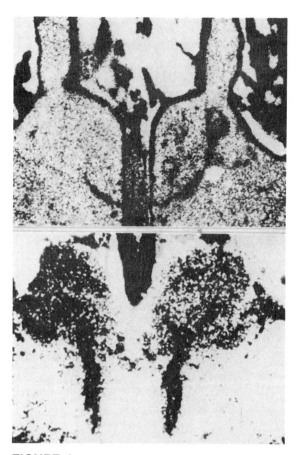

FIGURE 9 Deoxyglucose labeling of glomeruli. Top: Ipsilateral excitation by acetate. The contralateral branch of the axon is labeled, but not the glomerulus. Bottom: Bilateral excitation by benzaldehyde. Notice labeling on both sides. The left antenna was removed. The output axons are also labeled.

to be improved and made quantitative before a confident assessment of the invariance of activity patterns can be made.

Our knowledge of the circuitry of antennal lobes and the attendant processing of sensory signals is still very inadequate. Golgi preparations reveal a variety of interneurons (unpublished work by Fischbach, Borst, and Singh). These must integrate the incoming signals, but in the absence of detailed information about the network of relay neurons, not much can be said regarding the output signals that go to the mushroom body and dorsal protocerebrum. The input into glomeruli is convergent, but the output from the lobes diverges

OLFACTION IN DROSOPHILA

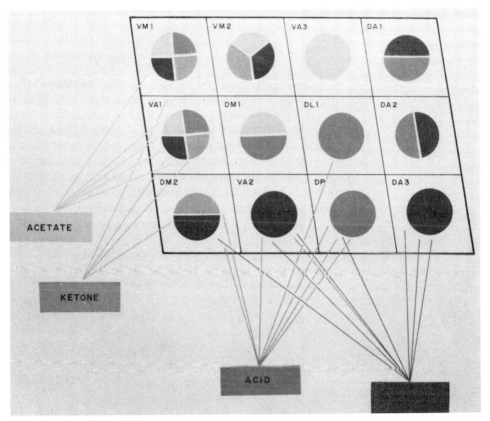

FIGURE 10 The pattern of projection of odors on a subset of 12 glomeruli (data from V. Rodrigues). Symbols D, V, A, and P refer to dorsal, ventral, anterior, and posterior locations.

to a few thousand Kenyon cells in the mushroom body. A major step forward in understanding odor perception will come from understanding the nature of mapping between the glomeruli and the Kenyon cells. In principle it should be possible to extend functional mapping beyond the antennal glomeruli.

V. OLFACTORY NEUROGENETICS

Mutations can alter the olfactory behavior of the fly in a variety of ways. Peripheral lesions may affect specific receptors, cellular signaling pathways, or electrical excitability of the membrane. In each case the mutation is expected to change the firing response of the neuron. Mutations affecting the central connections of the olfactory pathway are likely to change the osmotactic responses of the fly without changing the electrical output of the sensory neurons. One can use a number of diagnostic tests that should serve to distinguish peripheral from central defects.

In 1978 we described a set of four *olfactory* genes on the X chromosome of *Drosophila* (Rodrigues and Siddiqi, 1978). More mutants were subsequently isolated by Padhye and Ayyub. The *olf* mutants in our laboratory were induced by ethylmethanesulfonate and screened by testing mutagenized lines in the olfactory Y maze. John Carlson and his group at Yale have developed other methods of screening for olfatory mutants.

The six X-linked genes we studied are shown in Figure 11. Four of these, *olfA*, *olfB*, *olfE*, and *olfF*, reduce the response to aldehydes; *olfC* mutants are defective in the response to acetate esters. The gene *olfD* blocks responses to all odors that have been tested. Experiments in Carlson's laboratory show that *olfD* mutants are allelic to *smell blind*, a gene previously isolated by Aceves-Pina and Quinn (1979).

All the *olfactory* mutants examined so far are partial; absolute odor blocks have not been found. Some of the genes increase the threshold concentration

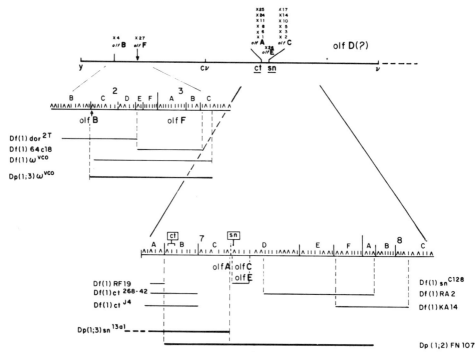

FIGURE 11 Cytogenetic localization of olfactory genes on the X chromosome.

of the response without reducing the maximal response; others cause a reduction in the maximal response (Figure 12). Mutations at all six loci affect the adults as well as the larvae, although the effect on the larval response is somewhat weaker.

The genes *olfA,* *olfC,* and *olfE* are closely linked to the genes *cut* and *singed* and have been investigated in some detail. Hasan and Ayyub have found

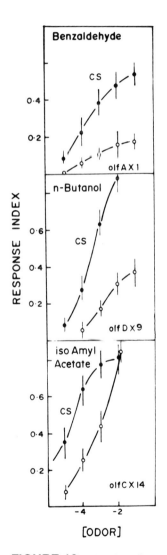

FIGURE 12 The larval responses of *olfactory* mutants to odors. (Siddiqi, 1987.)

that a fraction of P-induced mutations in the *singed* locus cause olfactory blocks. Several of these *sn olf* isolates fail to complement *olfA* and *olfE* alleles. The P element in these strains is inserted in the *singed* locus in only one of the two possible orientations. Hasan and Ayyub (in press) ascribe this to a polar effect of insertions on the neighboring *olf* loci. A chromosome walk to map the *olf* genes in this region has been carried out.

So far little is known about the primary functions of the *olf* genes. The most striking feature of the phenotype of these genes is the chemical specificity of the odor block. The six alleles of the *olfC* locus, for instance, belong to two sub groups. X2, X3, and X10 are defective in response to ethyl acetate as well as isoamyl acetate; X5, X14, and X17 are blocked to isoamyl acetate alone. This suggests that these mutations affect some step in reception. Venard and Pichon (1984) have reported that some of the *olfC* alleles have a defective electroantennogram (EAG) response to acetates. We are currently examining the single-unit responses of *olfactory* mutants.

It has been found that the *olfD* mutants are also defective in taste response (Mary Lilly, unpublished). Some of the new mutants isolated in Carlson's laboratory also show alterations in visual physiology. These results imply that there are common steps in the pathways of chemosensory and visual reception.

VI. SUMMARY

The most attractive feature of the olfactory pathway of *Drosophila* is its relative simplicity. Odorants are detected by an apparently small number of specific receptor sites. The presumptive receptors correspond to the functional groups in the volatiles of interest to the fly. The receptor sites are distributed on the sensory neurons in an overlapping fashion so that odors excite specific subsets of neuron types, each with a characteristic spectrum of sensitivity. The excitation patterns of the sensory neurons map on the glomeruli in the antennal lobes, where each odor is represented by a pattern of glomerular activity. The simplicity of glomerular organization in *Drosophila* makes this pattern recognizable.

A number of *olfactory* genes have been identified. Some of these give rise to odor-specific anosmias; others produce multiple blocks or smell blindness. The primary lesions in the *olfactory* mutants remain to be identified, but some of the *olf* mutations affect electrophysiologic responses, suggesting a defect in reception. There are mutations that simultaneously affect the sensory responses to olfactory, gustatory, and visual stimuli, showing that some of the steps in cellular signaling are common to the three modalities.

REFERENCES

Aceves-Pina, E. O., and Quinn, W. G. (1979). Learning in normal and mutant *Drosophila* larvae. *Science* **206**:93-95.
Barrows, W. (1907). The reactions of the pomace fly *Drosophila ampelophila* to odorous substances. *J. Exp. Zool.* **4**:515-537.
Begg, M., and Hogben, L. (1946). Chemoreceptivity of *Drosophila melanogaster*. *Proc. R. Soc. (Lond.) [Biol.]* **133**:1-19.
Borst, A. (1983). Computation of olfactory signals in *Drosophila melanogaster*. *J. Comp. Physiol.* **152**:373-383.
Borst, A., and Heisenberg, M. (1982). Osmotropotaxis in *Drosophila melanogaster*. *J. Comp. Physiol.* **132**:235-242.
Gailey, D. A., Lacaillade, R. C., and Hall, J. C. (1986). Chemosensory elements of courtship in normal and mutant, olfaction-deficient *Drosophila melanogaster*. *Behav. Genet.* **16**:375-405.
Hasan, G., and Ayyub, C. (in press). Olfactory behavior of P element-induced *singed* mutants of *Drosophila melanogaster*. (in preparation).
Jallon, J. M. (1984). A few chemical words exchanged by *Drosophila* during courtship and mating. *Behav. Genet.* **14**:483-520.
Pinto, L., Stocker, R. F., and Rodrigues, V. (1989). Anatomical and neurochemical classification of the antennal glomeruli in *Drosophila melanogaster*. *Int. J. Embryol. Exp. Morphol.* **17**:335-344.
Rodrigues, V. (1980). Olfactory behaviour of *Drosophila melanogaster*. In *Development and Neurobiology of* Drosophila, O. Siddiqi, P. Babu, L. Hall, and J. Hall (Eds.). Plenum, New York, pp. 361-371.
Rodrigues, V. (1988). Spatial coding of olfactory information in the antennal lobe of *Drosophila melanogaster*. *Brain Res.* **453**:299-307.
Rodrigues, V., and Buchner, E. (1984). [^{3}H]2-deoxyglucose mapping of odour-induced activity in the antennal lobes of *Drosophila melanogaster*. *Brain Res.* **324**:374-378.
Rodrigues, V., and Bulthoff, I. (1985). Freeze substitution of *Drosophila* heads for subsequent [^{3}H]2-deoxyglucose antoradiography. *J. Neurosci. Methods* **13**:183-190.
Rodrigues, V., and Siddiqi, O. (1978). Genetic analysis of chemosensory pathway. *Proc. Indian Acad. Sci. [B]* **87**:147-160.
Siddiqi, O. (1983). Olfactory neurogenetics of *Drosophila*. In *Genetics: New Frontiers*, Vol. III (Proceedings XV Int. Congress of Genetics, December 1983), V. L. Chopra, R. P. Sharma, B. C. Joshi, and H. C. Bansal (Eds.). Oxford I.B.L., New Delhi, pp. 243-261.
Siddiqi, O. (1987). Neurogenetics of olfaction in *Drosophila malanogaster*. *Trends Genet.* **3**:137-142.
Singh, R. N., and Nayak, S. (1985). Fine structure and primary sensory projections of sensilla on the maxillary palp of *Drosophila melanogaster*. *Int. J. Insect Morphol. Embryol.* **14**:291-306.
Singh, R. N., and Singh, K. (1984). Fine structure of the sensory organs of *Drosophila melanogaster* larva. *Int. J. Insect Morphol. Embryol.* **13**:255-273.

Stocker, R. F., Singh, R. N., Schorderet, M., and Siddiqi, O. (1983). Projection patterns of different types of antennal sensilla in the antennal glomeruli of *Drosophila melanogaster*. *Cell Tissue Res.* **232**:237-248.

Thorpe, W. H. (1939). Further studies on pre-imaginal olfactory conditioning in insects. *Proc. R. Soc. Lond.* [*Biol.*] **127**:424-433.

Tompkins, L. (1984). Genetic analysis of sex appeal in *Drosophila*. *Behav. Genet.* **14**:411-440.

Venard, R., and Pichon, Y. (1984). Electrophysiological analysis of peripheral response to odours in wild type and smell-deficient *olfC* mutants of *Drosophila melanogaster*. *J. Insect Physiol.* **30**:1-5.

Venkatesh, S., and Singh, R. N. (1983). Sensilla on the third antennal segment of *Drosophila melanogaster*. *Int. J. Insect Morphol. Embryol.* **13**:51-63.

8
A New Homoeotic Mutant with Defects in the *Drosophila* Olfactory System

Richard Ayer, Paula Monte, and John Carlson

Yale University, New Haven, Connecticut

I. INTRODUCTION

Drosophila is an appealing organism in which to address the function, organization, and development of the invertebrate olfactory system. The *Drosophila* olfactory system is very sensitive to a wide variety of chemicals, and it can be studied through several powerful approaches. Among these is genetics, which has been successfully brought to bear in analysis of other parts of the fly's nervous system, including the visual system.

Olfaction in *Drosophila* is mediated primarily through the antenna, which is composed of three main segments and a featherlike structure known as the arista. The first and second antennal segments bear bristles; the third segment bears no bristles but is covered with several hundred fine sensilla. The morphology and distribution of these sensilla have been described by Venkatesh and Singh (1984). Many of the sensilla contain pore tubules, which presumably allow passage of volatile chemicals into the interior of the sensilla, which contain dendrites of as many as four neurons. Electrophysiologic recordings from sensilla (Siddiqi, 1984) confirm that at least some of these neurons respond to chemical stimulation.

There are several ways in which genetics may be applied to the study of olfactory system development and function. One method is based on the measurement of olfactory-driven behavior. Precedent for this approach comes from the isolation and analysis of mutants defective in visually driven behavior, which has been a very successful approach to investigation of the *Drosophila* visual system (Benzer, 1967; Hall, 1982). Mutants containing a defect in olfactory response have been isolated by virtue of abnormal behavior in any of several olfactory paradigms, including an olfactory Y maze (Rodrigues and Siddiqi, 1978), a T maze (Helfand and Carlson, 1989), an olfactory trap assay (Woodard, Huang, Sun, Helfand, and Carlson, 1989), and a chemosensory jump response (McKenna, Monte, Helfand, Woodard, and Carlson, 1989). Such paradigms are especially useful in that they allow isolation of mutants with defects at any level of the response pathway, including olfactory transduction or processing.

A second genetic approach is based on the isolation of mutants by virtue of abnormal physiology. A variety of *Drosophila* visual system mutants have been isolated using the electroretinogram as a screen: such mutants include those with defects in receptor molecules (Pak, 1979; O'Tousa et al., 1985; Zuker et al., 1985) and in presumed components of second messenger pathways (Pak et al., 1970; Hotta and Benzer, 1970; Bloomquist et al., 1988). Measurement of antennal physiology is also feasible (Venard and Pichon, 1981), but its use as a genetic tool is still in its infancy.

A third approach, the focus of this chapter, seeks to isolate mutants bearing morphologic defects affecting the olfactory system. Such mutants may be useful both in studying the development and maintenance of the antenna and in studying the sensory functions it supports. We begin with a discussion of precedent for the use of morphologic mutants to study the olfactory system. Recent work designed to explore further the potential of this approach is then presented. We describe a simple method of screening for antennal mutants, the isolation of three mutants with abnormal antennal morphology, and characterization of one mutant whose antennae exhibit a homoeotic transformation of antennae toward legs.

II. PRECEDENT FOR THE USE OF ANTENNAL MUTANTS

Genetic analysis has already proven to be a powerful means of investigating pattern formation and determination in the antenna, as shown by studies of certain homoeotic mutations at the *Antennapedia* complex (Kaufman et al., 1980). These mutations cause the imaginal disks that normally give rise to the antenna during the course of development to undergo a change of fate: rather than developing into antennae, they develop into legs. A detailed understanding of the molecular mechanisms underlying antennal development requires a more complete description of the genes required in the process. The isolation and characterization of additional mutants bearing defects in antennal morphology may be a valuable step toward this end.

Mutants with abnormal antennal morphology may also be of great value in investigating the function of the olfactory system. Historically, the use of such mutants has played an important role in the study of *Drosophila* olfaction: localization of this sensory modality to the antenna was confirmed by early studies of the mutants *antennaless* and *aristapedia* (Begg and Hogben, 1943, 1946). The severe perturbations of antennae afforded by these mutations may still be useful as new means of measuring olfactory function become available. The fly contains a number of other sensory structures, some of which may also receive various forms of olfactory information (Harris, 1972; Singh and Nayak, 1985; Stocker and Gendre, 1989; Monte and Carlson, unpublished results). The roles of these other structures in response to various types of olfactory stimulation may in some cases be assessed more sensitively in mutants lacking antennal function.

Mutations disrupting antennal form at a finer level can be elegant tools for the study of the olfactory system, as shown by the work of Stocker and Gendre (1988) with *lozenge* mutants. These mutants lack one particular subset of antennal sensilla, the sensilla basiconica, and in addition are missing one subunit of the antennal lobes, glomerulus V, previously shown to be a target of neurons in the sensilla basiconica. Further analysis of these mutants has implied a role for sensory input in the development or maintenance of the antennal lobes. Moreover, a *lozenge* mutant has been studied behaviorally with the aim of investigating the functional role of the subset of the olfactory system affected by the mutation (Stocker and Gendre, 1989).

Visual system mutants analogous to these two classes of mutants have been characterized in detail. Gross perturbations of the eye are effected by *sine oculis*, which drastically reduces the size of the eye, and *glass*, which both reduces the area of the eye and eliminates its light-stimulated electrical potential (Pak et al., 1969). A finer perturbation is caused by the mutation *sevenless* (Harris et al., 1977; Campos-Ortega et al., 1979), in which one class of photoreceptor cells, the R7 cells, fails to develop, thereby eliminating one spectral component of the animal's visual response.

A third class of visual system mutant, exemplified by the mutant *rdgB* (Hotta and Benzer, 1970), appears to develop normally but exhibits a morphologic phenotype—degeneration of photoreceptor cells—following stimulation with light (Harris and Stark, 1977). Degeneration is postulated to result from the abnormal function of the light-stimulated transduction cascade. This class contains a number of visual system mutants that, like *rdgB*, show both defective visual system physiology and receptor cell degeneration. We do not know whether mutants defective in olfactory transduction may be identified by virtue of such a visible degenerative phenotype.

III. A SIMPLE SCREEN FOR ANTENNAL MUTANTS

The mutants described in this chapter were isolated from a defined parental strain of *Drosophila*, Canton-S-5 (CS-5), which is homozygous for a particular X chromosome. The CS-5 stock was mutagenized by means of the chemical mutagen ethyl methane sulfonate (EMS), x-irradiation, or hybrid dysgenesis (for details, see Ayer et al., 1989). A behavioral enrichment for mutants defective in olfactory function, described in Woodard et al., 1989, was used following mutagenesis with x-irradiation and hybrid dysgenesis.

A simple device was used to allow rapid screening for antennal mutants. A micropipette tip was severed so that the internal diameter of its small aperture, approximately 1 mm, was of a size such that the anterior surface of a fly head, including the antennae, could protrude slightly. The tip, fastened to a stand (Figure 1), was placed under a Zeiss light dissecting microscope with the tapered end upward in the beam of a light source positioned above. A fly, inserted into the base of the micropipette tip by means of a standard mouth aspirator, walked upward, presumably driven by positive phototaxis and negative geotaxis. Air pressure was then applied with the mouth aspirator to lodge the fly at the top of the pipette tip, its head protruding slightly through the opening of the tip and its thorax resting beneath the aperture. In this position, antennae were examined at ×128 magnification. The fly was removed from the apparatus without harm simply by applying gentle suction.

Flies were scored for the following characteristics: (1) morphology and relative positions of the second antennal segment, third antennal segment, and

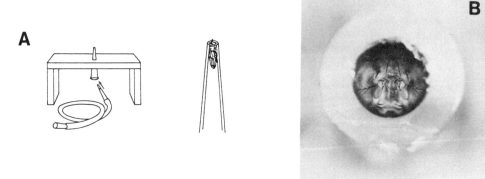

FIGURE 1 Device for examining fly antennal morphology. (A) Schematic of device showing plexiglass stand, pipette tip, and mouth aspirator used for inserting single flies into pipette tip. Inset shows close-up of tip with inserted fly. The stand is aligned below a ×128 dissecting microscope so that the anterior aspect of the fly head is centered in the field of view. (B) Photomicrograph through the dissecting microscope showing the fly's appearance in the device. Outside diameter of pipette tip, 1.5 mm. (From Ayer et al., 1989.)

arista; (2) "hairy" texture of the third segment; (3) position, number, size, and shape of bristles on the second antennal segment; (4) movement of antennae in response to air currents; and (5) antennal pigmentation. Flies were scored at a rate of approximately 100 per h.

IV. THE ISOLATION OF NEW ANTENNAL MUTANTS

Three antennal mutants, one induced by x-irradiation, one by EMS, and one by hybrid dysgenesis, were isolated from a screen of 1048 mutagenized X chromosomes: 335 mutagenized by EMS, 270 by x-irradiation, and 443 by hybrid dysgenesis.

The x-ray-induced mutant, X-M4, was isolated by virtue of its abnormally shaped third antennal segment, which was observed to bend laterally toward the eye. The EMS-induced mutant, EMS-M13, was isolated on account of lack of antennal movement upon stimulation with air currents. Closer examination revealed that antennal motions were hindered by the abnormal form of the head cuticle at the ptilinal suture: the prefrons was recessed with respect to the postfrons such that the antennae were locked in a fixed position. Although stable for many generations, neither of these mutants exhibited high penetrance, and neither has been maintained.

The dysgenic mutant was originally isolated on account of an increase ($\geq 35\%$) in the number of bristles on the anterior aspect of the second antennal segment (Figure 2). In approximately 15% of the population, the third segment was seen to be abnormally shaped. Moreover, the aristae of some flies were thickened and the long lateral hairs of the arista were visibly reduced in length and in regularity of pattern. Ectopic bristles appeared on the third antennal segment, although again with low penetrance. These recessive antennal phenotypes led us to name the mutant *extra antennal bristles* (*eab*) and to suggest the interpretation that it exhibited a weak homoeotic transformation of antennae toward legs (Ayer et al., 1989).

V. CHARACTERIZATION OF *eab*, A NEW DYSGENIC HOMOEOTIC MUTANT

Recent observations have confirmed the interpretation that *eab* is a homoeotic mutant. Dramatic transformations of antennae toward legs were produced in the course of an attempt to determine the genetic map position of the antennal phenotype. Briefly, *eab* males were mated to females homozygous for an X chromosome bearing the visible markers *yellow*, *crossveinless*, *vermilion*, and *forked* (*y cv v f*), and the female progeny were then crossed to males containing a balancer FM7a X chromosome. Among the F_2 male progeny that contained the parental *eab* X chromosome, an appreciable frac-

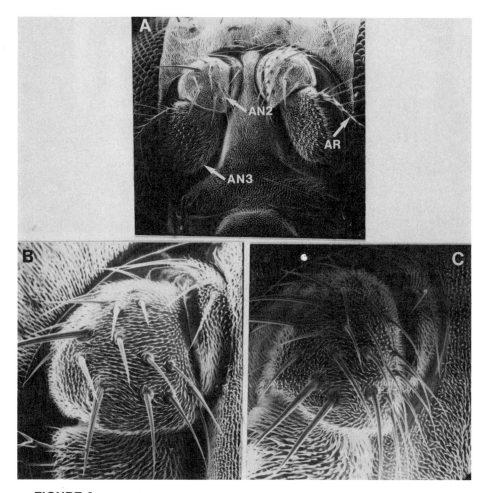

FIGURE 2 Second antennal segments of wild type (CS-5) and mutant (*eab*). (A) Anterior aspect of a CS-5 head: second segment (AN2), bearing large bristles; third segment (AN3), covered with sensilla; feathery arista (AR). (B) Second segment of wild type: ~15 bristles are visible on the anterior aspect. (C) Second segment of *eab*: ~27 bristles are visible on anterior aspect. (Taken from Ayer et al., 1989.)

tion (~10%) were found to exhibit severe phenotypes, such as those shown in Figure 3. These extreme transformations reveal a gross thickening of the arista into a leglike structure. In some cases (see Figure 3B and D) these transformed antennae bear a claw at the distal tip, reminiscent of the tarsal claw seen at the distal tip of the leg.

One simple interpretation of the increase in severity of the phenotype is that the *eab* defect is sensitive to the genetic background: perhaps the expres-

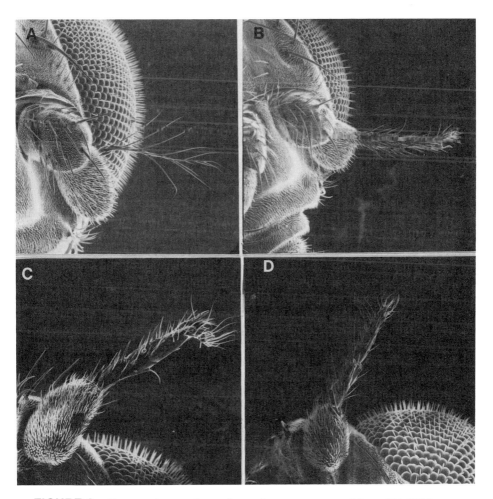

FIGURE 3 Homoeotic transformations of antennae toward legs. (A) Wild type (CS-5); (B, C, D) *eab*. Note the transformation of the feathery arista into a thickened leglike structure. In B and D a claw can be seen at the distal tip.

sivity of the defect is increased by the presence of an enhancing factor found in the autosomes of the *y cv v f* or balancer stocks but not in the autosomes of the original *eab* stock. Alternatively, it is possible that the more severely affected individuals have undergone a secondary dysgenic event. For example, if the original phenotype were caused by insertion of a transposable P element, then that element may have been mobilized by the cross to the *y cv v f* stock, which is of M cytotype and may therefore have allowed such secondary dysgenic events to occur at the observed frequency (Kidwell, 1987).

The *eab* mutant contains several other defects besides the antennal phenotype. These defects were all observed in the original *eab* stock, in which they were determined to be recessive; they also appear in those animals that subsequently acquired more severe antennal transformations. The eye appears roughened: the arrangement of ommatidia is less regular than in the wild type; the orientation of the hairs is less precise, and in some cases a pair of hairs exist immediately adjacent to each other (Figure 4). Irregular wing margins appear with high penetrance (Figure 5), and the wings are divergent. The metathoracic legs of a small fraction of flies are bent, such that the tarsal segments are at an angle to the tibia. Finally, viability is reduced among males carrying the *eab* X chromosome.

In addition to these morphologic phenotypes, *eab* has a behavioral phenotype. Unlike wild-type control males, which spend much of their time on the walls of the culture vials, *eab* males generally remain on the surface of the

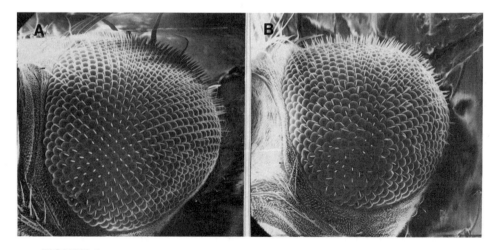

FIGURE 4 Eyes of wild type (CS-5) and mutants (*eab*). (A) Wild type; (B) *eab*. Note the disorganized facets of the mutant eye. Many hairs in the mutant eye are irregular in orientation, and in some cases two hairs exist as a pair, immediately adjacent to each other.

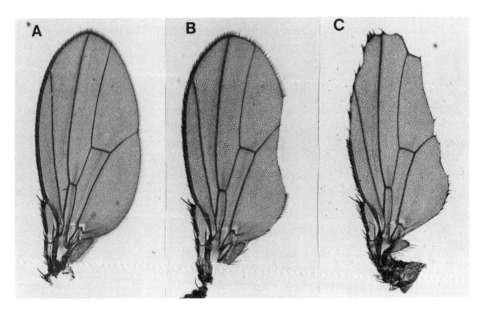

FIGURE 5 Wings of wild type (CS-5) and mutant (*eab*). (A) Wild type; (B) *eab*, low expressivity; (C) *eab*, high expressivity. Light micrographs. (From Ayer et al., 1989.)

food. The mutants appear to be defective in the ability to climb up the walls. The penetrance of this behavioral phenotype exceeds that of the leg phenotype just described. Mutant males also exhibit a greater tendency to become mired in the culture medium.

Analysis of recombinant chromosomes from the mapping cross suggests that both the antennal phenotype and the wing phenotype map to the *cv-v* interval of the X chromosome.

There are several homoeotic genes whose mutations produce transformations of antennae toward legs. Most but not all of these mutations are dominant, and most of the genes are autosomal. *Antennapedex* maps to the X chromosome but has been localized near the base of the chromosome, far from the *cv-v* interval (Ginter, 1969). The mutation *tumorous head-1* (*tuh-1*), which in combination with the mutation *tuh 3* produces transformations of antennae toward legs, is also X linked but again maps near the base of the X chromosome (Gardner, 1970). Further genetic analysis of the *eab* mutant, including determination of whether all its phenotypes are effects of a single mutation, is required before definitive interpretations can be drawn as to the nature of the defect. We note, however, that at least one of the mutants exhibiting transformations of antennae to legs, *Antennapedia of Le Calvez*, also has divergent wings (Le Calvez, 1948), and another, *Brista* (from *Bristle on arista*), has effects on the distal portions of the legs (Sunkel and Whittle, 1987).

VI. SUMMARY

A genetic approach to investigation of the *Drosophila* olfactory system, based on the use of mutations affecting antennal morphology, was discussed. We described a convenient means of screening for mutants with antennal defects and the isolation of three such mutants. A new X-linked recessive mutant, *eab* (*extra antennal bristles*), is found to exhibit a homoeotic transformation of antennae toward legs.

ACKNOWLEDGMENTS

We are grateful to Efrain Aceves-Pina for demonstrating the method of holding the fly during screening and to Henry Sun, Stephen Helfand, and Craig Woodard for supplying mutagenized material. Portions of the text of this manuscript are taken from Ayer et al., 1989. Grant support for this work was from the NIH. Carlson is an Alfred P. Sloan Research Fellow.

REFERENCES

Ayer, R., Monte, P., and Carlson, J. (1989). The isolation of antennal mutants and their use in *Drosophila* olfactory genetics. In *Neurobiology of Sensory Systems*, R. Singh and N. Strausfield (Eds.). Plenum Press, New York, pp. 411-418.

Begg, N., and Hogben, L. (1943). Localization of chemoreceptivity in *Drosophila*. *Nature* **152**:535.

Begg, N., and Hogben, L. (1946). Chemoreceptivity of *Drosophila melanogaster*. *Proc. R. Soc. (Lond.)* [*Biol.*] **133**:1-19.

Benzer, S. (1967). Behavioral mutants of *Drosophila* isolated by countercurrent distribution. *Proc. Natl. Acad. Sci. USA* **58**:1112-1119.

Bloomquist, B. T., Shortridge, R. D., Schneuwly, S., Perdew, M., Montell, C., Steller, H., Rubin, G., and Pak, W. L. (1988). Isolation of a putative phospholipase C gene of *Drosophila, norpA*, and its role in phototransduction. *Cell* **54**:723-733.

Campos-Ortega, J. A., Jurgens, G., and Hofbauer, A. (1979). Cell clones and pattern formation: Studies on *sevenless*, a mutant of *Drosophila melanogaster*. *Wilhelm Roux's Arch. Dev. Biol.* **186**:27-50.

Gardner, E. J. (1970). Tumorous head in *Drosophila*. *Adv. Genet.* **15**:116-146.

Ginter, E. K. (1969). *Dros. Info. Serv.* **44**:50-51.

Hall, J. (1982). Genetics of the nervous sytem in *Drosophila*. *Q. Rev. Biophys.* **15**: 223-479.

Harris, W. A. (1972). The maxillae of *Drosophila melanogaster* as revealed by scanning electron microscopy. *J. Morphol.* **138**:451-456.

Harris, W. A., and Stark, W. S. (1977). Hereditary retinal degeneration in *Drosophila melanogaster*. *J. Gen. Physiol.* **69**:261-291.

Harris, W. A., Stark, W. S., and Walker, J. A. (1977). Genetic dissection of the photoreceptor system in the compound eye of *Drosophila melanogaster*. *J. Physiol.* (*Lond.*) **256**:415-439.

Helfand, S. L., and Carlson, J. R. (1989). Isolation and characterization of an olfactory mutant in *Drosophila* with a chemically-specific defect. *Proc. Natl. Acad. Sci. USA* **86**:2908-2912.

Hotta, Y., and Benzer, S. (1970). Genetic Dissection of the *Drosophila* nervous system by means of mosaics. *Proc. Natl. Acad. Sci. USA* **67**:1156-1163.

Kaufman, T. C., Lewis, R., and Wakimoto, B. (1980). Cytogenetic analysis of chromosome 3 in *Drosophila melanogaster*: The homeotic gene complex in polytene chromosome interval 84A-B. *Genetics* **94**:115-133.

Kidwell, M. G. (1987). A survey of success rates using P element mutagenesis in *Drosophila melanogaster*. *Dros. Info. Serv.* **66**:81-86.

Le Calvez, J. (1948). In (3R) ss^{ar} : Mutation "*aristapedia*" heterozygote dominante, homozygote lethale chez *Drosophila melanogaster* (inversion dans le bras droit du chromosome III). *Bull. Biol. Fr. Belg.* **82**:97-113.

McKenna, M., Monte, P., Helfand, S., Woodard, C., and Carlson, J. (1989). A simple chemosensory response in *Drosophila* and the isolation of *acj* mutants in which is affected. *Proc. Natl. Acad. Sci. USA* **86**:8118-8122.

O'Tousa, J. E., Baehr, W., Martin, R. L., Hirsh, J., Pak, W. L., and Applebury, M. L. (1985). The *Drosophila ninaE* gene encodes an opsin. *Cell* **40**:839-850.

Pak, W. L. (1979). Study of photoreceptor function using *Drosophila* mutants. In *Neurogenetics, Genetic Approaches to the Nervous System*, X. D. Breakfield (Ed.). Elsevier, New York, pp. 67-99.

Pak, W. L., Grossfield, J., and White, N. V. (1969). Nonphototactic mutants in a study of vision of *Drosophila*. *Nature* **222**:351-354.

Pak, W. L., Grossfield, J., and Arnold, K. (1970). Mutants of the visual pathway of *Drosophila melanogaster*. *Nature* **227**:518-520.

Rodrigues, V., and Siddiqi, O. (1978). Genetic analysis of chemosensory pathway. *Proc. Indian Acad. Sci.* [B] **87**:147-160.

Siddiqi, O. (1984). Olfactory neurogenetics of *Drosophila*. In *Genetics: New Frontiers*, V. L. Chopra, B. C. Joshi, R. P. Sharma, and H. C. Bansal (Eds.). Oxford and IBH, New Delhi, pp. 243-261.

Singh, R. N., and Nayak, S. V. (1985). Fine structure and primary sensory projections of sensilla on the maxillary palp of *Drosophila melanogaster* Meigen (Diptera: Drosophilidae). *Int. J. Insect Morphol. Embryol.* **14**:291-306.

Stocker, R. F., and Gendre, N. (1988). Peripheral and central nervous effects of *lozenge*[3]: A *Drosophila* mutant lacking basiconic antennal sensilla. *Dev. Biol.* **127**: 12-24.

Stocker, R. F., and Gendre, N. (1989). Courtship behavior of *Drosophila*, genetically or surgically deprived of basiconic sensilla. *Behav. Genet.* **19**:371-385.

Sunkel, C. E., and Whittle, J. R. S. (1987). *Brista*: A gene involved in the specification and differentiation of distal cephalic and thoracic structures in *Drosophila melanogaster*. *Wilhelm Roux's Arch. Dev. Biol.* **196**:124-132.

Venard, R., and Pichon, Y. (1981). Etude electro-antennographique de la response peripherique de l'antenne de *Drosophila melanogaster* a des stimulations odorantes. *C.R. Acad. Sci. (Paris)* **293**:839-842.

Venkatesh, S., and Singh, R. N. (1984). Sensilla on the third antennal segment of *Drosophila melanogaster* Meigen (Diptera: Drosophilidae). *Int. J. Insect Morphol. Embryol.* **13**:51-63.

Woodard, C., Huang, T., Sun, H., Helfand, S., and Carlson, J. (1989). Genetic analysis of olfactory behavior in *Drosophila*: a new screen yields the *ota* mutants. *Genetics* **123**:315-326.

Zuker, C. S., Cowman, A. F., and Rubin, G. M. (1985). Isolation and structure of a rhodopsin gene from *D. melanogaster*. *Cell* **40**:851-858.

9
Genetics of a Moth Pheromone System

Wendell Roelofs

Cornell University, Geneva, New York

Thomas J. Glover

Hobart and William Smith Colleges, Geneva, New York

I. INTRODUCTION

How species are defined and the processes involved in speciation continue to be the subjects of debate. Unfortunately, most arguments supporting various concepts of speciation lack experimental approaches or are based on inferences drawn from events that occurred in the distant past. Experimental data, when obtained, are usually confounded by the complexities involved in attempting to discriminate among the various selective contexts operating during reproduction (Thornhill, 1979; Thornhill and Alcock, 1983). Factors involved in reproduction, however, appear to be key to any mode of speciation, and thus the long-distance sex pheromone communication system found in most moth species should be central to any speculative discussions on the evolution of these species. More information on the chemistry, biochemistry, physiology, and genetics of these tightly controlled mating communication systems should help to unravel some of the complexities inherent to the speciation processes operating in this particular group of insects. This chapter principally discusses data obtained on the sex pheromone system of one moth species and the implications that these findings may have in the speciation debate.

II. BACKGROUND

Commonly in moth species, the adult females release a sex pheromone composed of a very specific blend of components (usually one to seven compounds) that is highly attractive to conspecific males. The chemical compounds are only one element of the complex communication system that can have specificity because of a variety of factors, including release rates, blend ratios, seasonal and photoperiodic cycles, and habitat. It is the great specificity of these long-distance signals that makes them an important factor in defining and maintaining a group as a distinct species. An interesting example of this is found in some populations of New Zealand leafrollers (Foster and Dugdale, 1988). In a number of instances, moth populations from different locations on the two major islands were found to have distinctly different sex pheromone blends, even though the populations were previously synonymized on morphologic criteria. The differences were not minor but involved changes in chemical structure, such as in three sibling species in the *Planotortrix* genus that were each found to use one of the following combinations of monounsaturated 14-carbon acetates: (1) Z5-14:0Ac; (2) Z8-14:0Ac/Z10-14:0Ac; and (3) Z5-14:0Ac/Z7-14:0Ac/Z9-14:0Ac.

A study of the biosynthetic pathways for these compounds shows that they are produced by different routes, with the Δ^5, Δ^7, and Δ^9 components resulting from the action of the ubiquitous Δ^9-desaturase and the Δ^8 and Δ^{10} components coming from precursors produced with a rare Δ^{10} desaturase (Figure 1)

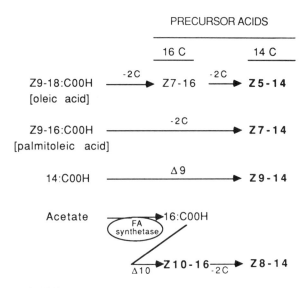

FIGURE 1 Biosynthetic pathways for pheromone components of plesiomorphic New Zealand and Australian leafroller species using ubiquitous Δ^9- or rare Δ^{10}-desaturases.

(Roelofs and Wolf, 1988). An interesting pattern has been found in New Zealand (Foster and Dugdale, 1988) and in Australia (Horak et al., 1988) in which leafroller species found to be pleiomorphic, or less evolved, according to morphologic criteria, all used pheromone components produced with either the Δ^9- or Δ^{10}-desaturase. More evolved leafroller species in these countries were found to use pheromone systems based on Δ^{11} desaturation, which has been found commonly for the pheromone systems of most other leafroller species around the world (Roelofs and Brown, 1982). What processes have occurred to produce these different pheromone communication systems? How plastic are the pheromone blends, and what are the modes of inheritance of the factors controlling the various elements of these communication systems? Although research has begun on some of the New Zealand leafroller species (Hansson et al., 1989), most of the information on these questions to date comes from other moth species.

III. STABILITY OF PHEROMONE SYSTEMS

The long-range sex pheromone communication system used by moth species is highly specific and must be subject to stabilizing selection, with a definite disadvantage to individuals that produce or respond to deviant signals. Support for this statement comes from selection studies with a North American leafroller species, the redbanded leafroller moth (RBLR, *Argyrotaenia velutinana*) (Du et al., 1984; Roelofs et al., 1986; Sreng et al., 1989), in which it was concluded that the specific pheromone blend is strongly canalized and that it would be very difficult to shift this intractable blend with direct selection. The two main pheromone components in this species are (Z)- and (E)-11-tetradecenyl acetate (Z- and E11-14:0Ac), which are common pheromone components produced by involvement of Δ^{11}-desaturases (Figure 2). A study

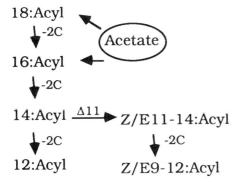

FIGURE 2 Biosynthetic pathway for pheromone components using Δ^{11}-desaturase, which is found commonly in lepidopteran pheromone glands.

on the individual variability of the E/Z11-14:0Ac ratio (Miller and Roelofs, 1980) showed that it had a low coefficient of variation (9.7%) in both a laboratory population that ranged from 4 to 14% and a field population that ranged from 4 to 16% E isomer. Analyses of airborne pheromone in an unrelated ermine moth, *Yponomeuta padellus*, showed that it also demonstrated a low coefficient of variation (15%) for these same two components (Du et al., 1987). Further research with the redbanded leafroller moth utilizing three selection protocols was carried out to determine if any of the observed individual variability had a genetic basis. In each protocol females producing the highest ratio of E isomer were selected. Although slight differences in the E/Z ratio could be achieved, the selection gain usually diminished after several generations and the mean ratio never moved beyond the range of the base population. In one particular line, the mean E/Z ratio never progressed beyond 11:89 in 25 generations under various selection pressures, and several times this line exhibited a mean ratio below the normal 9:91.

Since the E/Z ratio appeared to be quite intractable, it was thought that the higher variation found in the ratio of the 12-carbon minor components (E- and Z9-12:0Ac) to the 14-carbon compounds (E- and Z11-14:0Ac) would provide a better basis for shifting the pheromone blend. These 12-carbon compounds are produced by removal of two carbons from the unsaturated 14-carbon compounds through one cycle of β-oxidation (Figure 2). The 12-carbon to 14-carbon ratio responded quickly to direct selection and produced a high line with 42% and a low line with 14% 12-carbon acetates, but they stabilized at these ratios after several generations. Estimated heritability values of 0.416 and 0.644 for reciprocal crosses of these lines indicated some plasticity in this ratio, but in reality it was not possible to shift them beyond a limited range with high selection pressure.

In a similar study with the pink bollworm moth, Collins and Cardé (1985) found a low coefficient of variation of 5.3% in a laboratory colony for the two pheromone components (Z,Z)- and (Z,E)-7,11-hexadecadienyl acetates. Artificial selection for 12 generations changed the blend of the high line from 42.94 ± 0.97 to 48.23 ± 1.25% Z,E isomer. Although statistically significant, this change would not represent a shift outside the range of blend production and response of normal females and males, respectively, in the field.

IV. EUROPEAN CORN BORER PHEROMONE SYSTEM

Studies to determine the genetic basis of the tightly controlled sex pheromone systems within monomorphic populations have proved to be time consuming and unsuccessful in elucidating the underlying genetic systems. It therefore appears to be more promising to conduct genetic analyses with populations that utilize distinctly different pheromone systems but that have enough genetic compatibility to produce viable and fertile hybrids. The European corn borer (ECB, *Ostrinia nubilalis*) races in North America fulfill the criteria for such analyses. A bivoltine and a univoltine race utilize mainly the Z isomer in a two-component sex pheromone blend (E/Z11-14:0Ac ratio is 3:97), and another bivoltine race uses mainly the E isomer (E/Z ratio is 99:1) (Klun et al., 1973; Kochansky et al., 1975; Glover et al., 1987). Hybridization can be achieved in the laboratory and the field, and analyses of hybrid females from reciprocal crosses have shown that they produce an intermediate pheromone blend of approximately 68:32 E/Z isomers (Klun and Maini, 1979; Roelofs et al., 1987). Since F_2 and backcrossed progenies are viable and readily achieved in the laboratory, these races provided an excellent opportunity for a genetic analysis of the mate recognition system, including sex pheromone production in females, detection of the signals by the male moth's antennal olfactory cells, and upwind-oriented flight responses by the male moths to the pheromone (Roelofs et al., 1987).

A. Female Sex Pheromone Production

Laboratory crosses of the three ECB races were established in our laboratories and were analyzed periodically to verify the purity of the pheromone strains. Individual female pheromone glands were analyzed by capillary gas-liquid chromatography (GLC) for component ratio and amount. Reciprocal crosses were made among the three races, and 5-10 females were analyzed from each of 10 families from the F_1, F_2, paternal backcross, and maternal backcross progenies. The patterns found (Figure 3) fit the inheritance pattern expected for a single autosomal gene, with the two alleles exhibiting incomplete dominance (Klun and Maini, 1979; Roelofs et al., 1987). An in-depth analysis of the data (Bengtsson, unpublished) has demonstrated the occurrence of one (or more) genetic factors modifying the E/Z ratio produced by the females. Although there are many steps in the biosynthetic pathway for the production of the two acetate pheromone components, the autosomal factor could be controlling an isomer-specific enzyme in the last reduction step between the unsaturated 14-carbon precursor acids (Figure 2) and the corresponding acetates. An analysis of the fatty acyl intermediates from the parent populations and hybrids shows that the E/Z11-14:Acyl precursors are in the same 70:30 ratio in every case and that the specific E/Z ratios of the acetates must be effected in the reduction sequence.

B. Male Antennal Receptors for Pheromone

Electrophysiologic recordings from single olfactory sensilla on male ECB antennae were made with an antennal tip recording technique (Van Der Pers and den Otter, 1978). It was interesting to find that the parent populations

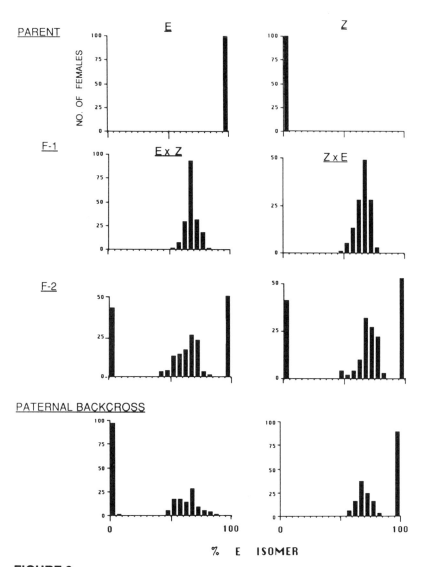

FIGURE 3 Data summaries for female pheromone blend production for parental ECB races and crosses. The maternal backcross data were similar to those of the reciprocal paternal backcross.

and the hybrids could be distinguished easily on the basis of the olfactory cell responses (Figure 4) (Hansson et al., 1987). Males of the Z races exhibited a high-amplitude spike response from an olfactory cell responding specifically to Z11-14:0Ac, whereas the minor pheromone component, E11-14:0Ac, elicited a low-amplitude spike from another olfactory cell. The second cell was specific for the minor component, but the amplitude was about 35% of the sum of the amplitudes of the Z and E cells. Males of the E race were distinguished by high-amplitude spike cells that responded specifically to the E isomer and a low-amplitude spike cell that specifically responded to the Z isomer. Males from all races possessed an additional third cell that responded with a very small amplitude spike to Z9-14:0Ac, a compound that has been found to be antagonistic to pheromone behavioral responses in ECB (Struble and Byers, 1987; Glover et al., 1989). Hybrid males were found to possess two olfactory cells that give intermediate spike amplitudes to either Z- or E11-14:0Ac.

The ability to distinguish easily the parent and hybrid males provided the opportunity to carry out a genetic analysis of this aspect of the communication system with the various crosses and backcrosses (Hansson et al., 1987). The study showed (Figure 4) that the distribution of males exhibiting the E, Z, and intermediate phenotypes was similar to that obtained in the previous analysis of female production and consistent with an inheritance scheme determined by a single autosomal gene with the two alleles showing incomplete dominance.

C. Linkage of Pheromone Production and Olfactory Cell Detection

The studies just described have shown that pheromone production in female ECB moths and pheromone detection on male ECB moth antennae are each controlled by an autosomal gene. An additional study (Löfstedt et al., 1989) was carried out to investigate the linkage relationship between these two autosomal factors. Males from the progeny obtained by backcrossing hybrid males with E females were analyzed by single-cell recordings to separate them by E type and hybrid antennae. Males with E-type antennae were mated with E females. If pheromone production is linked with the antennal cells, then all females in the progeny should be E producers. If there is no linkage, then one would expect the normal production distribution of 50% hybrid and 50% E females in the progeny. Female analyses of six families in this experiment showed that only one was all E females and the other five contained both E and hybrid females, indicating an absence of close linkage between genes controlling the pheromone component ratio and genes controlling olfactory cell detection.

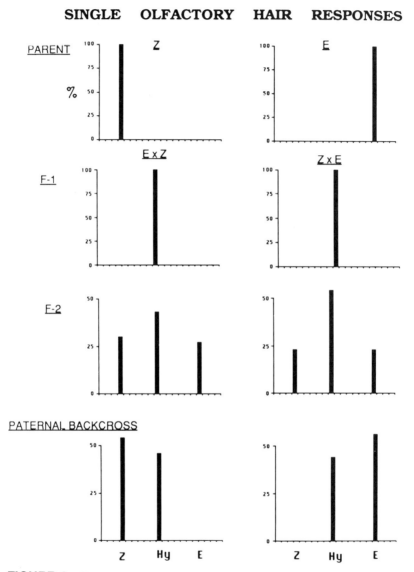

FIGURE 4 Data summaries of the male antennal olfactory cell response phenotypes for parental ECB races and crosses. The maternal backcross data were similar to those of the reciprocal paternal backcross.

D. Male Upwind Flight Responses

The upwind-oriented flight responses of male ECB moths can be assayed in a laboratory flight tunnel. With the appropriate light, temperature, airspeed, and humidity conditions, along with proper conditioning of the moths and release of pheromone, the assay can be very discriminating among the races (Glover et al., 1987). The behavioral analysis involves recording the responses of individual males flying to a particular pheromone blend. The elapsed time is also recorded as the male goes through the sequence of (1) wing fanning and walking, (2) taking flight, (3) orientated flight in the pheromone plume, (4) flight to within 30 cm of the source, (5) to within 5 cm of the source, (6) touching source, and (7) courtship display, including clasper extrusion with wings held vertically and abdomen waving slowly from side to side. Not only are the E and Z populations easily distinguished, but the assay is sensitive enough to discriminate between bivoltine (BZ) males and univoltine (UZ) males. Although males from both races respond extremely well to the natural blend containing 3% E isomer and very poorly to pure Z isomer, the males from the BZ race responded well (80% through flight and display) to a blend containing only 0.5% E isomer, whereas UZ males did not (17% display). The UZ males did respond well (73% display) to a 1% E isomer blend.

This discriminating flight-response assay provided the opportunity to carry out a genetic analysis of behavior with the various crosses and backcrosses of ECB. Individual males from the various progenies were flown to one of five different blends (30 µg on rubber septum), including 99:1, 65:35, 50:50, 35:65, and 3:97, until a total of 40 males from each progeny had flown to each treatment. The results show (Figure 5) that the parent populations are quite specific to their natural blend, although the E population has a broader pattern of response, with some males responding to the other blends, excluding the 3:97 E/Z blend. The males of both pheromone types typically did not even activate to the opposite pheromone blend. The profile of responses for hybrid males, which shows the percentage completing flights and touching the source, was very similar for all reciprocal crosses (data for both Z races are combined since the patterns appear to be very similar for all crosses and backcrosses). Surprisingly, the hybrid males responded equally to the blends containing from 65 to 3% E isomer, but they gave almost no response to the 99% E blend. These data indicate that hybrid males would be as responsive to the parent Z females as they are to the hybrid females.

The flight-response profiles for the F_2 and backcrossed males do not fit the patterns expected for control by an autosomal gene. The pattern for the (E ♀ × Z ♂) F_2 males is similar to that obtained by combining the hybrid profile with the parent Z profile, indicating that the progeny is composed of a 1:1 mix of hybrid and Z males, instead of the 1:2:1 composition expected

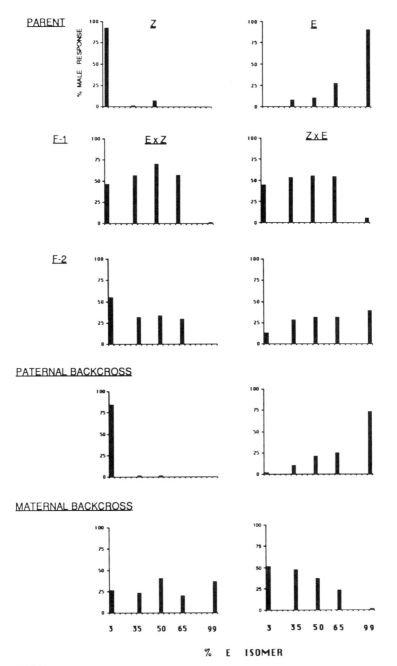

FIGURE 5 Data summaries for flight tunnel behavioral responses for parental ECB races and crosses. The profiles represent the percentages of males that flew upwind and touched the various pheromone sources.

for autosomal control. The response profile for F_2 males of the (Z♀ × E♂) cross, however, shows that this progeny is a 1:1 mix of hybrid and E males, again deviating from the 1:2:1 composition expected for autosomal control. The patterns are exactly those expected for sex-linked inheritance.

Control of the behavioral responses by a sex-linked gene was also confirmed by the response profiles of the paternal backcrosses. Calculation of the expected patterns using the sex determination system that predominates in Lepidoptera in which females are heterogametic (ZW) shows that the backcrossed males should all respond in the same way as the original male parent. In accordance with this prediction, the profile for the (E♀ × Z♂) paternal backcrossed males is the same as that for the Z populations, and the profile for the (Z♀ × E♂) paternal backcrossed males is the same as that for the E parents. The maternal backcrossed males exhibit response profiles that are composites of half-hybrid and half-maternal line responses, but this result is expected for both sex-linked and autosomal inheritance patterns. The good fit between the simple sex-linked chromosome model and the data suggests that there is minimum interaction of the major factor on the Z chromosome and modifying factors elsewhere in the genome. It is not known, however, if the major factor on the Z chromosome is a single gene or whether it consists of a set of closely linked genes. In any case, it is interesting to find that the upwind-oriented flight responses are controlled by a sex-linked gene that could not be linked to the autosomal gene controlling the male antennal pheromone-detecting olfactory cells.

E. TPI Enzyme, a Sex-linked Marker

In 1985, the three ECB races under culture were surveyed for allozyme differences using standard horizontal starch gel electrophoresis. At that time attempts were made to find allozyme differences among the cultures that could be utilized as genetic markers for investigations of inheritance patterns of the different pheromone communication system components. A total of 20 randomly chosen females and 20 randomly chosen males from each ECB culture were screened for 34 enzymes. Of the enzymes tested, 21 appeared to be monomorphic and of the remaining enzymes, only one exhibited sufficient differences in allele frequencies to distinguish among cultures. This one locus is the sex-linked gene for triose phosphate isomerase (TPI). Sex linkage was determined by the presence of a hybrid conditions for this gene only in the males. These males exhibited three separate bands for the TPI enzyme (Figure 6), indicating that the enzyme consisted of a randomly associating dimer. Since the males are the homogametic sex (ZZ), they can produce both forms of polypeptide and can therefore exhibit this hybrid phenotype. The heterogametic female ECB possess only a single copy of sex-linked genes. This copy is manifested as a single allozyme band on the starch gel.

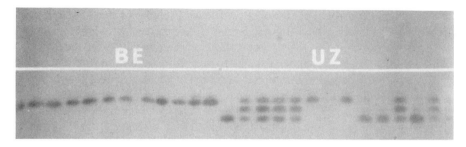

FIGURE 6 Analysis of the TPI gene by starch gel electrophoresis. Samples are individual male moths (12 each) from the bivoltine E race (BE) and from the univoltine Z (UZ) race. The BE race is fixed for allele; the UZ exhibits both homozygotes (1 and 2) as well as heterozygotes (three bands).

In 1985 the TPI banding patterns indicated that the UZ race had 98% allele 1 and 2% allele 2, whereas the BZ race had intermediate frequencies for both alleles and the BE race had 100% allele 2. Since the UZ and BE races are substantially different with respect to allele frequencies, it is anticipated that the TPI enzyme can be used as a sex-linked marker to confirm the composition of males in the flight tunnel-response profiles of the previous section (Figure 5). For this study, UZ females will be mated individually with BE males. The parent females will be analyzed for the TPI enzyme, and only families from females with allele 1 will be used. The hybrids will be used to generate F_2 and maternal backcross progeny to give behavioral response profiles that are either 1:1 E-hybrid males or Z-hybrid males, respectively (Figure 5). Individual males from these progenies will be assayed as before in the flight tunnel and all males preserved for TPI analysis. Assuming reasonably close linkage between the gene controlling male behavioral response and the TPI locus (both sex linked), it should be possible to assign various TPI genotypes and, therefore Z chromosome origin, to portions of the behavioral response profile. An interesting aspect of this project will be to determine the TPI genotype of nonresponders to each pheromone blend.

F. Recombinations of Olfactory Cell and Behavioral Response Genetic Factors

An examination of the different inheritance patterns exhibited for antennal olfactory cell responses and for male flight responses predicts the existence of "unusual" males that posses antennal receptors for one blend of pheromone components but exhibit upwind flight to a different blend. For example, in the F_2 populations, half the males respond as hybrids and half as Z males for the (E♀ × Z♂) cross and half as E males and half as hybrids in the (Z♀ × E♂)

cross (Figure 5). The profile (Figure 4) for male antennal olfactory cells, however, shows a 1:2:1 distribution of E-hybrid-Z males for the F_2 populations. Since the olfactory cells are controlled by an autosomal gene and the behavioral responses by a sex-linked gene, it is assumed that there is independent inheritance of these two genetic factors. Therefore, males responding to the various pheromone blends in the F_2 behavioral response profiles should have a 1:2:1 distribution of antennal types for each blend. If this is the case, then some males with an E type of antenna will be responding to the Z blend, and vice versa. Research on this subject will be carried out by assaying F_2 males in the flight tunnel to the five pheromone blends and analyzing each of them electrophysiologically for its antennal type.

The implications of these findings is that there is a control center in the central nervous system that is under different genetic control than the peripheral detection system, and the control center determines whether the male will respond to a particular pattern of impulses in the neural circuitry. It also implies that high or low spike amplitude in the peripheral cells may not be an important factor in the final analysis for motor output.

V. GENERALIZATIONS

Moth pheromone systems using specific mixtures of E and Z isomers have been found to be very tightly controlled and not subject to significant blend changes by high artificial selection pressure. It has also been found with the ECB moth that three different genetic factors control female pheromone blend, male pheromone-responding olfactory cells, and male upwind flight responses. This information suggests that changes in the pheromone system that could be important in the speciation process are probably made by major perturbations, rather than by gradual shifts to different blends. The existence of cryptic sibling species in New Zealand with quite different pheromone compositions in various geographic pockets seem to support this idea.

Speculation on this theme follows from the proposal of Carson (1975), who hypothesized that "genetic differences between newly evolved or closely related species consist principally of differences in obligatory internal genetic balances or coadaptations of the genetic system." The point is made that there are two systems of genetic variability in bisexual species: one that is "open" and the other "closed." Systems exhibiting open variability are subject to natural selection and random drift by the ordinary undirected forces of chance recombination. In the case of the ECB moths, the TPI alleles could be part of this open variability system. The genes involved in this open system would have a superficial or trivial effect on the speciation process.

In the closed system, however, a block of internally balanced genes is locked into strong epistases. Under normal circumstances, this system is not affected by gene flow. A speciation event involving these types of genetic factors, then, would not take place by the accumulation of slow shifts in gene frequencies but rather by a series of catastrophic, stochastic genetic events. The closed variability system must undergo an unusual forced reorganization. Carson suggests that a flush-crash-founder cycle could be important in this type of speciation process. The moth sex pheromone system appears to be a good candidate for this closed variability system. Any catastrophic event that forces a change in this system and yields survivors with a different mate recognition system should generate a new species. Perhaps the heat from volcanic actions in New Zealand played a role in the evolution of those cryptic leafroller species.

In accordance with this discussion, it seems logical that incipient species produced by significant changes in their sex pheromone systems would be defined better by the recognition concept of species (Paterson, 1985) than by one that involves isolation from other species (Mayr, 1963). In the recognition concept, sex pheromones are part of the specific mate recognition system, which is described as the central defining property of a species. "It is mate recognition which defines reproductive lineages, which are incidentally isolated from other lineages" (Lambert et al., 1987). This view contrasts with the isolation concept, in which "species are more unequivocally defined by their relation to nonconspecific populations ('isolation') than by the relation of conspecific individuals to each other. The decisive criterion is not the fertility of individuals, but the reproductive isolation of populations" (Mayr, 1963). The isolation concept, therefore, does not seem as pertinent in cases in which the mate recognition system is changed by a catastrophic effect because the effect on speciation should be independent of other species.

Although these concluding speculative remarks are made from a small amount of experimental data, they are made only to provoke more genetic research in this interesting area of moth sex pheromone systems, which are central to the reproductive process.

REFERENCES

Carson, H. L. (1975). The genetics of speciation at the diploid level. *Am. Naturalist* **109**:83-92.

Collins, R. D., and Cardé, R. T. (1985). Variation in and heritability of aspects of pheromone production in the pink bollworm moth, *Pectinophora gossypiella* (Lepidoptera: Gelechiidae). *Ann. Entomol. Soc. Am.* **78**:229-234.

Du, J.-W., Linn, C. E., Jr., and Roelofs, W. L. (1984). Artificial selection for new pheromone strains of redbanded leafroller moths, *Argyrotaenia velutinana*. *Contr. Shanghai Inst. Entomol.* **4**:21-30.

Du, J.-W., Löfstedt, C., and Löfqvist, J. (1987). Repeatability of pheromone emissions from individual female ermine moths, *Yponomeuta padellus* and *Yponomeuta rorellus*. *J. Chem. Ecol.* **13**:1431-1441.

Foster, S. P., and Dugdale, J. S. (1988). A comparison of morphological and sex pheromone differences in some New Zealand tortricinae moths. *Biochem. Syst. Ecol.* **16**:227-232.
Glover, T. J., Tang, X.-H., and Roelofs, W. L. (1987). Sex pheromone blend discrimination by male moths from E and Z strains of European corn borer. *J. Chem. Ecol.* **13**:143-151.
Glover, T. J., Perez, N., and Roelofs, W. L. (1989). Comparative analysis of sex pheromone-response antagonists in three races of European corn borer. *J. Chem. Ecol.* **15**:863-873.
Hansson, B. S., Löfstedt, C., and Roelofs, W. L. (1987). Mendelian inheritance of olfactory response to sex pheromone components in male European corn borer, *Ostrinia nubilalis*. *Naturwissenschaffen* **74**:497-499.
Hansson, B. S., Löfstedt, C., and Foster, S. P. (1989). Z-linked inheritance of male olfactory response to sex pheromone components in two species of tortricid moths, *Ctenopseustis obliquana* and *Ctenopseustis sp*. *Entomol. Exp. Appl.* **53**:137-145.
Horak, M., Whittle, C. P., Bellas, T. E., and Rumbo, E. R. (1988). Pheromone gland components of some Australian tortricids in relation to their taxonomy. *J. Chem. Ecol.* **14**:1163-1175.
Klun, J. A., and Maini, S. (1979). Genetic basis of an insect chemical communication system: The European corn borer. *Environ. Entomol.* **8**:423-426.
Klun, J. A., Chapman, D. L., Mattes, K. C., Wojtkowski, P. W., Beroza, M., and Sonnet, P. E. (1973). Insect sex pheromones: Minor amount of opposite geometrical isomer critical to attraction. *Science* **181**:661-663.
Kochansky, J., Cardé, R. T., Liebherr, J., and Roelofs, W. L. (1975). Sex pheromones of the European corn borer in New York. *J. Chem. Ecol.* **1**:225-231.
Lambert, D. M., Michaux, B., and White, C. S. (1987). Are species self-defining? *Syst. Zool.* **36**:196-205.
Löfstedt, C., Hansson, B. S., Roelofs, W. L., and Bengtsson, B. O. (1989). The linkage relationship between factors controlling female pheromone production and male pheromone response in the European corn borer, *Ostrinia nubilalis*. *Genetics* **123**:553-556.
Mayr, E. (1963). *Animal Species and Evolution*. Harvard University Press, Cambridge, Massachusetts, p. 20.
Miller, J. R., and Roelofs, W. L. (1980). Individual variation in sex pheromone component ratios in two populations of the redbanded leafroller moth, *Argyrotaenia velutinana*. *Environ. Entomol.* **9**:359-363.
Paterson, H. E. H. (1985). The recognition concept of species. In *Species and Speciation*, E. S. Vrba (Ed.). Transvaal Museum Monograph No. 4, Transvaal Museum, Pretoria, pp. 21-29.
Roelofs, W. L., and Brown, R. T. (1982). Pheromones and evolutionary relationships of tortricidae. *Annu. Rev. Ecol. Syst.* **13**:395-422.
Roelofs, W. L., and Wolf, W. A. (1988). Pheromone biosynthesis in the lepidoptera. *J. Chem. Ecol.* **14**:2017-2029.
Roelofs, W. L., Du. J.-W., Linn, C. E., Jr., Glover, T. J., and Bjostad, L. B. (1986). The potential for genetic manipulation of the redbanded leafroller moth sex phero-

mone gland. In *Evolutionary Genetics of Invertebrate Behavior*, M. D. Huettel (Ed.). Plenum, New York, pp. 263-272.

Roelofs, W. L., Glover, T. J., Tang, X.-H., Sreng, I., Robbins, P., Eckenrode, C. J., Löfstedt, C., Hansson, B. S., and Bengtsson, B. O. (1987). Sex pheromone production and perception in European corn borer moths requires both autosomal and sex-linked genes. *Proc. Natl. Acad. Sci. USA* **84**:7585-7589.

Sreng, I., Glover, T., and Roelofs, W. L. (1989). Canalization of the redbanded leafroller moth sex pheromone blend. *Arch. Insect Biochem. Physiol.* **10**:73-82.

Struble, D. L., and Byers, J. R. (1987). Sex pheromone components of an Alberta population of European corn borer, *Ostrinia nubilalis* (Hbn.) (Lepidoptera: Pyralidae). *Can. Entomol.* **119**:291-299.

Thornhill, R. (1979). Male and female sexual selection and the evolution of mating strategies in insects. In *Sexual Selection and Reproductive Competition in Insects*, M. S. Blum and N. A. Blum (Eds.). Academic Press, New York, pp. 81-121.

Thornhill, R., and Alcock, J. (1983). *The Evolution of Insect Mating Systems*. Harvard University Press, Cambridge, Massachusetts.

Van Der Pers, J. N. C., and den Otter, C. J. (1978). Single cell responses from olfactory receptors of small ermine moths to sex attractants. *J. Insect Physiol.* **24**:337-343.

10
Genetic Alteration of the Multiple Taste Receptor Sites for Sugars in *Drosophila*

Teiichi Tanimura

Kyushu University,, Fukuoka, Japan

I. INTRODUCTION

Recent molecular studies of vision are increasing our understanding of the molecules involved in the visual excitation and the subsequent transduction pathways. In contrast, the molecular mechanisms of chemoreception remain unsolved in both vertebrates and invertebrates. Although earlier studies reported the isolation and characterization of the sweet-sensitive protein from bovine and rat taste buds (Dastoli and Price, 1966; Hiji and Sato, 1973), the direct evidence for their real function as the taste receptor is lacking (Cagan and Morris, 1979). There are still arguments about whether the taste reception is mediated by a specific receptor protein (Teeter et al., 1987). Since specific inhibitors or antagonists for the taste receptors are not available, it is difficult to isolate the putative receptor molecule solely using biochemical methods. Genetic approaches should provide an alternative way to solve this problem. If a gene coding proteins involved in the taste receptor functions could be identified, the molecular cloning of the gene would clarify the nature of the receptor substance.

The fruit fly, *Drosophila melanogaster*, is a favorable material to genetically dissect a variety of neurobiologic problems (Hall, 1985). The method of inducing mutations is established. If an appropriate screening method is available, mutants can be readily isolated from thousands of flies. Genetic and cytologic analysis allow us to finely locate the genes on the chromosomes.

The presence of the polytene chromosomes in *Drosophila* enable us to map the absolute location of genes in chromosomal bands. Recent progress in the gene-cloning methodology in *Drosophila* also make this fly an excellent material to study gene expression at the molecular level. Particularly, P element mediated germline transformation allows us to reintroduce a piece of foreign DNA into the genome, and thus the function of a gene can be tested *in vivo* (Rubin and Spradling, 1982). Using these genetic techniques, mutants for membrane excitability, the circadian rhythm, and learning have been isolated and investigated at a molecular level. These approaches can also be applied to study the nature of the taste receptor. This chapter summarizes the current knowledge of the multiple receptor sites for sugars in *Drosophila*, mainly obtained through genetic analyses.

II. TASTE RECEPTORS IN *DROSOPHILA*

Much of the study of insect chemoreception has been done in larger flies, the blowfly and the flesh fly, by Dethier and his coworkers (Dethier, 1976) and also by the Morita's group (Morita and Shiraishi, 1985). The basic structural and functional organization of chemoreceptors does not differ between in the larger flies and *Drosophila* (Falk et al., 1976). Chemosensilla are located on the tarsi and the labellum. The interpseudotracheal papillar receptors are located inside the labellar lobes. A labellar chemosensillum usually contains four sensory cells. Electrophysiologic experiments showed that each sensory cell specifically responds to sugar, salt, or water. The sensory cells that respond to sugars are called the sugar receptor cells. In *Drosophila*, effective stimulants for the sugar receptor cell are monosaccharides, such as glucose, fructose, and di- or trisaccharides. In the fleshfly, some amino acids and fatty acids stimulate the sugar receptor cells (Shimada, 1987); in *Drosophila* these substances are not effective stimulants. Sensory cells send axons to the central nervous system without synapses. This feature makes the fly excellent material for electrophysiologic studies. By stimulating a single chemosensillum using a tip recording or a side wall recording method, we can record the nerve response originating from a single sugar receptor cell (Morita and Yamashita, 1959; Fujishiro et al., 1984). Since a sugar solution stimulates both the water and the sugar cells, two kinds of nerve impulses are elicited. The size and the shape of impulses originating from the two sensory cells are different, and the concentration-response relationship also differs. Two kinds of impulses can be discriminated.

III. MULTIPLE RECEPTOR SITES FOR SUGARS

To identify the putative taste receptor molecule, we should know whether a single species of receptor molecules is involved in all kinds of sweet substances

GENETIC ALTERATION OF THE MULTIPLE TASTE RECEPTOR SITES 127

or separate multiple receptors are involved in a given group of sugars. In insects, the presence of multiple sugar receptor sites has been well documented (Shimada, 1987). Pharmacologic treatments of a chemosensillum with chemicals or enzymes that modify proteins has differential effects on the nerve response to sugars. In the flesh fly, four kinds of sugar receptor sites are hypothesized: the G site for pyranose and pyranosides, the F site for furanoses, the Ar site for aromatic amino acids and dipeptides, and the R site for aliphatic amino acids and fatty acids (Shimada, 1987). In *Drosophila*, experiments using protease treatment showed separation of the receptor sites for furanose and for pyranose and pyranosides (Tanimura and Shimada, 1981). Treatment of a labellar chemosensillum with papain or trypsin specifically eliminated the nerve response to fructose without affecting the responses to other sugars. When the concentration-response curves for glucose and fructose were recorded before and after protease treatment, the response to fructose was completely depressed but the concentration-response relationship of glucose did not change after treatment. Since the elimination occurred using trypsin with a strict substrate specificity, limited proteolysis may occur on proteins involved in fructose reception. These results clearly showed that separate proteins are involved in the recognition of furanose and pyranoses. Treatment with a higher concentration of trypsin eliminated the response to all sugars, but the water and the salt receptor cells are unaffected, showing the proteinaceous nature of all the sugar receptor sites. *Drosophila* chemosensilla seem to be more susceptible to modification treatment than flesh fly chemosensilla. Further pharmacologic experiments in *Drosophila* might reveal the organization of multiple receptor sites for sugars.

IV. GENETIC VARIATION OF THE TASTE SENSITIVITY TO TREHALOSE

Our first observation is that the commonly used laboratory wild-type strains, Canton-S and Oregon-R, respond differently to trehalose (Tanimura et al., 1982). A convenient behavioral assay to measure taste sensitivity was developed and used to measure the sensitivity difference. Agar gels containing two kinds of sugar solutions, each colored with a different food color, were alternately placed in wells of the microtiter plate. Starved flies were introduced into the plate and allowed to feed. The ingested sugar solution in the crop and the gut of the flies can be easily observed through the body wall. In the two-choice preference test between 20 mM trehalose and 2 mM sucrose, Canton-S preferred trehalose and Oregon-R preferred sucrose. Figure 1 shows the sensitivity curves for trehalose. Two-choice preference tests were performed at different concentrations of trehalose. In Canton-S, at the trehalose concentration of 30 mM, the preference completely turned to the trehalose side; in Oregon-R the concentration of trehalose must be raised to over 200 mM to

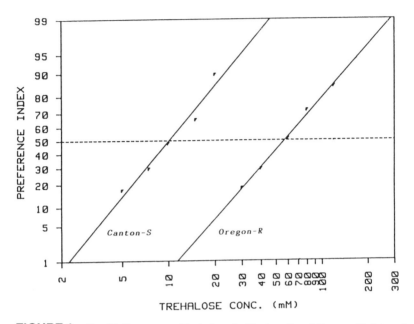

FIGURE 1 Sensitivity curves of trehalose in Canton-S and Oregon-R determined by the two-choice preference tests at five different concentrations of trehalose; 2 mM sucrose was used as a control. Preference index values for trehalose were plotted against the logarithmic concentration of trehalose.

turn the preference. Most of the laboratory strains show either high sensitivity or low sensitivity to trehalose. We surveyed isofemale lines derived from flies collected in the natural population. The results indicated that the dimorphism exists in wild populations (Tanimura, unpublished results).

The strain difference observed is due to the difference in the taste sensitivity to trehalose, not in that to sucrose. When the amount of intake of sucrose, trehalose, glucose, and fructose was measured using a colorimetric method after feeding with the sugar solution mixed with a food color, a difference was observed only in the amount of trehalose between the two strains (Tanimura et al., 1982). Electrophysiologic recordings from the labellar and tarsal chemosensilla also shown that the nerve response to trehalose markedly decreased in the low-sensitivity strain. These results indicate that the mutation alters the sensitivity of the receptors for trehalose located on both the labellum and the tarsi in a similar way. Since the sensitivity to glucose, fructose, sucrose, and other pyranoses did not change between the high- and low-sensitivity strains, the sensitivity difference must be in the receptor site for trehalose in the sugar receptor cell. We do not yet have enough data on the second messengers involved in the transduction mechanisms in insects. If we assume that there are separate second messenger pathways for sugars, there remains a possibility that the difference in the trehalose sensitivity is in the second messenger system specific to trehalose.

V. *Tre* GENE CONTROLS THE DIFFERENTIAL TASTE SENSITIVITY TO TREHALOSE

The gene controlling the sensitivity to trehalose was mapped at 13.6 on the X chromosome by recombination analysis (Tanimura et al., 1982). The gene was designated *Tre*. The high-sensitivity allele is designated *Tre*$^+$ and the low-sensitivity allele as *Tre*. Using deficiency and translocation chromosomes, the *Tre* gene was cytologically located between 5A10 and 5B1-2 on the X chromosome (Tanimura et al., 1988).

Low-sensitivity mutants can be isolated through mutagenesis of the high-sensitivity strain Canton-S with ethylmethanesulfonate. Several low-sensitivity mutants were isolated. Genetic analysis showed that the *Tre* gene is also involved in the decrease in trehalose sensitivity. This result shows that *Tre* is the only gene on the X chromosome controlling trehalose sensitivity. Electrophysiologic studies indicated that *Tre* gene is likely to be functioning in the primary step of chemoreception. If the *Tre* gene controls trehalose sensitivity, we can expect that a mutation within the *Tre* gene results in a decrease in sensitivity to a different degree. Through mutagenesis of low-sensitivity flies, an allele designated *Tre*^{222} was recovered. This mutant shows a super low sensitivity to trehalose but is not totally taste blind to trehalose. These results suggest that the *Tre* gene codes a protein involved in the recognition of trehalose in the sugar receptor cell.

VI. GENE DOSAGE ALTERS TREHALOSE SENSITIVITY

In *Drosophila*, the gene dosage of most parts of the genome can be manipulated using segmental aneuploidy techniques. We examined the effects of gene dosage alteration of *Tre* alleles on the sensitivity to trehalose (Tanimura et al., 1988). In *Drosophila*, a female has two X chromosomes and a male has one X chromosome. Each X chromosome in a female produces half the total gene product produced by the single X chromosome in a male. This phenomenon is called "dosage compensation" (Baker and Belote, 1983). When a small region of the X chromosome is duplicated in one of the X chromosomes in a female, the genes in the duplicated region produced 150% products, if we assume that a normal female or male produces 100% products. In a male, 200% products are produced in a duplicated region. Most enzymatic proteins in *Drosophila* are known to be produced according to the dose of the coding genes (O'Brien and MacIntyre, 1978; Stewart and Merriam, 1974).

Knowing the cytologic location of the *Tre* gene, we can manipulate the number of gene dosages of the *Tre* gene using deletion and translocation chromosomes and examine how the composition of *Tre* alleles affects the sensitivity. Table 1 shows the expected amount of gene product produced in the euploids and aneuploids with a different dosage and composition of *Tre* alleles. To measure the taste sensitivity to trehalose, two-choice behavioral tests were done between 2 mM sucrose and three different concentrations of trehalose. Figure 2 summarizes the results. Trehalose sensitivity is represented by the concentrations of trehalose that have stimulating effect equivalent to that of 2 mM sucrose. For simplicity, the *Tre* and *Tre*$^+$ alleles are shown here as *H* and *L*, respectively. *L/H* heterozygous females showed intermediate sensitivity compared to the respective homozygotes. Since the trehalose sensitivity of *H/L* heterozygous flies did not differ significantly from that of females with a haploid dosage of *Tre*, *Tre* does not seem to inhibit or interfere with the function of the *Tre*$^+$ allele. This result implies that *Tre* could

TABLE 1 Gene Dosages in Euploid and Aneuploid Flies [a]

		Compensated gene dosage of	
		Tre^+	Tre
	Genotype	(H)	(L)
Females			
Euploid	H/H	100	0
	L/L	0	100
Deletion	H/O	50	0
	L/O	0	50
Duplication	H/H/H	150	0
	H/H/L	100	50
	H/L/L	50	100
	L/L/L	0	150
Males			
Euploid	H/Y	100	0
	L/Y	0	100
Duplication	H/H/Y	200	0
	L/L/Y	0	200

[a] Compensated gene dosage: the relative gene activity, assuming that the total gene activity of a normal diploid genome is 100. *H* and *L* represent the *Tre*$^+$ and the *Tre* alleles, respectively.

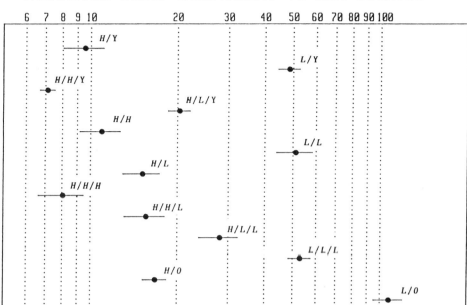

FIGURE 2 Effects of the gene dosage alterations and the composition of *Tre* on trehalose sensitivity. See Table 1 for explanation of abbreviation.

be an amorphic allele; that is, it is genetically inert. However, females with one dose of the *Tre* allele showed even lower sensitivity than euploid L/L females. The *Tre* allele therefore contributes in some way to trehalose sensitivity. Consequently, one would expect $H/L/Y$ duplicated males to be at least as sensitive as normal males. The results, however, show that the duplicated males are half as sensitive. Assuming that $H/L/Y$ duplicated males produce twice as much of the Tre^+ gene product as normal males, the fact that their sensitivity did not increase significantly suggests that there is a limit to the amount of product the receptor membrane can accommodate. The failure of trehalose sensitivity to respond to increasing gene dosages is also observed in cases of duplication-bearing females of Tre^+ and those of *Tre*. That H/L duplicated males showed a similar sensitivity to L/H females suggests that the products of the two alleles compete for a limited number of sites on the receptor membrane. $H/H/L$ females exhibited a lower sensitivity to trehalose than did H/H females, despite there being the same gene dosage

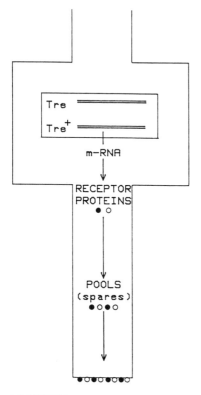

FIGURE 3 A model for the synthesis, storage, and integration of the putative receptor molecule for trehalose. This model assumes that there is no preference for the integration of products of the *Tre* and *Tre*$^+$ alleles and that there is a limit to the number of receptor molecules that the receptor membrane can accommodate.

of *Tre*$^+$ alleles in each of these two genotypes. If there is no upper limit to the incorporation of the gene products, these flies should have a sensitivity comparable to that of euploid *Tre*$^+$ flies. Females of the genotype $H/L/L$ also showed lower sensitivity than females of the genotype H/L. The differences in the sensitivity of the various duplication-bearing females suggest that there is no preference for incorporation of *Tre* and *Tre*$^+$ products into the receptor membrane. In females of the genotype $H/L/L$, for example, the total amount of both products incorporated is the same as in the euploid female, and the ratio of *Tre* and *Tre*$^+$ products in the membrane would be 2:1. Since the ratio of defective to functional molecules is more than H/L flies (1:1) and less than in L/L flies (2:0), they should show a sensitivity intermediate between those of the two euploid genotypes. These results further support the view that there is a limit to the number of molecules that the receptor membrane can accommodate and that Tre^+ and Tre products are equally incorporated.

Figure 3 is a model of the production and integration of the receptor molecule. We assume that there is a pool for the spare receptor molecules in the taste cells. We found that the response eliminated by papain treatment gradually recovered in 2 h (Tanimura and Shimada, 1981). When the elimination is not complete, the recovery time is much shorter. These results suggest that there are spare receptor molecules in the receptor cell that can immediately replace the damaged receptor. The model we present here supposes that the Tre gene product is incorporated into a limited number of sites in the receptor cell membrane. The Tre^+ and Tre products are equally able to be integrated and therefore compete for sites. We were measuring behaviorally the activity of the intact receptor system. Then, even if the Tre-encoded products are expressed depending on enhanced gene dosage, only those molecules incorporated into the membrane should contribute to the activity we measured.

Our results suggest that the Tre gene is a structural gene for the receptor molecule of trehalose and that different Tre alleles may code the receptor proteins with different affinities to trehalose. Direct evidence that the Tre locus is the structural gene for a trehalose receptor protein has not been demonstrated. However, it should be possible to identify and isolate the Tre gene by microcloning dissected polytene chromosomes or by P element transposon tagging. Molecular techniques can then be applied to characterize this putative taste receptor molecule.

VII. MUTATIONS AFFECTING THE PYRANOSE SITE

So far two kinds of mutants have been reported in which the responses to pyranose sugars are reduced. An autosomal mutant was isolated (Isono and Kikuchi, 1974), but further genetic study was not done. Rodrigues and Siddiqi reported *gustA* mutants on the X chromosome in which the response to pyranose and pyranosides is drastically depressed (Rodrigues and Siddiqi, 1981). However, further studies on the two mutants have not been published.

I have screened mutants with a reduced sensitivity to glucose. The two-choice test was done with a combination of glucose and trehalose, since the receptor sites for these two sugars should be separate. From about 1000 mutagenized autosomes, several glucose low-sensitive mutants have been isolated. Complementation tests revealed that all the mutations belong to a single complementation group. The gene controlling the glucose sensitivity was mapped on the third chromosome. Electrophysiologic analysis showed that the nerve responses to glucose and sucrose are reduced. These results provide genetic evidence for the separation of the receptor site for furanose and for pyranose as shown in the protease treatment experiments.

VIII. SUMMARY AND PERSPECTIVES

Genetic analysis of the taste receptor in *Drosophila* showed the presence of three kinds of receptor sites for sugars. Additional isolations of mutants affecting taste receptor functions are necessary. One noticeable fact in these studies is that no mutants completely abolish the sensitivity to sugars belonging to one receptor site. We can speculate about the reason. One possibility is that one kind of sugar may react with more than one kind of receptor site. The second possibility is that a receptor site is composed of multiple receptors with different affinities to sugar. These possibilities can be ascertained by examining a double mutant affecting taste receptor sites.

To reveal the nature of the taste receptor molecule, cloning of the gene should provide conclusive evidence. Such work would be possible through studies of taste mutants of *Drosophila*. Do studies of *Drosophila* contribute to our knowledge about mammalian systems? Although *Drosophila* and mammals belong to divergent phylogenies, recent molecular studies prove that many proteins are homologous. In this regard we suspect that studies in *Drosophila* could provide a basic principle common to both mammals and the fly.

REFERENCES

Baker, B. S., and Belote, J. M. (1983). Sex determination and dosage compensation in *Drosophila melanogaster*. *Annu. Rev. Genet.* **17**:345-393.

Cagan, R. H., and Morris, R. W. (1979). Biochemical studies of taste sensation: Binding to taste tissue of ^{3}H-labeled monellin, a sweet-tasting protein. *Proc. Natl. Acad. Sci. USA* **76**:1692-1996.

Dastoli, F. R., and Price, S. (1966). Sweet-sensitive protein from bovine taste buds: Isolation and assay. *Science* **154**:905-908.

Dethier, V. G. (1976). *The Hungry Fly*. Harvard University Press, Cambridge, Massachusetts.

Falk, R., Bleiser-Avivi, N., and Atidia, J. (1976). Labellar taste organs of *Drosophila melanogaster*. *J. Morphol.* **150**:327-342.

Fujishiro, N., Kijima, H., and Morita, H. (1984). Impulse frequency and action potential amplitude in labellar chemosensory neurons of *Drosophila melanogaster*. *J. Insect Physiol.* **30**:317-325.

Hall, J. C. (1985). Genetic analysis of behavior in insects. In *Comprehensive Insect Physiology Biochemistry and Pharmacology*, Vol. 9, G. A. Kerkut and L. I. Gilbert (Eds.). Pergamon Press, Oxford, pp. 287-373.

Hiji, Y., and Sato, M. (1973). Isolation of the sugar-binding protein from rat taste buds. *Nature (New Biol.)* **244**:91-93.

Isono, K., and Kikuchi, T. (1974). Autosomal recessive mutation in sugar response of *Drosophila*. *Nature* **248**:243-244.

Morita, H., and Shiraishi, A. (1985). Chemoreception physiology. In *Comprehensive Insect Physiology Biochemistry and Pharmacology*, Vol. 6, G. A. Kerkut and L. I. Gilbert (Eds.). Pergamon Press, Oxford, pp. 133-170.

Morita, H., and Yamashita, S. (1959). Generator potential of insect chemoreceptor. *Science* **130**:922-923.

O'Brien, S. G., and MacIntyre, R. J. (1978). Genetics and biochemistry of enzymes and specific proteins of *Drosophila*. In *Genetics and Biology of Drosophila*, Vol. 2a, M. Ashburner and T. R. F. Wright (Eds.). Academic Press, New York, pp. 396-551.

Rodrigues, V., and Siddiqi, O. (1981). A gustatory mutant of *Drosophila* defective in pyranose receptors. *Mol. Gen. Genet.* **181**:406-408.

Rubin, G. M., and Spradling, A. C. (1982). Genetic transformation of *Drosophila* with transposable element vectors. *Science* **218**:348-353.

Shimada, I. (1987). Stereospecificity of the multiple receptor sites in the sugar receptor cell of the fleshfly. *Chem. Senses* **12**:235-244.

Stewart, B. R., and Merriam, J. R. (1974). Segmental aneuploidy and enzyme activity as a method for cytogenetic localization in *Drosophila melanogaster*. *Genetics* **76**:301-309.

Tanimura, T., and Shimada, I. (1981). Multiple receptor proteins for sweet taste in *Drosophila* discriminated by papain treatment. *J. Comp. Physiol.* **141**:265-269.

Tanimura, T., Isono, K., Takamura, T., and Shimada, I. (1982). Genetic dimorphism in the taste sensitivity to trehalose in *Drosophila melanogaster*. *J. Comp. Physiol.* **147**:433-437.

Tanimura, T., Isono, K., and Yamamoto, M.-T. (1988). Taste sensitivity to trehalose and its alteration by gene dosage in *Drosophila melanogaster*. *Genetics* **119**:399-406.

Teeter, J., Funakoshi, M., Kurihara, K., Roper, S., Sato, T., and Tonosaki, K. (1987). Generation of the taste cell potential. *Chem. Senses* **12**:217-234.

11
A Possible Mechanism of the High Differential Sensitivity of Taste in *Drosophila*

Ichiro Shimada

Tohoku University, Sendai, Japan

I. INTRODUCTION

Every animal is always exposed to continuous fluctuations of the environment, and the response of each sensory cell is usually accompanied by a considerable, inherent variability. How can the animal, then, gain reliable information? Not only qualitative but also quantitative discrimination of stimuli is indispensable for survival. Quantitative discrimination involves two types of thresholds: absolute and differential (the resolving power). Surprisingly, the differential threshold ΔI has scarcely been studied in invertebrates (see Dethier and Bowdan, 1984; Dethier and Rhoades, 1954; Richter and Campbell, 1940; Frings, 1946; von Frisch, 1934).

Recent studies of differential taste thresholds in the blowfly in response to sucrose have been conducted at both the behavioral and neurophysiologic levels (Detheir and Bowdan, 1984), the lowest values (expressed as the Weber fraction $\Delta I/I$) of which were 0.1 and 0.06, respectively. Based on the close correspondance between the results from these two methods, it was suggested that the coding mechanism for differential thresholds in the fly derives directly from the mechanism of receptor transduction and that no complex intervening mechanism is involved.

Previously, Smith et al. (1983) first evaluated the capability of information transmission for sucrose concentration in a tarsal sugar receptor of the blowfly. According to their results, a single sugar receptor can encode the sucrose concentration for at most 5.3 ($= 2^{2.4}$, or 2.4 bits) discriminable levels within the stimulus range from 10 to 1000 mmol/liter. This value, suggesting a low differential sensitivity for a single sugar receptor, is contrasted by the high sensitivity in the feeding behavior of the blowfly (Dethier and Bowdan, 1984).

Here we report a significantly lower value for the Weber fraction, 0.025, in *Drosophila melanogaster*. This value was obtained by colorimetrically determining the amount of intake after homogenization of 30-40 flies, following ingestion of sucrose solution containing dye (Shimada et al., 1987). A simple model is proposed in which a single central nervous system (CNS) neuron integrates the input from many peripheral receptors, suggesting the importance of the central nervous system in feeding behavior. According to our numerical results (Nakao et al., 1987), the model can provide sufficient discrimination at the behavioral level by integrating the responses of a realistic number of receptors and so explain the apparent contradiction with the results at the receptor level.

II. FEEDING PROCEDURES

The fruit fly, *D. melanogaster*, was reared on a sucrose-cornmeal-yeast medium at 25 °C under constant illumination. Wild-type (Canton-S) flies were used 1 day after eclosion. Before the tests they were starved, usually for 20 h unless otherwise stated, but supplied with distilled water.

The blue food dye, brilliant blue FCF ($C_{37}H_{34}O_9SNa$), was used as an internal marker of the intake at a concentration of 0.125 mg/ml. The red food dye, acid red, was not used because it was shown to inhibit ingestion slightly (data not shown). Tests were carried out using microtest plates with 60 small wells (10 μl each, Nunc, Denmark) filled with sucrose solution, mixed with the food dye and made up in 1% agar (Difco, Noble). Male flies (30-40) were usually introduced onto the plate unless otherwise stated. After being left in the dark at 25 °C to feed on the sucrose-dye mixture for 1 h at a fixed time of day (4-6 p.m.), the flies were sacrificed by freezing at -80 °C. For convenience, in this chapter the various concentrations and combinations of sucrose solutions are abbreviated: 10S, all 60 wells in each microtest plate are filled with 10 mmol/liter of sucrose; 10S-5S, two-choice tests in which alternate wells are filled with 10 and 5 mmol/liter of sucrose, and so on. Sucrose intake was determined colorimetrically. The flies were homogenized at 25,000 rpm for 30 s in 1.5 ml of 100 mmol/liter of potassium phosphate buffer, pH 7.2, containing 50% ethanol (Physcotron Homogenizer, Niti-on Medical & Physical Instruments Mfg. Co., Ltd.). After centrifugation at 12,000 $\times g$ for 30 minutes, the absorbance of the supernatant was measured at 630 nm (the absorbance maximum of brilliant blue). The absolute amount

POSSIBLE MECHANISM OF SENSITIVITY OF TASTE IN DROSOPHILA

of intake was calculated from the calibration curves of the dye after subtracting the background absorbance of the fly homogenate alone.

Two-choice tests were performed using the microtest plates symmetrically filled with sucrose solutions of differing concentrations (Tanimura et al., 1982). One concentration solution was marked with the blue food dye. This simple quantitative method confines the sensory inputs of feeding behavior to taste alone.

III. CHOICE

On average, each fly ingested 0.11 µl of 10 mmol/liter sucrose per h for 10S, but only 0.07 µl of 5 mmol/liter of sucrose was taken in the 5S condition (these values are the means of 13 experiments). The extent of ingestion of sucrose solution clearly depends on concentration, that is, taste intensity. Although the flies evidently preferred 10S to 5S, they also ingested 5S to a considerable extent. In the 10S-5S situation, however, the 5 mmol/liter sucrose wells were almost ignored. In the three-choice 10S-7.5S-5S, the highest concentration of sucrose (10 mmol/liter) was exclusively preferred, with the ingested volume of the other two solutions significantly smaller than for the 7.5S or 5S tests. This clear choice behavior suggests that the CNS may play an important role in feeding, but there are various problems concerning the mechanism involved. To what extent can the fly discriminate between different concentrations of sucrose solution? What factors affect discrimination ability? Where does discrimination actually take place, at the level of the central nervous system or at the peripheral sensory level?

IV. DISCRIMINABILITY

The discriminability of *D. melanogaster* was evaluated at the standard concentration of 10 mmol/liter which lies just within the dynamic range of the labellar sugar receptor of the fruitfly: 10-1000 mmol/liter (Fujishiro et al., 1984). The Weber fraction $\Delta I/I$ for fluid intake I was obtained by setting up a concentration pair I and $I + \Delta I$ and gradually closing the gap (i.e., decreasing ΔI) until there was no significant difference in fluid intake (see Dethier and Rhoades, 1954). The difference in the ingested volumes between 10.25 and 10.00 mmol/liter was statistically significant according to Wilcoxon's test ($P < 0.02$, $N = 8$, 30-40 flies for each experiment). Using the same procedure, discriminability was evaluated at other standard concentrations. The ability of the flies to discriminate, however, was highest at 10 mmol/liter standard concentration, as shown in Figure 1. The preference index is expressed as the ratio $(V_h - V_l)/(V_h + V_l)$, where V_h and V_l indicate the ingested volumes of sucrose solutions of high and low concentrations, respectively. The ratio

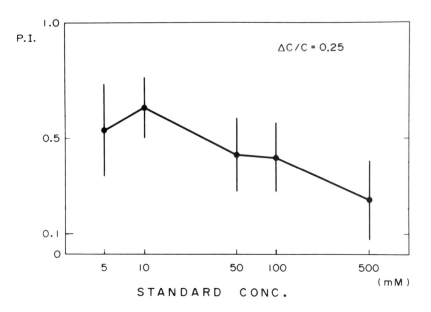

FIGURE 1 Preference index (PI, defined in the text) at each standard concentration with the constant ratio $\Delta C/C$ of 0.25 (mean values for eight experiments, with bars representing ± SD.

of concentrations $(C + \Delta C)/C$ was fixed at 1.25. The difference between the largest value of the PI at 10 mmol/liter and those at other concentrations was statistically significant ($P < 0.02$). This is one order of magnitude different from the concentration of highest sensitivity for the blowfly (100 mmol/liter) (Dethier and Bowdan, 1984; Dethier and Rhoades, 1954). The value for $\Delta I/I$ of 0.025 [i.e., $(10.25 - 10)/10$] reported here is the lowest so far for insects.

V. FACTORS AFFECTING DISCRIMINABILITY

The significant difference between the present data and those of Dethier and Bowdan (1984) on the feeding behavior of blowflies seems partly due to species differences, since the response characteristics of a fruit fly sugar receptor shows a dynamic range of 2 log units but that of the blowfly is extended for 3 log units (according to the electrophysiologic results of Fujishiro et al., 1984, and Shiraishi and Tanabe, 1974). The main reason for the difference, however, is probably due to differences in experimental procedure. There are at least five such differences between the present study and that of Dethier and Bowdan (1984).

First, with regard to the feeding times, various hormonal, metabolic, and environmental changes may more heavily affect the ingested volume during a feeding time of 24 h, for example in two-choice tests on *Phormia* (Dethier and Bowdan, 1984), compared with 1 h in our experiment for *D. melanogaster*. According to our preliminary experiments, the time course of ingested volume (for 10S) increases almost linearly up to 1 h. Within 30 minutes, the concentration-ingestion curves in a series of single drinks resemble the dose-response curves of the labellar sugar receptor of *Drosophila* in a similar dynamic range of 10-1000 mmol/liter (Fujishiro et al., 1984). The ingested volumes therefore show very close correspondence to the neural responses when measured over the much shorter period of 1 h.

A second difference in technique is in the distribution of sucrose solutions. In the *Phormia* experiments there were only two capillary tubes containing sucrose solutions, arranged in a Plexiglas cylinder, in contrast to our preparation, in which 60 wells are set in a microtest plate. Discrimination clearly depends on the distribution of sucrose solutions, as shown in Figure 2: the sparser the distribution of foods (sucrose solutions), the poorer the discrimination results.

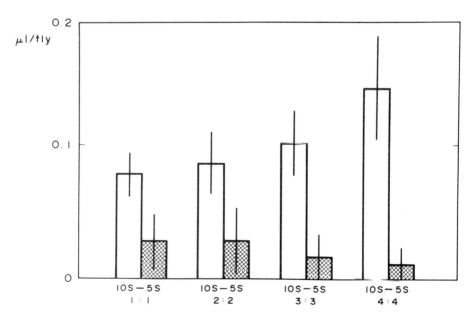

FIGURE 2 The effect of the distribution of sucrose solutions on ingestion choice. Open and shadowed blocks show results for ingested volumes of 10S and 5S, respectively, for different distributions. The 1:1, for example, is a sparse distribution in which only 2 of 60 wells are filled with 10S and 5S, leaving the other 58 wells filled with water.

A third difference is in the method of measuring the ingested volume. Although this method does not need any special skills, we have confirmed that it provides reasonable reproducibility. Besides, our procedure confines the sensory input to taste alone. Olfactory input can be excluded as far as the nonvolatile sugar solutions are concerned, and visual input can be neglected because the flies are left to feed in the dark. Mechanical input can also be neglected since all the wells are exactly the same shape and therefore indistinguishable.

Fourth, there is the effect of starvation. Our flies were usually deprived of food for 20 h before the 1 h experiment, but the experiment with *Phormia* was performed over 24 h after 48 h deprivation (Dethier and Bowdan, 1984). Starvation markedly affected the discrimination sensitivity in two-choice tests with a smaller difference between sucrose concentrations, as shown in Figure 3. However, starvation level is not significant with large differences in concentrations, such as 10S-5S (see Shimada et al., 1987; Campbell, 1958). Complete starvation (0S) enhanced discrimination significantly compared with the various hunger levels 10S, 30S, and 50S.

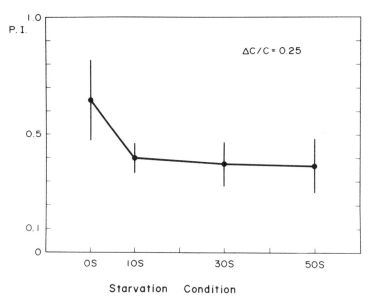

FIGURE 3 The effect of hunger on PI. A 100 mmol/liter sucrose solution is usually supplied when the fruit fly is reared on sucrose solution alone. 0S, 10S, 30S, and 50S on the abscissa indicate sucrose concentrations of 0 mmol/liter (i.e., complete starvation) and 10, 30, and 50 mmol/liter applied to the flies during starvation period (20 h). The PI was determined at the standard concentration of 200 mmol/liter (mean ± SD).

The last difference is in the method of determining the Weber fraction. According to Dethier and Bowdan (1984), ΔI and the Weber fraction $\Delta I/I$ for each standard concentration I were calculated from the equation $\Delta I =$ (75% concentration - 25% concentration)/2, where 75% concentration means the concentration at which 75% of the flies consumed more than the median of the standard and, similarly, the concentration at which 25% consumed more, were calculated. These percentage values are rather arbitrary: a value of 50% is commonly used in measurements of odor detection (e.g., Engen, 1971). Dethier and Bowdan's calculations were based on a series of percentage values obtained for 10-20 flies by measuring the volume ingested by each fly. The lowest ΔI obtained, expressed as the Weber fraction, was 0.1 at 100 mmol/liter sucrose solution. In our study, the volume ingested was measured colorimetrically after homogenization of 30-40 flies because of the difficulty of determining volume intake of such a tiny fly as *Drosophila*. The data were analyzed after using Wilcoxon's test to find any statistical differences from the standard solutions. The smallest difference in the concentration detected by the flies was defined as ΔI, and the Weber fraction $\Delta I/I$ was calculated (where I is the standard concentration of sucrose).

VI. RESOLVING POWER OF ONE SUGAR RECEPTOR

Smith et al. (1983) evaluated the capability of information transmission in terms of sucrose concentration in a tarsal sugar receptor of the blowfly using the methods and concepts of information theory. They logarithmically divided the stimulus range from 10 to 1000 mmol/liter into 11 concentration levels. Even if each concentration level could be distinguished from the others, the estimated Weber fraction is around 0.4. According to their results based on electrophysiologic data, a single tarsal sugar receptor can encode the sucrose concentration levels into at most 3.8 or 5.3 discriminable levels (1.9 or 2.4 bits, respectively) within the stimulus range applied, depending upon the variability of responses. However, the behaviorally obtained values expressed as the Weber fraction, mentioned earlier are much smaller than this estimate for a single sugar receptor.

To seek a biologic mechanism to explain this improvement in discrimination, Nakao et al. (1987) proposed a simple model with one CNS neuron integrating inputs from many peripheral taste receptors (Figure 4).

VII. ESTIMATION OF CONVERGENCE IN THE PROPOSED MULTIRECEPTOR MODEL

Histologically, axons of each taste receptor of the fly ascend directly into the CNS without synaptic interconnections (Dethier, 1976), suggesting that the responses of receptors are transmitted to the CNS in parallel. Shiraishi and Tanabe (1974) revealed that the threshold for the proboscis extension response (part of the feeding behavior) could be explained by the central summation of peripheral receptor responses of the blowfly. Shiraishi and Miyachi (1976) reported that the acceptance threshold for the proboscis extension response in sucrose sensing with two legs is clearly lower than with one leg. A plausible model for the central information processing of taste in the fly should therefore consist basically of the convergence of many receptor cells onto one CNS neuron. The CNS neuron integrates all inputs from the receptors, as shown in Figure 4. Following Smith et al. (1983), the magnitude of nth receptor response $x_n(s)$ to the stimulus intensity level (concentration) of sucrose s is defined as the count of spikes firing in the first second just after the onset of stimulation. The output of the integrating CNS neuron, $u(s)$, can therefore be expressed as follows:

$$u(s) = W^x X(s)$$

where:

$$W^x = [w_1, w_2, \ldots, w_N]$$
$$X(s)^x = [x_1(s), x_2(s), \ldots, x_N(s)]$$
$$x = 1, 2, \ldots, L_R$$
$$s = 1, 2, \ldots, L_S$$

where W is the vector of synaptic weight, N the number of receptors, $X(s)$ the vector of the receptor response, L_R the number of response levels of a receptor, L_S that of stimulus intensity levels experimentally applied, and the subscript t a transpose of a vector. Since the receptor response and the output of the CNS neuron are expected to behave stochastically, the standard

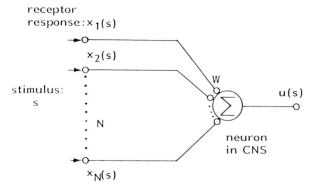

FIGURE 4 A multireceptor model. A single CNS neuron integrates the responses of several taste receptors by linear summation.

information theoretical tools are adopted to estimate the mutual information entropy $I(u, s)$ between the stimuli s and the output of the CNS neuron $u(s)$. The value of $I(u, s)$ quantifies exactly how the CNS neuron transmits the information concerning the stimulus intensity level against the uncertainty of receptor responses. Based on the electrophysiologic data for individual tarsal sugar receptors (Smith et al., 1983), I(u, s), was computed for various number of receptors. Mutual information entropy approaches a limit of 3.46 ($\log_2$ 11) bits, as the number of receptors N increases. At $N = 33$, it reaches the maximal value, the entropy of the stimuli, when the CNS neuron can discriminate completely all 11 intensity levels of stimuli applied. This number of receptors, 33, is expected since it lies within the realistic range of the number of taste hairs of the fly that usually participate in feeding behavior (Dethier, 1976; Shiraishi and Tanabe, 1974). Such an improvement of discrimination in the model (from 2.4 to 3.46 bits) may explain the difference between the low differential sensitivity of a single sugar receptor of the blowfly (Smith et al., 1983) and the high differential sensitivity in the feeding behavior of the blowfly (Dethier and Bowdan, 1984) and the fruit fly (Shimada et al., 1987). Integration in the CNS neurons can make allowance for individual variations in the responses of population of peripheral taste receptors and thereby improve the overall differential sensitivity.

REFERENCES

Campbell, B. A. (1958). Absolute and relative sucrose preference thresholds for hungry and satiated rats. *J. Comp. Physiol. Psychol.* **51**:795-800.

Dethier, V. G. (1976). *The Hungry Fly*. Harvard University Press, Cambridge, Massachusetts.

Dethier, V. G., and Rhoades, M. V. (1954). Sugar preference-aversion functions for the blowfly. *J. Exp. Zool.* **126**:177-204.

Dethier, V. G., and Bowdan, E. (1984). Relations between differential threshold and sugar receptor mechanism in the blowfly. *Behav. Neurosci.* **98**:791-803.

Engen, T. (1971). Olfactory psychophysics. In *Handbook of Sensory Physiology*, Vol. IV, *Chemical Senses* Part 1, *Olfaction*, L. M. Beidler (Ed.). Springer-Verlag, Berlin, pp. 216-244.

Frings, H. (1946). Gustatory thresholds for sucrose and electrolytes for the cockroach *Periplaneta americana* (Linn.). *J. Exp. Zool.* **102**:23-50.

Fujishiro, N., Kijima, H., and Morita, H. (1984). Impulse frequency and action potential amplitude in labellar chemosensory neurons of *Drosophila melanogaster*. *J. Insect. Physiol.* **30**:317-325.

Nakao, M., Kawazoe, Y., and Shimada, I. (1987). Intensity discrimination in a model of the sense of taste in insect. Proc. 9th Ann. Conf. IEEE Engin. Med. Biol. Soc., 14-15.

Richter, C. P., and Campbell, K. H. (1940). Sucrose taste thresholds of rats and humans. *Am. J. Physiol.* **128**:291-297.

Shimada, I., Nakao, M., and Kawazoe, I. (1987). Acute differential sensitivity and role of the central nervous system in the feeding behavior of *Drosophila melanogaster*. *Chem. Senses* **12**:481-490

Shiraishi, A., and Miyachi, N. (1976). The peripheral inhibition of the tarsal sugar receptor by sodium chloride in the proboscis extension response of the blowfly, *Phormia regina* M. *J. Comp. Physiol.* **110**:97-109.

Shiraishi, A., and Tanabe, Y. (1974). The proboscis extension response and tarsal and labellar chemosensory hairs in the blowfly. *J. Comp. Physiol.* **92**:161-179.

Smith, D. V., Bowdn, E., and Dethier, V. G. (1983). Information transmission in tarsal sugar receptors of the blowfly. *Chem. Senses* **8**:81-101.

Tanimura, T., Isono, K., Takamura, T., and Shimada, I. (1982). Genetic dimorphism in the taste sensitivity to trehalose in *Drosophila melanogaster*. *J. Comp. Physiol.* **147**:433-437.

von Frisch, K. (1934). Uber den Geschmackssinn der Biene. *Z. Vergl. Physiol.* **21**:1-156.

12
Chemical Cues from Conspecifics and Resource Response Variation in *Drosophila*

Ary A. Hoffmann

La Trobe University, Bundoora, Victoria, Australia

I. INTRODUCTION

A number of behavioral events are involved in the process that leads an insect to feed and oviposit. These can be broadly defined under the terms "finding" and "accepting" (Miller and Strickler, 1984), the first term encompassing behaviors that occur before arrival at a resource and the second term encompassing behaviors that determine whether an insect feeds or oviposits once it gets there. As in many other insects, research in *Drosophila* has shown that chemical cues from food are important in finding and accepting hosts. For example, species using fermenting resources, such as *Drosophila melanogaster* and the cactophilic *Drosophila*, are strongly attracted to some yeasts and rotting plant tissue in olfactometers (Kellogg et al., 1962; Fogleman, 1982) and females preferentially oviposit on certain yeasts (Vacek et al., 1985). Chemicals that are fermentation products act as adult and larval attractants (Kellogg et al., 1962; Parsons, 1979; Hoffmann and Parsons, 1984) and may stimulate oviposition (McKenzie and Parsons, 1972).

Much of the *Drosophila* work on responses to food cues has focused on the demonstration of genetic variation within populations. Reviews of some of this work can be found in Grossfield (1978), Futuyma and Peterson (1985), Visser (1986), and Taylor (1987). In general, laboratory studies have demonstrated genetic variation for olfactory responses, chemoreception, and oviposition. Most of these investigations have used fermentation products or cues from artificial media. However, several recent studies have demonstrated genetic variation for finding and accepting natural food resources (Hoffmann et al., 1984; Barker et al., 1986; Jaenike, 1986; Klazcko et al., 1986).

Far less attention has been directed at the impact of cues from conspecifics on resource response. Such cues are often chemical and may influence a variety of behaviors in insects. Well-known examples include the marking pheromones of tephritid fruit flies, which inhibit further oviposition (Propoky, 1981), and the aggregation pheromones of bark beetles (Birch, 1984). In *Drosophila*, chemical cues from conspecifics have been examined mainly in the context of courtship (Jallon, 1984; Tompkins, 1984). Courtship behaviors in *Drosophila* species utilizing fermenting fruit may be associated with breeding and feeding resources (Spieth, 1974), but they are clearly distinct from processes involved in finding and accepting food resources.

In this chapter I review chemical cues from conspecifics that influence resource response in *Drosophila* and consider the evidence for heritable variation in these behaviors. Most relevant studies have been done in the laboratory, although some inferences about the importance of conspecific cues have been made from the distribution of *Drosophila* species on natural resources. I also examine the potential fitness consequences of responding to conspecifics and whether there may be an advantage to a signaler or receiver as postulated in broad definitions of insect communication systems (Holldobler, 1984). Finally I discuss fitness and heritable variation in responses to conspecifics.

II. CHEMICAL CUES FROM CONSPECIFICS AFFECTING RESOURCE RESPONSE

Cases in which chemical cues associated with conspecifics have been implicated in resource response are listed in Table 1.

A. Adult Aggregation

Aggregation pheromones that attract both sexes and are produced by sexually mature males have been isolated in 17 *Drosophila* species by Bartelt, Jackson, and Schaner and coworkers (Bartelt and Jackson, 1984; Bartelt et al., 1985a, b, 1986b, 1989; Moats et al., 1987; Schaner et al., 1987; Bartelt, personal communication). In their assay, approximately 1000 starved adults are placed in a horizontal wind tunnel and a pheromone sample is placed on the lip of a glass vial. The vials are sealed after 3 minutes, and the number of trapped flies indicates attractivity. This assay probably tests for long-distance resource finding because *Drosophila* locate distant odor sources by upwind movement (Kellogg et al., 1962). The assay was initially used to isolate a male-produced

TABLE 1 Chemical Cues from Conspecifics Affecting Resource Response in Drosophilids

Chemical cue	Behavior	Species	Consequence of response	Reference
Male pheromones	Olfaction (adults)	D. virilis group	Adult aggregation	Bartelt and Jackson, 1984; Bartelt et al., 1985a, 1989
		D. melanogaster group		Bartelt et al., 1985b; Schaner et al., 1987
		D. hydei		Moats et al., 1987
Female pheromones derived from males	Olfaction (adults)	D. melanogaster group	Adult aggregation	Bartelt et al., 1985b; Schaner et al., 1987
Residual female odors	Olfaction (adults)	D. melanogaster, D. simulans	Adult aggregation	Spence et al., 1984; Parsons and Hoffmann, 1985
Larval cues	Olfaction (larvae)	D. melanogaster	Larval attraction/repulsion	Pruzan and Bush, 1977
Larval cues	Olfaction (adults)	D. melanogaster, D. simulans	Adult aggregation	Hoffmann and Parsons, 1986
Residual male cues	Oviposition	D. melanogaster, Zaprionus tuberculatus	Increase fecundity, egg aggregation	Hoffmann and Harshman, 1985; Harshman and Hoffmann, 1987
		D. funebris	Decrease fecundity	Harshman and Hoffmann, 1987
Larval cues	Oviposition	D. melanogaster, D. simulans	Inhibit oviposition	Chess and Ringo, 1987
Extract from medium with larvae	Oviposition	D. melanogaster, D. simulans	Inhibit oviposition	Aiken and Gibo, 1979
Male peptides transferred during mating	Oviposition	D. melanogaster D. funebris	Stimulate oviposition	Chen and Buhler, 1970 Baumann, 1975

pheromone in *Drosophila virilis* consisting of a cuticular hydrocarbon that was synergistic with five male-derived esters (Bartelt et al., 1985a). The number of flies attracted to the pheromone was seven times greater than attracted to fermented rearing medium. Attractivity of the pheromone was doubled when it was coupled with rearing medium (Bartelt et al., 1985a).

Additional pheromones have been isolated in six other species of the *D. virilis* group (Bartelt et al., 1986b, 1989). All species responded to extracts from males of their own species, and there was also some attraction to extracts from other species. Of particular interest for field behavior is the finding that these male-derived pheromones were synergistic with bark extracts on which this group feeds and oviposits. There was little response to either male extract or bark in two species (*Drosophila borealis* and *Drosophila littoralis*) unless these odor sources were combined (Table 2) (Bartelt et al., 1988).

Bartelt et al. (1985b) and Schaner et al. (1987) isolated aggregation pheromones in *D. melanogaster* and *Drosophila simulans*. They found that a male extract was particularly attractive to flies from both sexes when added to laboratory rearing medium (Table 2). One of the active components was *cis*-vaccenyl acetate (cVA), which may act as a close-range pheromone in *D. melanogaster* by inhibiting male courtship (Zawitowski and Richmond, 1986). Females did not produce cVA, but males transferred this pheromone to females during mating and females (unlike males) released it rapidly after mating. The transfer of pheromone from mature males to females does not occur in *D. virilis*; the *D. virilis* pheromones are cuticular, whereas cVA is found in the reproductive tract of *D. melanogaster* males (Bartelt, personal communication).

These experiments demonstrate that chemicals from *Drosophila* males attract conspecifics and are probably active over relatively long distances.

Table 2 Attraction of *Drosophila* Species to Male Extract and Extract in Combination with Food Odors[a]

Species	Treatment	Mean catch in bioassay
D. melanogaster	Male extract	0.9*
	Rearing medium (with yeast)	7.5†
	Male extract + medium	21.5‡
D. borealis	Male extract	2.2*
	Aspen bark extract	1.5*
	Male + aspen extracts	9.5†
D. littoralis	Male extract	1.8*
	Aspen bark extract	1.3*
	Male + aspen extracts	11.3†

[a]Means followed by the same symbol are not significantly different by the $P < 0.05$ criterion.
Source: Adapted from Bartelt et al. (1985b) and Bartelt et al. (1989) with permission.

The synergistic effects of the pheromones with feeding and breeding resources and the release of large amounts of pheromone onto surfaces (Schaner et al., 1987; Bartelt et al., 1989) suggest that these aggregation pheromones are important in responses to food resources, although field trials have not been undertaken.

Experiments on adult aggregation in *D. melanogaster* and *D. simulans* were carried out by Spence et al. (1984). Glass cylinders were marked with odors by placing 600-700 *D. melanogaster* and *D. simulans* adults in them for 8 h. The attractiveness of these cylinders was tested in a wind tunnel. Cylinders not marked by residues catch less than 1% of the released flies. Residual odors from nonvirgin females caught 18% of the released males and 6% of the released females. Residues from virgin females and males were less attractive, although both increased the attractiveness of cylinders. Males tended to be attracted to conspecific odors. The strength of this effect can be evaluated by an attraction index Δ, defined as

$$\Delta = P_{1,1} - P_{1,2}$$

where the probability estimate $P_{1,1}$ is the number of flies of species 1 attracted to odors of species 1, divided by the total number of species 1 flies that were captured, and $P_{1,2}$ is the number of species 2 flies attracted to species 1 odors divided by the total number of species 2 flies captured. This index has values from -1 to 1, with values > 0 indicating a positive association. The mean attraction index value in these experiments was 0.38 (Parsons and Hoffmann, 1986).

These effects are fairly weak compared to the attraction of flies to odors from natural resources in the same wind tunnel. For example, the odors from one fruit item catch 60-80% of the released flies in the tunnel if the same protocol is used. The experiments also require the use of several hundred adults to obtain residues; only a few ($<5\%$) of the released flies are captured when cylinders are marked with less than 100 adults (Hoffmann and Parsons, unpublished).

The residues in these experiments may be related to hydrocarbons, particularly as Bartelt et al. (1985b) have shown that active components can be isolated by hexane rinsing of rearing vials. We have recently shown that male extract combined with laboratory medium is more attractive for *D. melanogaster* than medium alone in our wind tunnel (O'Donnell and Hoffmann, unpublished). The greater attractiveness of nonvirgin females compared to virgin females in the Spence et al. (1984) experiments is consistent with a pheromone like cVA that is transferred during mating. However, this acetate is probably not solely responsible for the behavioral differences between *D. simulans* and *D. melanogaster* because there was preferential attraction to odors from conspecific virgin females.

B. Nonmating Effects of Males on Oviposition

Mainardi (1968) found that *D. melanogaster* females tended to oviposit in vials previously exposed to males rather than in vials that had not been exposed and argued that a male pheromone had induced oviposition. The experiment used a "choice" assay, in which both types of vials were available to females for oviposition. This work has been widely cited as an example of conspecific chemical cues affecting oviposition, although Mainardi's interpretation was questioned by David (1970) and Atkinson (1983) on the grounds that surface texture was not controlled (males walking over medium could alter texture). Hoffmann and Harshman (1985) replicated Mainardi's results in a nonchoice test (females were held in vials that had been either exposed or not exposed to males), with a 67% increase in fecundity over a 7 day period. However, they found that a similar increase could be obtained by exposing vials to virgin females. They suggested that the increase in fecundity was associated with the growth of microorganisms transmitted by adult flies rather than a male pheromone.

Hoffmann and Harshman (1985) isolated a male-specific effect by holding flies in small bottles containing only agar (which does not support microbial growth but provides moisture). Bottles were attached to vials with laboratory medium after flies had been removed. Females laid more eggs when they were exposed to bottles that had held males than when they were exposed to bottles that had held virgin females or no flies. A 22% increase in fecundity was found when females were exposed to bottles with residues from 20 males. The equivalent increase was only 6% when females were held in vials with live yeast as well as residues (Harshman and Hoffmann, 1987), perhaps because cues from yeast override the stimulating effect of male residues. This response is probably independent of aggregation because females would have been attracted away from the egg-laying surface.

Harshman and Hoffmann (1987) tested for residual male and female effects in other drosophilids. Both male and female residues reduced fecundity by up to 30% in *Drosophila funebris*, and male residues increased fecundity by up to 23% in *Zaprionus tuberculatus*. The *D. funebris* result suggests that marking pheromones inhibiting oviposition occur in *Drosophila* as in many other insects (Propoky, 1981), although reduced fecundity and attraction could be confounded. The increased fecundity in *D. melanogaster* and *Z. tuberculatus* seems to be a novel observation, although this was not a strong effect.

The fecundity increase in *D. melanogaster* may be associated with hydrocarbons. To test this, we used the method of Antony and Jallon (1982) to extract hydrocarbons with hexane (Hoffmann and Cacoyianni, unpublished). The extract of 50 flies was applied to food medium in a vial. *D. melanogaster* females were placed individually into vials and left to oviposit for 24 h. The number of eggs deposited in vials with male extract ($\bar{x} = 25.5$, SD = 21.2, $N = 36$) was significantly higher (Kruskal-Wallis test, $P < 0.01$) than the number in vials with only hexane ($\bar{x} = 14.0$, SD = 18.3, $N = 36$) and was also significantly higher ($P < 0.05$) than the number in vials with extract from nonvirgin females ($\bar{x} = 17.6$, SD = 17.7, $N = 36$). Hydrocarbons deposited by males in vials could mediate the male effect described by Hoffmann and Harshman (1985).

C. Olfactory Attraction to Larval Cues

Pruzan and Bush (1977) examined the olfactory response of larvae from two *D. melanogaster* strains in a U tube. Larvae were relatively more attracted to food with larvae than to food without them; 45% more larvae were attracted to tubes with food and larvae in one strain, and 27% in another. The number of larvae not moving into either of the tubes increased with larval density, suggesting that larvae could also be repelled by conspecific odors.

Hoffmann and Parsons (1986) examined the attraction of starved *D. melanogaster* and *D. simulans* adults to larvae and laboratory medium in a wind tunnel olfactometer. Few flies (<5%) were attracted to flasks with only labtory medium over the 8 h that the experiment was run. Flasks became more attractive as the larvae developed, attracting up to 86% of the released adults.

Males and females tended to be preferentially attracted to flasks with larvae from their own species at all stages of larval development, although there was considerable variation between experiments. There was even a significant tendency of adults to avoid their own species in some cases, suggesting that larvae may have variable effects on adult attraction. Attraction index values (as defined earlier) ranged from 0.63 to -0.40 for males and from 0.42 to -0.33 for females. Some of the variation between experiments may have been related to larval density because females were preferentially attracted to their own species when flasks contained 50 larvae per 300 ml but not when larval density was 10-fold higher.

D. Oviposition in Sites with Larvae or Eggs

Del Solar and Palomino (1966) found that *D. melanogaster* females tend to lay eggs on medium containing larvae of the same or a different species (*D. funebris*) rather than on uninhabited medium in a choice test. This tendency was strong; on average, females laid 87% of their eggs on medium with larvae. Larval density was low (15 larvae in 5 ml medium). The behavioral basis of this response was not determined, so it is not clear if chemical cues were involved.

Atkinson (1983) argued that these findings reflected a response to surface texture rather than larvae. He found that oviposition by *D. melanogaster* increased when a surface was scarified and that the presence of eggs on a scarified surface did not affect oviposition. A problem with Atkinson's experiments is that total fecundity increases on a rough surface. Thus in three experiments with a scarified surface the mean fecundities per food cup were 110.6, 86.0, and 84.9, whereas mean fecundity in one experiment with a smooth surface carrying eggs was only 50.6. It is possible that there is a response to *Drosophila* eggs independent of surface texture that disappears when food odors are released by scarification. Atkinson does not consider responses to larval cues, so his findings do not direct address the experiments of Del Solar and Palomino (1966). Sokolowski (unpublished) has found that *D. melanogaster* aggregates eggs even on a rough surface.

Aiken and Gibo (1979) examined the effect of larvae on the fecundity of *D. melanogaster* and *D. simulans* in a nonchoice test by confining 50 larvae (mixed instars) on 6 g medium. Females of both species showed lower fecundity in the presence of larvae of either species, as previously found by Chiang and Hodson (1950) and Moth and Barker (1976). To examine the effect of larval residues, agar plugs were soaked in a water extract from larval cultures that had been passed through a plankton filter. Residue of either species reduced the number of eggs laid by females; *D. melanogaster* laid 63% fewer eggs on plugs with residue than on plugs with water after 48 h, and the equivalent figure for *D. simulans* was 55%. Chemical cues associated with larval activity can therefore inhibit oviposition.

Chess and Ringo (1985) considered the oviposition response of *D. melanogaster* and *D. simulans* to laboratory media that had been exposed to a high density of larvae and then frozen and sieved to kill and remove the larvae and to control for surface texture. Adults were exposed to sieved medium that had contained larvae or no larvae in choice tests. *D. melanogaster* and *D. simulans* laid 1.4 and 1.8 times more eggs than expected on the unconditioned medium, respectively. *D. simulans* females tended to lay more of their eggs on unconditioned medium than *D. melanogaster*, but neither species discriminated between medium with *D. melanogaster* and *D. simulans* larvae. These results provide additional evidence for chemical cues with an inhibitory effect on oviposition.

In summary, oviposition may be stimulated at low larval densities and inhibited at high larval densities. The inhibitory effects involve chemical cues that do not require the presence of live larvae.

E. Male Peptides That Stimulate Oviposition

The aggregation pheromones discussed earlier indicate that male cues transferred during mating can affect resource finding in species of the *D. melanogaster* group. Male substances can also affect resource acceptance by stimulating females to oviposit. Chen and Buhler (1970) found that a peptide from the accessory glands of *D. melanogaster* males injected into virgin females caused a two- to threefold increase in oviposition. This peptide was transferred to females during copulation. Baumann (1975) isolated a similar oviposition-stimulating peptide from male *D. funebris*. Electrophoretic patterns of accessory gland proteins differ between species, and their stimulating effects on oviposition are species specific (Chen et al., 1985).

F. Concluding Remarks

This survey indicates that conspecific chemical cues can influence a range of behaviors in *Drosophila* that are involved in resource response, although some of the behavioral changes mediated by these cues are small. The cues are associated with larvae as well as adults and are involved in accepting as well as finding food resources. Many of the responses show some degree of species specificity. All the studies discussed here hve been carried out in the laboratory using artificial resources. Attempts to extrapolate findings to natural resources have so far only been made for adult aggregation pheromones. Specific pheromones have been demonstrated in only a few cases. Although many of the cues may be related to hydrocarbons, some may represent resource modification rather than the release of specific signals.

III. EVIDENCE FOR HERITABLE VARIATION

Genetic variation may involve variation in the emission of chemicals and/or responses to them. Variation in both these components leads to the possibility of a genetic correlation between emission and response. This may result in individuals being preferentially attracted to (or repulsed by) cues from their own strain.

Bartelt et al. (1986b) demonstrated variation in the aggregation response of two *D. virilis* strains from different geographic locations. One strain showed a fourfold increase in attraction to the hydrocarbon (Z)-10-heneicosene when it was combined with the ester ethyl tiglate, but the response of another strain to the hydrocarbon was not significantly affected by ethyl tiglate. Males from the strains produced a similar amount of the ester. Response variation to aggregation pheromones has not been investigated in other species. There is evidence for strain differences in the production of cuticular hydrocarbons in *D. americana* and *D. lummei* (Bartelt et al., 1986a) as well as in *D. melanogaster* (Jallon, 1984; van den Berg et al., 1984). This variation has not been related to aggregation behavior, although the active hydrocarbon of *D. americana* [(Z)-9-heneicosene] was predominant in both strains examined and none of the *D. lummei* hydrocarbons showed strong activity (Bartelt et al., 1986b).

Parsons and Hoffmann (1985) examined the attraction of *D. melanogaster* and *D. simulans* populations originating from Melbourne and Townsville to cylinders containing the residues of nonvirgin females. Males tended to be attracted to cylinders with residues from their own population. Attraction index values were somewhat lower than in the species comparisons (see earlier), with a mean value of 0.26 for *D. melanogaster* and 0.22 for *D. simulans*. Parsons and Hoffmann (1985) found a similar trend for males from two isofemale lines originating from the same population, with a mean Δ value of 0.16. The number of males attracted to the cylinders was fairly low, the average for the *D. melanogaster* and *D. simulans* population comparisons being 30 and 26%, respectively.

These results suggest an association between responses to odors and odor emission, although more strains need to be tested. This association has not been found for *D. melanogaster* mating behavior; males from the Canton-S and Oregon-R laboratory strains court each other less in the presence of volatile compounds extracted from their own strain than from the other strain (Tompkins and Hall, 1984).

Pruzan and Bush (1977) found genetic variation for responses to larval odors in two laboratory *D. melanogaster* strains. Larvae were relatively more attracted to odors from their own strain when arms of a U tube olfactometer contained larvae from different strains. This suggests heritable variation for attraction to larval odors and the emission of these odors. The effect was not strong, however, and was apparent at only one of four larval densities examined.

Hoffmann and Parsons (1986) considered the attraction of Melbourne and Townsville *D. melanogaster* and *D. simulans* females to larval odors from these populations when flasks contained 50 larvae per 300 ml medium. Attraction index values were positive in these experiments ($\Delta = 0.15$), indicating differential attraction of females to odors from their own populations. Variation in odor response and odor emission are again suggested by these results.

Both the direct and the indirect effects of conspecifics may contribute to this variation (Parsons and Hoffmann, 1986). For example, variation in the production of larval cues may involve pheromones, genetic differences in the products released by larvae in laboratory medium (e.g., Weisbrot, 1966; Dawood and Strickberger, 1969), or simply genetic variation for larval foraging behavior (Sokolowski et al., 1986) that affects the release of food odors.

Other cases of heritable variation in resource response may also involve conspecific chemical cues. Del Solar (1968) successfully selected for increased and decreased tendencies of *D. pseudoobscura* females to aggregate eggs. Sokolowski (1986, unpublished) showed that females from two *D. melanogaster* strains tended to oviposit on media plugs occupied by larvae of their own strain. There is genetic variation in natural populations for male effects on female fertility that may involve chemical cues. For example, female *D. melanogaster* mated to males from lines homozygous at different enzyme loci have different fecundities (Serradilla and Ayala, 1983).

In summary, the few studies that have been carried out to date indicate heritable variation in responses to conspecific odors and their emission. Much of this evidence involves variation at the interpopulation level or differences between laboratory strains. There is almost no information on variation within populations, and there has been no attempt to assess levels of variation in emission and response separately. It seems likely that response variation to conspecific chemical cues will be found within populations, given the evidence for genetic variation in olfactory response and chemoreception in stocks from natural populations (Becker, 1970; Fuyama, 1978; Hoffmann, 1983; Alcorta and Rubio, 1988).

IV. CONSPECIFIC CUES AND RESPONSES TO FIELD RESOURCES

Field studies on conspecific cues have not been carried out with *Drosophila*, although Atkinson and Shorrocks and coworkers have made inferences about the importance of such cues from the distribution of *Drosophila* emerging from individual fruit collected in nature (Atkinson and Shorrocks, 1984; Atkinson, 1985; Shorrocks et al., 1984). Their studies indicate that individuals emerging from individual resources have an aggregated distribution that fits the negative binomial distribution and that species abundance shows only weak positive correlations across resources. Atkinson and Shorrocks (1984) argue that the aggregated distributions arise solely because individual females leave clusters of eggs at equally attractive breeding sites. They consider that the number of eggs laid by females is not affected by the presence of other individuals, citing the absence of strong correlations and Atkinson's (1983) laboratory results in support.

These arguments are not convincing. As already discussed, Atkinson (1983) considered only eggs (not larvae, as stated in Atkinson and Shorrocks, 1984), and other laboratory studies indicate an effect of larvae on oviposition and attraction when surface texture is not a confounding factor (see earlier). Moreover, Atkinson and Shorrocks used a statistical test with low power for detecting interspecific interactions (Worthen and McGuire, 1988), and processes other than independent aggregation could result in the absence of strong correlations across breeding sites in any case. For example, weak positive correlations could arise between species even if adults of one species show a tendency to avoid cohabiting with another species when resources vary overall in their suitability as a breeding site.

Two preliminary experiments on the response of *D. melanogaster* to natural resources (Hoffmann and O'Donnell, unpublished) illustrate the potential impact of cues from other individuals on resource response. We used ripe apples that had been damaged by extracting a small core from two sides of each apple. In the first experiment, half the apples were exposed to five males and five females for 1 day, left another day, and then placed in containers in a cage. The distribution of 30 males and 30 females on these resources was investigated after 1 day. Almost all adults were attracted to apples previously exposed to adults, indicating that cues associated with adults or larvae dramatically increased the attractiveness of fruit (Table 3). In the second experiment, cuticular components extracted with hexane from 50 males was applied to one side of the apple (hexane alone did not attract more adults). Apples were placed in canisters, and 10 flies were released in each canister. The number of flies alighting on each side of the apple was counted at regular intervals. On average, there were more than twice as many adults on the damaged area with extract than on the area without extract (Table 3). We obtained similar results with *D. simulans*.

The attractiveness of field resources for feeding, breeding, or mating may depend on several interacting factors that include conspecific cues. These

TABLE 3 Effects of Conspecific Cues on the Attraction of *D. melanogaster* Adults to Apples[a]

	Number of adults at cored sites	
Experiment	Conspecific cues	No cues
1: Adult or larval cues		
Trial 1	53	3
Trial 2	36	11
Trial 3	57	1
Total	146	15
2: Male extract		
Trial 1	27	12
Trial 2	27	13
Trial 3	29	16
Total	83	41

[a]Sites attractive to adults were obtained by taking cores (4-6 mm deep) from two sides of an apple. Counts of adults on sites were made 15 h after adults were released. In the first experiment, five apples that had been previously exposed to adults were placed in a cage along with five apples that had not been exposed. A total of 60 adults were released into a cage. In the second experiment, extracts from 50 males were placed on one side of an apple. Five males and five females were released in a canister with one apple, and counts were pooled over 10 canisters.

interactions are summarized in Figure 1 for such species as *D. melanogaster* that may feed, breed, and mate on the same resource. Most fruit or other suitable plant tissue must be damaged before it can support adult feeding and larval growth. Adult *D. melanogaster* die when they are confined with intact apples and oranges in a cage. The initial attraction of adults to a resource depends on the damaged plant substrate and the microorganisms that colonize it. Once adults arrive they may release aggregation pheromones and affect the microorganism community by feeding selectively on yeasts (e.g., Vacek et al., 1979, 1985) and by bringing in new microorganisms. *Drosophila* may be important vectors for microorganisms. For example, Gilbert 1980) found that *Drosophila* were notable agents of dispersal for yeasts and bacteria in a deciduous forest. The types of microorganisms that colonize a damaged area affect its suitability for feeding and breeding, as documented in the cactophilic *Drosophila* (Vacek, 1982). Larvae may modify the attractiveness of a resource in various ways. Larval activity may increase the release

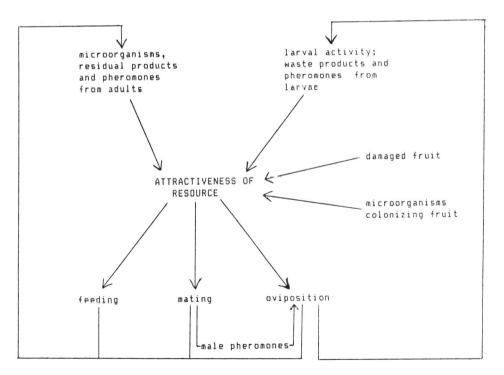

FIGURE 1 Interactions between factors affecting the attractiveness of a fruit resource for feeding, mating, and oviposition.

of fruit odors and alter the surface area available for the growth of yeasts and bacteria. Larvae may also influence attractiveness more directly by releasing waste products or other chemicals.

These considerations emphasize that the attractiveness of natural resources is determined by a combination of cues from plant substrates, microorganisms, and conspecifics. The nature and importance of these cues change over time as the plant substrate is modified by microorganism growth and the activity of adults and immatures.

V. FITNESS AND RESPONSES TO CONSPECIFIC CUES

Responses to conspecific cues alter the density of adults or immatures on a resource. Numerous laboratory studies indicate that density changes affect fitness-related traits in *Drosophila*. In *D. melanogaster*, for example, larval density alters viability, adult body size, and development time (e.g., Chiang and Hodson, 1950; Barker and Podger, 1970; Moth and Barker, 1976), and adult density alters fecundity, egg hatchability, longevity, and mating behavior (e.g., Chiang and Hodson, 1950; Moth, 1977; Moth and Barker, 1977; Harshman et al., 1988). A few density effects have been described for field resources. For example, high larval densities decreased adult body size in *Drosophila* species breeding in fruit (Atkinson, 1979).

These density effects may generate fitness advantages for a signaler or receiver. Adult aggregation may help in mate location, particularly when species densities are low relative to available resources. *Drosophila* females may gain carbohydrates and protein as well as sperm from males (Markow and Ankey, 1984; Steele, 1986) and may therefore benefit nutritionally from attracting males or responding to male cues.

Larvae may benefit by attracting females for oviposition because larval viability is maximized at intermediate densities, at least on laboratory media (e.g., Lewontin, 1955). Larvae may benefit from releasing cues that inhibit further oviposition because their viability may be decreased when other individuals use the same resource as a breeding site (Moth and Barker, 1976). Respondents may also benefit because of the reduced viability of eggs deposited on medium with a high larval density (Moth and Barker, 1976).

Adult females may also benefit nutritionally by finding sites occupied by conspecific immatures and adults. This was suggested by an experiment in which *D. melanogaster* and *D. simulans* females were held for 1 day on cored apples before their fecundity was tested (Hoffmann, unpublished). Some apples had been previously exposed to other adults for 3 days and were inhabited by larvae; others had not been previously exposed. The females held on the exposed apples had more than twice the fecundity of the other females (Table 4), even though other experiments had indicated that oviposition was not inhibited on occupied sites.

TABLE 4 Fecundity of *D. melanogaster* and *D. simulans* Females Held with Apples Previously Exposed or Unexposed to Adults and Immature Stages[a]

	Exposed		Unexposed	
	$\bar{x}$	SD	$\bar{x}$	SD
D. melanogaster				
Trial 1	55.8	20.8	22.0	12.3
Trial 2	63.5	26.7	31.5	17.4
Trial 3	41.7	15.3	14.6	11.8
D. simulans				
Trial 1	40.0	17.9	23.6	14.3
Trial 2	39.3	21.7	16.5	11.3
Trial 3	33.3	15.3	19.3	16.4

[a]Cored apples were held in containers with five flies of each sex for 3 days or were not exposed to adults. Flies were cleared, and five females (3-4 days old) were held for 1 day in these containers before their fecundity was measured by holding them individually in vials for 24 h. Counts were pooled over the five females from the same container. Means and standard deviations are based on 10 containers.

In summary, there are many consequences of altered density that could favor responses to conspecific cues and their emission. Some of the behaviors described here may therefore fall within broad definitions of communication that require fitness advantages for a signaler or a respondent (Holldobler, 1984). However, most fitness studies in *Drosophila* have been carried out on artificial media under laboratory conditions, and the fitness consequences of responses to conspecific cues with field resources need to be investigated.

VI. FITNESS AND HERITABLE VARIATION

The preceding discussion indicates several cases of interspecific variation in chemical emission and production, such as the release of male-derived aggregation pheromones by *D. melanogaster* group females but not by *D. virilis* group females (Bartelt, personal communication) and the diverse effects of male residue on oviposition in drosophilids (Harshman and Hoffmann, 1987). These differences may be the result of past selection for different responses to conspecific cues associated with ecologic differences between species.

Ecologic correlates have been hypothesized and tested for insect species that release adult pheromones inhibiting oviposition; these species tend to have a narrow host range, low larval mobility, and relatively permanent hosts (Roitberg and Propoky, 1987). Adaptive arguments can also be used to postulate ecologic correlates for many of the behaviors described here. For example, selection for adult aggregation pheromones may be stronger in multiple-mating species that are rare than in common species that mate only once. There is currently insufficient behavioral and ecologic information to test such predictions for drosophilids.

Ecologic heterogeneity may also affect the fitness of genotypes with different responses or emission rates. Most theory on genetic variation in resource use considers the situation in which genotypes favored in one habitat are at a disadvantage in another habitat (Hedrick, 1986). This may be relevant to the situation in which there is genetic variation either for responses to conspecific cues or for their emission. Seasonal and spatial variation in the availability of resources and in the abundance of many *Drosophila* species ensures that there is considerable variation in the number of adults and larvae relative to available resources. Genotypes with high responses or emission rates may therefore be favoured under one set of conditions but selected against under another set of conditions.

The fitness of these genotypes becomes more difficult to predict when there is variation in signal production and emission. The relative fitness of individuals producing signals may then depend on the frequency of individuals responding to them (and vice versa). For example, females that release aggregation pheromones may have a fitness advantage when flies responding to these pheromones are common in the population but not when they are rare. The possibility that signal emission and response may be genetically correlated also needs to be considered.

Genetic variation in behaviors associated with conspecific cues may be under selection for sexual communication as well as resource response. For example, the cuticular hydrocarbons of *D. melanogaster* and *D. simulans* affect courtship behaviors (Jallon, 1984) as well as adult aggregation and oviposition. Explaining patterns of genetic variation in these cues will therefore require a consideration of different types of selection pressures.

There is currently enough laboratory evidence to suggest that conspecific chemical cues may be an important component of resource response in many *Drosophila* species and that heritable variation exists for behaviors associated with these cues. There are numerous ways in which the behaviors can interact with density-dependent or frequency-dependent fitness traits to account for heritable variation within and between species but insufficient empirical data to test the possibilities.

VII. SUMMARY

In *Drosophila*, laboratory studies indicate that chemical cues associated with conspecifics can have marked effects on behaviors involved with utilization of food resources by larvae and adults. These include adult aggregation in response to chemicals released by males and females, stimulation and inhibition of oviposition by larval and male cues, and attraction of larvae and adults to cues associated with larvae. There is evidence for differences between related species and between strains of the same species for many of these behaviors. Several of the cues are pheromones associated with hydrocarbons, whereas others probably reflect the chemical modification of food resources because of larval activity. Responses to conspecific cues alter the density of adults and larvae on a food resource. Density changes can influence numerous fitness traits in the laboratory and can affect the suitability of natural resources for feeding and breeding. This leads to the potential for adaptive chemical signaling and responses by *Drosophila*. The relative fitness of individuals responding to conspecifics or emitting signals varies with environmental conditions. The importance of conspecific cues have not been considered in field studies of resource response variation in *Drosophila*.

ACKNOWLEDGMENTS

I am extremely grateful to Robert Bartelt and Marla Sokolowski for copies of unpublished manuscripts. Stuart Barker, Robert Bartelt, Larry Harshman, and Peter Parsons provided many useful comments. My experimental work on habitat response in *Drosophila* is supported by the Australian Research Council.

REFERENCES

Aiken, R. B., and Gibo, D. L. (1979). Changes in fecundity of *Drosophila melanogaster* and *D. simulans* in response to selection for competitive ability. *Oecologia* **43**:63-77.

Alcorta, E., and Rubio, J. (1988). Genetical analysis of intrapopulational variation in olfactory response in *Drosophila melanogaster*. *Heredity* **60**:7-14.

Antony, C., and Jallon, J.-M. (1982). The chemical basis for sex recognition in *Drosophila melanogaster*. *J. Insect Physiol.* **28**:239-242.

Atkinson, W. D. (1979). A field investigation of larval competition in domestic *Drosophila*. *J. Anim. Ecol.* **48**:91-102.

Atkinson, W. D. (1983). Gregarious oviposition in *Drosophila melanogaster* is explained by surface texture. *Aust. J. Zool.* **31**:925-929.

Atkinson, W. D. (1985). Coexistence of Australian rainforest diptera breeding in fallen fruit. *J. Anim. Ecol.* **54**:507-518.

Atkinson, W. D., and Shorrocks, B. (1984). Aggregation of larval Diptera over discrete and ephemeral breeding sites: The implications for coexistence. *Am. Naturalist* **124**:336-351.

Barker, J. S. F., and Podger, R. N. (1970). Interspecific competition between *Drosophila melanogaster* and *Drosophila simulans*: Effects of larval density on viability, development period and adult body weight. *Ecology* **51**:170-189.

Barker, J. S. F., Vacek, D. C., East, P. D., and Starmer, W. T. (1986). Allozyme genotypes of *Drosophila buzzatii*: Feeding and oviposition preferences for microbial species, and habitat selection. *Aust. J. Biol. Sci.* **39**:47-58.

Bartelt, R. J., and Jackson, L. L. (1984). Hydrocarbon component of the *Drosophila virilis* (Diptera: Drosophilidae) aggregation pheromone: (Z)-10-Heneicosene. *Ann. Entomol. Soc. Am.* **77**:364-371.

Bartelt, R. J., Jackson, L. L., and Schaner, A. M. (1985a). Ester components of the aggregation pheromone of *Drosophila virilis* (Diptera: Drosophilidae). *J. Chem. Ecol.* **11**:1197-1208.

Bartelt, R. J., Schaner, A. M., and Jackson, L. L. (1985b). cis-Vaccenyl acetate as an aggregation pheromone in *Drosophila melanogaster*. *J. Chem. Ecol.* **11**:1747-1756.

Bartelt, R. J., Armold, M. T., Schaner, A. M., and Jackson, L. L. (1986a). Comparative analysis of cuticular hydrocarbons in the *Drosophila virilis* species group. *Comp. Biochem. Physiol.* **83B**:731-742.

Bartelt, R. J., Schaner, A. M., and Jackson, L. L. (1986b). Aggregation pheromones in five taxa of the *Drosophila virilis* species group. *Physiol. Entomol.* **11**:367-376.

Bartelt, R. J., Schaner, A. M., and Jackson, L. L. (1989). Aggregation pheromone components in *Drosophila mulleri*: A chiral ester and an unsaturated ketone. *J. Chem. Ecol.* **15**:399-412.

Baumann, H. (1975). Biological effects of paragonial substances PS-1 and PS-2 in females of *Drosophila funebris*. *J. Insect Physiol.* **20**:2347-2362.

Becker, H. J. (1970). The genetics of chemotaxis in *Drosophila melanogaster*. Selection for repellant insensitivity. *Mol. Gen. Genet.* **107**:194-200.

Birch, M. C. (1984). Aggregation in bark beetles. In *Chemical Ecology of Insects*, W. J. Bell and R. T. Carde (Eds.). Chapman and Hall, London, pp. 331-354.

Chen, P. S., and Buhler, R. (1970). Paragonial substance and other free ninhidrin-positive components in male and female adults of *Drosophila melanogaster*. *J. Insect Physiol.* **16**:615-627.

Chen, P. S., Stumm-Zellinger, E., and Caldelari, M. (1985). Protein metabolism of *Drosophila* male accessory glands. II. Species-specifity of secretion proteins. *Insect Biochem.* **15**:385-390.

Chess, K. F., and Ringo, J. M. (1985). Oviposition site selection by *Drosophila melanogaster* and *D. simulans*. *Evolution* **39**:869-877.

Chiang, H. C., and Hodson, A. C. (1950). An analytical study of population growth in *Drosophila melanogaster*. *Ecol. Monogr.* **20**:173-206.

David, J. (1970). Oviposition chez *Drosophila melanogaster*. Importance des caracteristiques physiques de la surface de ponte. *Rev. Comp. Anim.* **4**:70-72.

Dawood, M. M., and Strickberger, M. W. (1969). The effect of larval interaction on viability in *Drosophila melanogaster*. III. Effects of biotic residues. *Genetics* **63**:213-220.

Del Solar, E. (1968). Selection for and against gregariousness in the choice of oviposition sites by *Drosophila pseudoobscura*. *Genetics* **58**:257-282.

Del Solar, E., and Palomino, H. (1966). Choice of oviposition in *Drosophila melanogaster*. *Am. Naturalist* **100**:127-133.
Fogleman, J. C. (1982). The role of volatiles in the ecology of cactophilic *Drosophila*. In *Ecological Genetics and Evolution. The Cactus-Yeast-Drosophila Model System*, J. S. F. Barker and W. T. Starmer (Eds.). Academic Press, Sydney, pp. 191-206.
Futuyma, D. J., and Peterson, S. C. (1985). Genetic variation in the use of resources by insects. *Annu. Rev. Entomol.* **30**:217-238.
Fuyama, Y. (1978). Behavior genetics of olfactory response in *Drosophila*. II. An odorant-specific variant in a natural population of *Drosophila melanogaster*. *Behav. Genet.* **8**:399-414.
Gilbert, D. G. (1980). Dispersal of yeasts and bacteria by *Drosophila* in a temperate forest. *Oecologia* **46**:135-137.
Grossfield, J. (1978). Non-sexual behavior of *Drosophila*. In *The Genetics and Biology of Drosophila*, Vol. 2b, M. Ashburner and T. R. F. Wright (Eds.). Academic Press, London, pp. 3-126.
Harshman, L. G., and Hoffmann, A. A. (1987). Residual influences on fecundity in drosophilid species. *Experientia* **43**:213-215.
Harshman, L. G., Hoffmann, A. A., and Prout, T. (1988). Environmental effects on remating in *Drosophila melanogaster*. *Evolution* **42**:312-321.
Hedrick, P. W. (1986). Genetic polymorphism in heterogeneous environments: A decade later. *Annu. Rev. Ecol. Syst.* **17**:535-566.
Hoffmann, A. A. (1983). Bidirectional selection for olfactory response to acetaldehyde and ethanol in *Drosophila melanogaster*. *Genet. Select. Evol.* **15**:501-518.
Hoffmann, A. A., and Harshman, L. G. (1985). Male effects on fecundity in *Drosophila melanogaster*. *Evolution* **39**:638-644.
Hoffmann, A. A., and Parsons, P. A. (1984). Olfactory response and resource utilization in *Drosophila*: Interspecific comparisons. *Biol. J. Linnean Soc.* **22**:43-53.
Hoffmann, A. A., and Parsons, P. A. (1986). Inter- and intraspecific variation in the response of *Drosophila melanogaster* and *D. simulans* to larval cues. *Behav. Genet.* **16**:295-306.
Hoffmann, A. A., Parsons, P. A., and Nielsen, K. M. (1984). Habitat selection: Olfactory response of *Drosophila melanogaster* depends on resources. *Heredity* **53**:139-143.
Holldobler, B. (1984). Evolution of insect communication. In *Insect Communication*, T. Lewis (Ed.). Academic Press, Orlando, Florida, pp. 349-377.
Jaenike, J. (1986). Genetic complexity of host-selection behavior in *Drosophila*. *Proc. Natl. Acad. Sci. USA* **83**:2148-2151.
Jallon, J.-M. (1984). A few chemical words exchanged by *Drosophila* during courtship and mating. *Behav. Genet.* **14**:441-477.
Kellogg, F. E., Frizel, D. E., and Wright, R. H. (1962). The olfactory guidance of flying insects. IV. *Drosophila*. *Can. Entomol.* **94**:884-888.
Klazcko, L. B., Taylor, C. E., and Powell, J. R. (1986). Genetic variation for dispersal by *Drosophila pseudoobscura* and *D. persimilis*. *Genetics* **112**:229-235.
Lewontin, R. C. (1955). The effects of population density and composition on viability in *Drosophila melanogaster*. *Evolution* **9**:27-41.

Mainardi, M. (1968). Gregarious oviposition and pheromones in *Drosophila melanogaster*. *Boll. Zool.* **35**:135-136.

Markow, T. A., and Ankey, P. F. (1984). *Drosophila* males contribute to oogenesis in a multiple mating species. *Science* **224**:302-303.

McKenzie, J. A., and Parsons, P. A. (1972). Alcohol tolerance: An ecological parameter in the relative success of *Drosophila melanogaster* and *D. simulans*. *Oecologia* **10**:373-388.

Miller, J. R., and Strickler, K. L. (1984). Finding and accepting host plants. In *Chemical Ecology of Insects*, W. J. Bell and R. T. Carde (Eds.). Chapman and Hall, London, pp. 127-157.

Moats, R. A., Bartelt, R. J., Jackson, L. L., and Schaner, A. M. (1987). Ester and ketone components of aggregation pheromone of *Drosophila hydei* (Diptera: Drosophilidae). *J. Chem. Ecol.* **13**:451-462.

Moth, J. J. (1977). Interspecific competition between *Drosophila melanogaster* and *D. simulans*: Effects of adult density, species frequency and dietary ^{32}P on egg hatchability. *Aust. J. Zool.* **25**:699-709.

Moth, J. J., and Barker, J. S. F. (1976). Interspecific competition between *Drosophila melanogaster* and *Drosophila simulans*. *Oecologia* **23**:151-164.

Moth, J. J., and Barker, J. S. F. (1977). Interspecific competition between *Drosophila melanogaster* and *Drosophila simulans*: Effects of adult density on adult viability. *Genetics* **47**:203-218.

Parsons, P. A. (1979). Larval reactions to possible resources in three species of *Drosophila* as indicators of ecological diversity. *Aust. J. Zool.* **27**:413-419.

Parsons, P. A., and Hoffmann, A. A. (1985). Habitat marking: Parallel divergence in two *Drosophila* populations. *Heredity* **54**:203-207.

Parsons, P. A., and Hoffmann, A. A. (1986). Ecobehavioral genetics: Habitat preference in *Drosophila*. In *Evolutinary Processes and Theory*, S. Karlin and E. Nevo (Eds.). Academic Press, Orlando, Florida, pp. 535-559.

Propoky, R. J. (1981). Epidiectic pheromones that influence spacing patterns of phytophagous insects. In *Semiochemicals: Their Role in Pest Control*, D. A. Norland, R. L. Jones, and W. J. Lewis (Eds.). Wiley, New York, pp. 181-213.

Pruzan, A., and Bush, G. (1977). Genotypic differences in larval olfactory discrimination in two *Drosophila melanogaster* strains. *Behav. Genet.* **7**:457-464.

Roitberg, B. D., and Propoky, R. J. (1987). Insects that mark host plants. *BioScience* **37**:400-406.

Schaner, A. M., Bartelt, R. J., and Jackson, L. L. (1987). (Z)-11-octadecenyl acetate, an aggregation pheromone in *Drosophila simulans*. *J. Chem. Ecol.* **13**:1777-1786.

Serradilla, J. M., and Ayala, F. J. (1983). Alloprocoptic selection: A mode of natural selection promoting polymorphism. *Proc. Natl. Acad. USA* **80**:2022-2025.

Shorrocks, B., Rosewell, J., Edwards, K., and Atkinson, W. D. (1984). Interspecific competition is not a major organizing force in many insect communities. *Nature* **310**:310-312.

Sokolowski, M. B. (1986). *Drosophila* larval foraging behavior and correlated behaviors. In *Evolutionary Genetics of Invertebrate Behavior*, M. Heuttel (Ed.). Plenum Press, New York, pp. 197-214.

Sokolowski, M. B., Bauer, S. J., Wai-Ping, V., Rodriguez, L., Wong, J. L., and Kent, C. (1986). Ecological genetics and behavior of *Drosophila melanogaster* larvae in nature. *Anim. Behav.* **34**:403-408.

Spence, G. E., Hoffmann, A. A., and Parsons, P. A. (1984). Habitat marking: Males attracted to residual odours of two *Drosophila* species. *Experientia* **40**:763-764.

Spieth, H. T. (1974). Courtship behavior in *Drosophila*. *Annu. Rev. Entomol.* **19**: 385-405.

Steele, R. H. (1986). Courtship feeding in *Drosophila subobscura* I. The nutritional significance of courtship feeding. *Anim. Behav.* **34**:1087-1098.

Taylor, C. E. (1987). Habitat selection within species of *Drosophila*: A review of the experimental findings. *Evol. Ecol.* **1**:389-400.

Tompkins, L. (1984). Genetic analysis of sex appeal in *Drosophila*. *Behav. Genet.* **14**:411-440.

Tompkins, L., and Hall, J. C. (1984). Sex pheromones enable *Drosophila* males to discriminate between conspecific females from different laboratory stocks. *Anim. Behav.* **32**:349-352.

Vacek, D. C. (1982). Interactions between microorganisms and cactophilic *Drosophila* in Australia. In *Ecological Genetics and Evolution. The Cactus-Yeast-Drosophila Model System*, J. S. F. Barker and W. T. Starmer (Eds.). Academic Press, Sydney, pp. 175-190.

Vacek, D. C., Starmer, W. T., and Heed, W. B. (1979). Relevance of the ecology of *Citrus* yeasts to the diet of *Drosophila*. *Microb. Ecol.* **5**:43-49.

Vacek, D. C., East, P. D., Barker, J. S. F., and Soliman, M. H. (1985). Feeding and oviposition preferences of *Drosophila buzzatii* for microbial species isolated from its natural environment. *Biol. J. Linnean Soc.* **24**:175-187.

van den Berg, M. J., Thomas, G., Hendriks, H., and van Delden, W. (1984). A re examination of the negative assortative mating phenomenon and its underlying mechanism in *Drosophila melanogaster*. *Behav. Genet.* **14**:45-61.

Visser, J. H. (1986). Host odor perception in phytophagous insects. *Annu. Rev. Entomol.* **31**:121-144.

Weisbrot, R. D. (1966). Genotypic interactions among competing strains and species of Drosophila. *Genetics* **53**:427-435.

Worthen, W. B., and McGuire, T. R. (1988). A criticism of the aggregation model of coexistence: Non-independent distribution of dipteran species on ephemeral resources. *Am. Naturalist* **131**:453-458.

Zawitowski, S., and Richmond, R. C. (1986). Inhibition of courtship and mating of *Drosophila melanogaster* by the male-produced lipid, *cis*-vaccenyl acetate. *J. Insect Physiol.* **32**:189-192.

Part III
Genetics in Chemical Communication

13
A Hamster Macromolecular Pheromone Belongs to a Family of Transport and Odorant-Binding Proteins

Foteos Macrides

Worcester Foundation for Experimental Biology, Shrewsbury, Massachusetts

Alan G. Singer

Monell Chemical Senses Center, Philadelphia, Pennsylvania

I. INTRODUCTION

Hamster vaginal discharge contains a multipheromonal system in which a highly volatile compound, dimethyl disulfide, attracts male conspecifics and promotes contact investigation of the female's genitals, thereby permitting a sexually arousing macromolecular pheromone to be detected via the vomeronasal organ (VNO) of the male's accessory olfactory system (Singer et al., 1976, 1980, 1984a,b; Clancy et al., 1984b; Macrides et al., 1984a,b). We review here work on the isolation and characterization of the aphrodisiac pheromone, which proved to be a new member of the α_{2u}-globulin superfamily of extracellular proteins and has been given the functional name aphrodisin (Singer et al., 1986; Henzel et al., 1988). Several proteins in this superfamily have known or suspected transport functions, including retinol binding protein, β-lactoglobulin, the major urinary proteins, and recently characterized odorant binding proteins that are secreted abundantly in nasal mucus (Pervaiz and Brew, 1985; Pevsner et al., 1985, 1988; Cavaggioni et al., 1987; Sawyer, 1987; Singer et al., 1988; Clancy et al., 1988). We discuss some of the structural and functional attributes of proteins in this superfamily and the implications of these attributes for the evolution and possible mechanisms of pheromonal action of aphrodisin.

II. ISOLATION AND CHARACTERIZATION OF APHRODISIN

An unusual and experimentally advantageous behavioral characteristic of male hamsters is that they will attempt to copulate with an anesthetized male conspecific ("surrogate female") whose hindquarters have been anointed with vaginal discharge, thus providing a reliable means to assay the aphrodisiac activity of fractions or compounds from the discharge under conditions in which no other female-derived stimuli are present (Macrides et al., 1984a). The surrogate female is placed in the subject male's home cage and positioned on a mound of bedding to simulate a lordotic posture. Active chemosensory stimuli on average elicit two to three intromission attempts (mounts accompanied by repetitive pelvic thrusting) during observation periods lasting 5 minutes. Assays typically employ 20 subject males, and each male in an assay is tested with each of up to five experimental and control stimuli presented in counterbalanced orders on different days at 2-4 day intervals to minimize habituation or extinction. Differences in response levels are analyzed nonparametrically with the Wilcoxon matched-pairs signed-rank test (two-tailed, within-subject comparisons) to minimize effects of extreme values obtained when some subjects occasionally exhibit much higher levels of response to active stimuli.

Our initial studies with this assay (Macrides et al., 1977; Singer et al., 1980) indicated that, although dimethyl disulfide accounts for most of the attraction to vaginal discharge under other testing conditions in which the males cannot physically contact the discharge with their snouts, the volatile attractant pheromone does not also elicit copulatory behavior. Low-molecular-mass fractions from gel permeation chromatography of the discharge similarly were devoid of aphrodisiac activity. High activity, equivalent to that of unfractionated discharge, was recovered consistently in macromolecular fractions containing soluble proteins with masses between approximately 10 and 50 kD. Subsequent studies (Macrides et al., 1984b; Singer et al., 1984a) revealed that the concentrations of these proteins in the discharge varied with the estrous cycle (highest during estrus) and were substantially reduced by ovariectomy or hypophysectomy with corresponding reductions in the aphrodisiac activity of the macromolecular fractions. Furthermore, the activity in the fractions was destroyed by proteolysis, indicating that the presence of intact protein is necessary. All fractions with proteins of mass greater than 50 kD or less than 10 kD were inactive. The aphrodisiac activity thus appeared to be attributable to one or more endocrine-regulated proteins and/or ligand(s) selectively bound to the protein(s) for transport to the male's chemosensory system during contact genital investigation of the female.

Further chromatographic separation of the endocrine-regulated proteins yielded a macromolecular fraction that remained fully active and contained the most abundant and second most abundant soluble proteins in estrous vaginal discharge and trace quantitites of a minor soluble protein. This fraction remained odorous to the human nose and thus clearly contained bound volatile compounds. It was "deodorized" by passage through a column of the absorbent XAD-4 (Rohm and Haas), a neutral polystyrene resin that is generally effective in separating lipophilic ligands from carrier proteins. This deodorized high-molecular-mass fraction (HMF) elicited significant investigatory behavior in male hamsters only under testing conditions that permitted physical contact with the stimulus, but it showed high aphrodisiac activity in the surrogate female assay (Singer et al., 1984b). These results made it likely that the aphrodisiac pheromone is introduced into the nasal cavity in a nonvolatile state and may be detected through vomeronasal sampling of compounds dissolved in mucus of the nasal vestibule (Meredith and O'Connell, 1979). Peripheral deafferentation studies verified that the active material in the HMF is detected via the VNO (Clancy et al., 1984b).

The major soluble protein in estrous vaginal discharge, aphrodisin, was purified to homogeneity and was found to account for most of the activity in the HMF (Singer et al., 1986). The purification includes XAD-4 adsorption chromatography of the isolated protein to remove odorous volatile compounds. The second most abundant vaginal discharge protein (SMVDP) and the minor soluble protein in the HMF have recently been isolated and are devoid of aphrodisiac activity when tested in the absence of aphrodisin (Singer et al., 1989). Aphrodisin thus appears to be unique among hamster vaginal discharge proteins in its ability to facilitate copulatory behavior in male hamsters. It is an acidic protein, with an isoelectric point at approximately pH 5, and has a molecular mass of 17 kD (Figure 1). The SMVDP is comparably acidic and behaves on chromatography as a dimer of two identical 17 kD polypeptide chains, suggesting that it may be an inactive dimer of aphrodisin. However, parallel amino acid analyses of the SMVDP and aphrodisin have shown that this is not the case (Singer et al., 1989).

The average wet weight of vaginal discharge extruded by estrous hamsters on tactile stimulation of the genital region is 20 mg, and this quantity of estrous discharge contains approximately 100 μg aphrodisin (Singer et al., 1986). For purposes of bioassay, we refer to the amount of a compound normally contained in 20 mg estrous discharge or the amount of a fraction obtained from 20 mg as a female equivalent (FE). The dose of HMF that achieves maximal activity, statistically indistinguishable from that with an equal or higher dose of unfractionated discharge, is 0.1 FE. Figure 2 shows the dose-response curve for aphrodisin and compares it with the response to 0.1 FE of the HMF (active control). In assays with proteinaceous stimuli, the samples are lyophilized in coded vials (for subsequent testing blind) and are reconstituted with distilled water such that the intended dose is presented in 50 μl solvent. Because the protein is likely to be adsorbed to the fur and

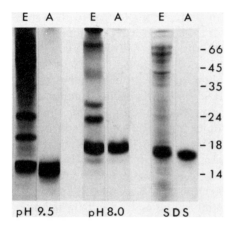

FIGURE 1 Electrophoresis of vaginal discharge extract and aphrodisin. Aqueous extracts of vaginal discharge (E) compared with isolated aphrodisin (A) by polyacrylamide gel electrophoresis in three different systems: at pH 9.5, at pH 8.0, and in 0.1% sodium dodecyl sulfate (SDS). The sample loads in the three systems were 0.2, 0.2, and 0.05 female equivalents (see text) of aqueous extract and 48, 20, and 17 μg isolated protein, respectively. The scale on the right indicates, for the SDS system, the mobilities and the molecular mass (kD) of marker proteins, which were bovine serum albumin (66 kD), hen ovalbumin (45 kD), porcine pepsin (35 kD), bovine pancreas trypsinogen (24 kD), bovine β-lactoglobulin (18 kD), and hen egg lysozyme (14 kD). (Reproduced from Singer et al., 1986.)

skin of the surrogate, the actual amount of protein that enters the subject male's nasal cavity and can reach the VNO may be considerably less than the specified dose (see later) or the amount that enters under natural conditions in which the female extrudes a bolus of discharge. The applied dose of aphrodisin that elicits maximal response levels in the surrogate female assay is 0.1-0.4 FE (10-40 μg).

As is the case for the HMF, the activity of aphrodisin is destroyed by proteolysis (Singer et al., 1986). More restricted cleavage of aphrodisin with a variety of endopeptidases in the hope of generating an active fragment also have consistently resulted in complete loss of activity. The activity similarly is eliminated if the protein is denatured by boiling. It thus appears that the structural integrity and solubility of the protein are critically important for aphrodisiac activity.

Although dialysis, gel filtration, and adsorption chromatography with polystyrene resin are generally effective in separating low-molecular-mass compounds from proteins, that the pheromonal activity of aphrodisin survived all these separation techniques does not rigorously exclude the possibility

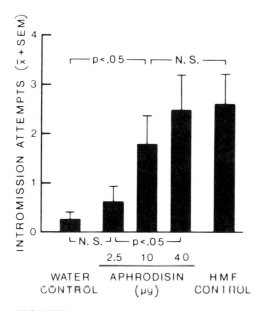

FIGURE 2 Behavioral assay of aphrodisin on 20 male hamsters. See text for details of sample preparations and behavioral testing. The HMF was tested at a dose of 0.1 female equivalents to contain approximately 10 μg aphrodisin, one-tenth the amount normally extruded by an estrous female and deposited on the male's snout during contact investigation. (Reproduced from Singer et al., 1986.)

that aphrodisin serves to transport a ligand, such as a volatile odorant, steroid, or peptide, to receptors in the VNO. To further assess the possible existence of such a ligand, aphrodisin was dissolved in 25% acetic acid and subjected to reversed-phase high-pressure liquid chromatography (HPLC), and the recovered protein versus side fractions were tested in the surrogate female assay (Figure 3). This is a denaturing procedure, and the recovered protein was insoluble in water but fortunately could be returned to its water-soluble, presumably native, state by dialysis of the acetic acid solution in water and then in 0.05 M NH_4HCO_3. The protein was verified to be aphrodisin by electrophoresis and was found to have retained its activity, whereas the side fractions were not significantly active. These results seem to rule out a peptide or other acid-soluble ligand. In the event that a ligand co-migrated with aphrodisin on HPLC, the 25% acetic acid solution of aphrodisin was also subjected to gel permeation chromatography and eluted with 50% acetic acid, and the restored (solubilized) aphrodisin fraction versus lower molecular mass side fraction was tested for aphrodisiac activity (Figure 4). Again, aphrodisin retained its activity and the side fraction was inactive. Although the

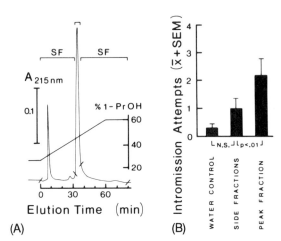

FIGURE 3 Reversed-phase HPLC and behavioral activity of acid-treated aphrodisin. (A) A representative chromatogram obtained from the analysis of 200 μg aphrodisin in 25% acetic acid. HPLC was performed on a 4.6 × 250 mm Proteosil 300 octyl column (Whatman). Adsorbed compounds were eluted at 0.5 ml/minute with a 60 minute linear gradient from 24 to 60% l-propanol in 0.15% trifluoroacetic acid and then with the final eluant for an additional 20 minutes. The peak at 8 minutes is from the acetic acid solvent. The peak fraction and side fractions (SF) were collected, as indicated by the slashes and brackets, from a series of such HPLC runs for the behavioral assay (B). The early and late side fractions were combined before lyophilization. They were not dialyzed to minimize loss of any low-molecular-weight compounds dissociated from the protein by the acid treatment. Samples were prepared for behavioral testing to contain the amount of combined side fraction obtained from 50 μg acid-treated protein and were assayed for aphrodisiac activity on 20 male hamsters. The peak fraction was identified as aphrodisin (see text) and was tested at a dose of 40 μg in the behavioral assay. (Reproduced from Singer et al., 1986.)

possibility that aphrodisin transports an active ligand still cannot be completely excluded, the alternative possibility, that its polypeptide chain contains an active site, seems to merit serious consideration. This would not be unprecedented in the vertebrate chemosenses. Proteins can be effective gustatory stimuli (Cagan, 1973).

The complete primary structure of aphrodisin has been determined by sequence analysis (Figure 5), together with analyses of its posttranslational modifications (Henzel et al., 1988). The protein contains 151 amino acid residues, with the amino-terminal glutamine (Q) cyclized to pyroglutamate. It has two glycosylation sites, at asparagine (N) residues in positions 41 and

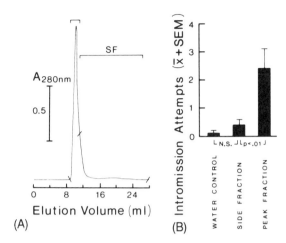

FIGURE 1 Gel permeation chromatography and behavioral activity of acid-treated aphrodisin. (A) A solution of aphrodisin (1.8 mg in 25% acetic acid) was chromatographed on a 1.5 × 11.5 cm column of Sephadex G-25 (void volume = 9 ml; total column volume = 20 ml) eluted with 50% acetic acid at 8.1 ml/h. The eluate was collected in two fractions as indicated by slashes and brackets. The peak fraction was identified as aphrodisin and was prepared for assay with the same procedures as for the peak fraction in Figure 3. It was tested in the behavioral assay (B) at a dose of 40 µg on 20 male hamsters. The side fraction (SF), which should contain any peptides or other acid-soluble small (molecular mass < 5 kD) compounds dissociated from the protein, was not submitted to dialysis. After lyophilization, samples were prepared for the behavioral assay to contain the amount of side fraction obtained from 40 µg acid-treated protein. (Reproduced from Singer et al., 1986.)

69, which each have a single residue of *N*-acetylglucosamine. It has 4 cysteine (C) residues, which form two disulfide bridges between positions 38 and 42 and between positions 57 and 149.

A computer search of the National Biomedical Research Foundation Library protein data bank revealed homology between the primary sequence of aphrodisin and a set of rodent proteins with the potential for a pheromonal function (see later). These consist of α_{2u}-globulin (UART), which is secreted in the urine of male rats, and the major urinary proteins of mice (MUP), a multigene family secreted by both males and females. These other rodent proteins appear to be synthesized in the liver under the control of gonadal hormones (Finlayson et al., 1963, 1965; Kuhn et al., 1984). The homology of aphrodisin to UART and a representative member of the mouse protein family (MUSMUPA) is illustrated in Figure 6. The urinary proteins are prototypical of the recently proposed α_{2u}-globulin superfamily (Pervaiz

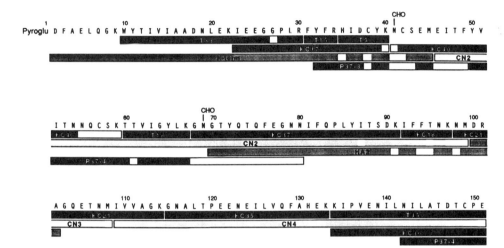

FIGURE 5 Amino acid sequence of aphrodisin. The alignment of peptides used for determining the sequence is represented by boxes. The shading indicates residues identified by Edman degredation. The designations T, KC, CN, HA, and P represent peptides derived by trypsin, lysine-C, cyanogen bromide, hydroxylamine, and pepsin digestions, respectively. Treatment of intact aphrodisin with pyroglutamate aminopeptidase unblocked the amino terminus, resulting in the identification of residues 1-9. FAB mass spectrometry showed that the asparagine residues at positions 41 and 69 each have a single residue of N-acetylglucosamine (CHO) and showed disulfide bonds (not illustrated) between the cysteine residues at positions 38 and 42 and at positions 57 and 149. (Reproduced from Henzel et al., 1988.)

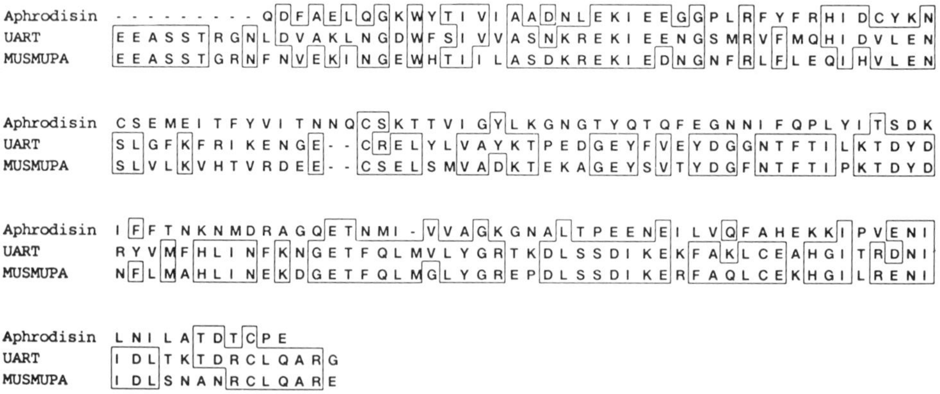

FIGURE 6 Homology of aphrodisin with rat α_{2u}-globulin (UART) and mouse major urinary protein (MUSMUPA). The sequence for UART is from Unterman et al. (1981). That for MUSMUPA is from Kuhn et al. (1984). (Reproduced from Henzel et al., 1988.)

and Brew, 1985), which appears to be ancient in that it includes proteins from insects as well as vertebrates. Distinct members of the superfamily show rather limited similarity in their primary sequences, rarely more than 25% identical residues (Pervaiz and Brew, 1987). They are all relatively small (150-190 amino acid residues), and sequences appear to be best conserved in the amino-terminal region. As shown in Figure 7 for 10 members, this region contains the characteristic motif -O-X-Gly-X-Trp-Y-X-O-, where O is hydrophobic, X unrestricted, and Y aromatic or basic. Another characteristic is a disulfide bridge between a cysteine residue near the carboxyl terminus and one about 100 residues toward the amino terminus. Figure 6 shows the alignment of the appropriate cysteine residues and the lower sequence homology of aphrodisin with the urinary proteins in the region between these residues relative to that near the amino terminus. Based on gene sequences determined for five distinct members of the superfamily (UART, β-lactoglobulin, retinol binding protein, apolipoprotein D, and α_1-acid glycoprotein), another characteristic that relates the proteins is a similar organization of introns and exons (Ali and Clark, 1988).

The three-dimensional structure has been determined by x-ray crystallography for four members of the superfamily: β-lactoglobulin, retinol binding protein, and two insect bilin binding proteins. Despite the low homologies in the primary sequences of these proteins, they share a unique three-dimensional organization. All of them consist of two sets of four antiparallel β-strands, arranged in two orthogonally stranded β-sheets that form a flattened cone flanked by a "handlelike" section of α-helix containing about 10 amino acid residues near the carboxyl terminus (Sawyer, 1987). The interior

```
Aphrodisin   - - - - - - - - - - - - - - - - - Q D F A E L Q G K W Y T I V I A A D N L E K I E E G G P L R
BLAC         - - - - - - - L I V T Q T M K G L D I Q K V A G T W Y C L A M A A S D I S I I D A Q S A P L
INSEC        G D I F Y P G Y C P D V K P V N D F D L S A F A G A W H E I A K L P L E N E N Q G K C T I A E
MUSMUPA      - - - - - - - E E A S S T G R N F N V E K I N G E W H T I I L A S D K R E K I E D N G N F R
BG           - - - - - Q C Q A D L P P V M K G L E E N K V T G V W Y G I A A A S N C K Q F L Q M K S D N M
RATANDRO     - - - - - - - - - - - A V V K D F D I S K F L G F W Y E I A F A S K M G T P G L A H K E E K
RRBP         - - - E R D C R V S S F R V K E N F D K A R F S G L W Y A I A K K D P E G L F L Q D N I I A E
UART         - - - - - - - E E A S S T R G N L D V A K L N G D W F S I V V A S N K R E K I E E N G S M R
APO-D        - Q A F H L G K C P N P P V Q E N F D V N K Y L G R W Y E I E K I P T T F E N G R C I Q A N Y
HC           - - G P V P T P P D N I Q V Q E N F N I S R I Y G K W Y N L A I G S T C P L K I M D R M T V S
```

FIGURE 7 Consensus sequence of aphrodisin with the amino-terminal region of members of the α_{2u}-globulin protein superfamily. Proteins are bovine β-lactoglobulin (BLAC), tobacco hornworm insecticyanin protein (INSEC), mouse major urinary protein (MUSMUPA), Bowman's gland protein (BG), rat androgen-dependent protein (RATANDRO), rat retinol binding protein (RRBP), rat α_{2u}-globulin (UART), apolipoprotein D (APO-D), and α_1-microglobulin (HC). The motif -O-X-Gly-X-Trp-Y-X-O (O is hydrophobic, X unrestricted, and Y aromatic or basic) is shown shaded. (Reproduced from Henzel et al., 1988.)

of the cone contains retinol (vitamin A) in retinol binding protein, and in the insect bilin binding proteins contains the pigment biliverdin IX.

The recently characterized proteins secreted in rat and bovine nasal mucus that bind a variety of pyrazines, thiazoles, and other odorous compounds have been categorized in this superfamily (Cavaggioni et al., 1987; Pevsner et al., 1988). Their homology with aphrodisin is shown in Figure 8. Complete sequence data are available for rat odorant binding protein (OBP). The analyses in rat indicate that this is a member of a multigene family (Pevsner et al., 1988). The primary sequence homology between aphrodisin and the rat OBP is approximately 40%, unusually high for any two distinct members of the superfamily. In addition to cysteine residues in positions that characteristically form a disulfide bridge between the carboxyl-terminal region and a more central region of the polypeptide backbone, the rat OBP has cysteine residues that line up with those that form a second disulfide bridge in aphrodisin. The OBP does not, however, have corresponding glycosylation sites. The bracket in Figure 8 indicates the region that corresponds with the α-helical "handle" in the four proteins analyzed by x-ray crystallography. For both aphrodisin and the rat OBP, Chou-Fasman (Chou and Fasman, 1978) and Robson (Garnier et al., 1978) analyses predict that this region is α-helical and flanked by extended regions. There is, however, only one instance of

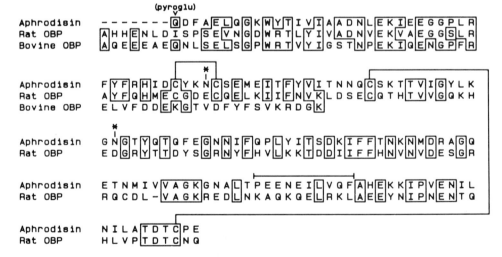

FIGURE 8 Homology of aphrodisin with odorant binding protein. The sequence for rat odorant binding protein (OBP) is from Pevsner et al. (1988). That for bovine OBP is from Cavaggioni et al. (1987). The disulfide bridges of aphrodisin are indicated by connecting lines, and the N-acetylglucosamine residues are represented by asterisks. The bracket indicates the region predicted to form an α-helical "handle," with the remainder of the protein forming a cuplike pair of β-sheets (see text).

acidic identical amino acid residues in this region for the two proteins, and this region is highly acidic in aphrodisin, with three glutamic acid (E) residues. This is also a region of low homology between aphrodisin and the urinary proteins. They have one glutamic acid residue in identical positions, and aphrodisin has a glutamine residue in alignment with MUSMUPA (Figure 6). Overall, aphrodisin and rat OBP are distinct proteins but appear to be close relatives in the α_{2u}-globulin superfamily and are considerably closer relatives than either is to the urinary proteins.

It is interesting in this regard that rat OBP appears to be synthesized in the lateral nasal glands and secreted via Stenson's duct into the vestibule of the nasal cavity (Pevsner et al., 1986). Although speculation about the function of OBP has centered on a possible role in facilitating transport of inhaled hydrophobic volatiles through the nasal mucous layer that coats the receptor neurons of the main olfactory system, based on anatomic considerations it seems just as, if not more, likely that it serves to trap and transport small molecules into the lumen of the VNO for sampling by the vomeronasal receptor neurons of the main olfactory system, based on anatomical considerationship between aphrodisin and OBP thus may not be surprising. Although we have not yet found any direct evidence that a ligand is critical for the activity of aphrodisin or is present after XAD-4 adsorption chromatography, the obvious relationship of aphrodisin with OBP keeps us wondering, and still looking for one (see later).

III. POSSIBLE MECHANISMS OF PHEROMONAL ACTION

Mice and rats generally utilize urine for pheromonal communication, and thus, like aphrodisin in hamsters, the urinary proteins have the potential for a pheromonal function. The milk protein β-lactoglobulin similarly has the potential for a pheromonal function in mother-offspring interactions of ungulates. The possibility of a pheromonal function also exists for OBP. Stenson's duct is contiguous with the nasolacrimal duct, which carries secretions of the lacrimal and Harderian glands, and OBP is present in tears from the inner canthus of rats (Pevsner et al., 1986). The secretions of the lateral nasal glands thus appear to be combined with those of the lacrimal and Harderian glands. The secretions to the inner canthus and nares in rodents are smeared onto the snout during grooming and have been shown to be involved in chemical communication during conspecific social interactions (Thiessen et al., 1976). Interestingly, in the surrogate female assay, if the surrogate is anointed with an inactive compound or fraction the subject male invests much more time in contact investigation of the surrogate's snout and eyes than of the genitals (Macrides et al., 1984a).

These considerations, and the observation discussed earlier that distinct members of the α_{2u}-globulin superfamily have an extremely similar three-dimensional organization, led us to test heterospecific members of the superfamily for aphrodisiac activity in male hamsters. Neither female mouse MUP nor bovine β-lactoglobulin had any aphrodisiac activity in the surrogate female assay (Singer et al., 1989). We have not yet tested OBP, but it appears that more than the overall shape of the protein is important for activity. Although this may not seem surprising, it should be borne in mind that under natural conditions the opportunity for contact investigation of another individual's secretions is behaviorally regulated, and pheromonal communication via a nonvolatile chemosignal may not have required as high a degree of chemical or species specificity as expected for volatile pheromones.

We have also begun to test the possibility that other members of the α_{2u}-globulin superfamily have pheromonal activity in conspecifics. Although we do not have a comparably efficient and reliable behavioral assay for aphrodisiac activity in male mice or rats (precopulatory courtship behaviors appear to be more important for the copulatory behavior of males in these species than in hamsters), female mouse urine reliably elicits an episode of luteinizing hormone (LH) secretion in male mice, the urinary pheromone is detected via the VNO, and the neuroendocrine response appears to be a correlate of sexual arousal (Coquelin et al., 1984; Clancy et al., 1984a). Our biochemical analyses (Singer et al., 1988) indicate that this pheromone is not a protein but does exhibit binding to the MUP. The pheromone can be separated from the MUP by routine chromatographic procedures without appreciable loss of its ability to elicit an LH secretion episode, and the MUP subjected to XAD-4 adsorption chromagraphy is devoid of this activity. To the extent that the MUP is involved in this neuroendocrine response, its mechanism of pheromonal action appears to be the most obvious that might be suspected for proteins in this superfamily: that it is a "carrier protein" for an "active" ligand.

Aside from OBP, whose function as we have discussed is still open to speculation, the only other member of this superfamily that "transports" a ligand with a known function to known targets is retinol binding protein. Although several additional members have been shown to bind and "carry" a ligand, biologically important transport functions are merely suspected and have yet to be rigorously demonstrated for them. In the case of retinol binding protein, the target cells have receptors for the protein, and binding of the protein-ligand complex to the target cell membrane is necessary for transport of the vitamin into the cell (Maraini and Gozzoli, 1975; Rask and Peterson, 1976). The retinol appears to induce an important conformational change in the retinol binding protein because the binding affinity of the protein to transthyretin is reduced when retinol is not present in the β-pleated cone, and such a change appears to regulate the binding of the protein-retinol complex versus apoprotein to the membrane receptors (Rask and Peterson, 1976).

Binding of the complex to the receptors similarly appears to yield a conformational change or other effect that serves to release retinol from its binding protein appropriately for transport of the vitamin through the membrane into the cell. These findings have noteworthy implications for any consideration of possible pheromonal functions of proteins in this superfamily. If the proteins can have receptors in target cell membranes, such receptors in turn could have evolved coupling to an electrogenic process in chemosensory receptor neurons. Although the proteins may have a characteristic ligand, the pheromonal function of the ligand, if any, could be like that of an allosteric effector in enzyme action, not that of the "active" (electrogenic) stimulus itself. Another possibility is distinct receptor molecules for both the protein and the ligand, with coordinate or serial binding to the receptors required for a maximal electrogenic response.

Let us first consider these possibilities in relation to what is known at present about OBP. The reported binding affinities of odorous volatile compounds to OBP are relatively low (Pevsner et al., 1986), considerably lower than the affinity of retinol for retinol binding protein (Cogan et al., 1976). Relatively low affinity binding would be appropriate for the hypothesized function of facilitating transport of lipophilic volatiles through an aqueous phase (mucus) to presumably higher affinity receptor molecules of the chemosensory receptor neurons. It is possible, however, that the OBP-ligand complex becomes a more effective or specific stimulus for the receptor neurons. To carry this thought one step further, if OBP is an exocrine secretory product, as suggested by its presence in tears, it may have a very high affinity ligand like other members of the superfamily and the complex might become a potent stimulus for conspecifics. Note in Figure 8 that in the region where sequence data are available for both the rat and bovine OBP, the homology between them is no greater than that between rat OBP and hamster aphrodisin. Under these circumstances, a hypothetical high-affinity ligand for OBP could be a reproductively or socially relevant metabolite common to different species, and the hypothetical pheromone complex could have species specificity due to heterospecific structural differences in OBP.

Recent findings in our attempts to characterize the structural basis for aphrodisin's pheromonal activity suggest that these possibilities and speculations may be more than idle thoughts. This work is still preliminary, but some of our findings have been replicated sufficiently to merit discussion here. Colleagues at Genentech, Inc., under the direction of Drs. Hugh Niall and Patrick Gray, have been successful in cloning and expressing the complete polypeptide backbone of aphrodisin in *Escherichia coli*. The bacteria properly trim the leader sequence and secrete the protein into the periplasmic space, which should favor formation of the disulfide bonds. However, the amino-terminal glutamine is not cyclized, and the bacteria do not glycosylate the protein so that the N-acetylglucosamine residues are absent at positions Asn_{41} and Asn_{69} (see Figures 5 and 8). Because it is extremely unlikely that the bacteria would also be expressing a hypothetical high-affinity ligand for aphrodisin, we are in a position to assess whether the polypeptide backbone is sufficient for aphrodisiac activity and whether post-translational modifications contribute to the activity. Also, we anticipated, if a hamster-specific ligand were required for activity, the bacterially expressed protein (ECAPH) could serve as a binding protein for testing low-molecular-mass fractions or compounds in the surrogate female assay.

Across the dose range from 10 to 160 μg (0.1-1.6 FE), the ECAPH generally elicits somewhat higher mean response levels than the solvent or heterospecific protein (β-lactoglobulin) control stimuli, but the differences in response levels are not statistically significant in our standard assays with 20 subject males. The amino acid backbone, by itself, thus cannot account for the activity of hamster aphrodisin. We are able to impart significant activity to the ECAPH by mixing it with organic (ethanol, butanol, or hexane) extracts of hamster vaginal discharge. The ECAPH may be adsorbing a volatile attractant that elicits more vigorous contact investigation and in effect increases the dose, but if so, the adsorbed attractant does not appear to be dimethyl disulfide. Mixing the ECAPH with dimethyl disulfide did not make it significantly active.

A possible explanation for these results is that there is indeed a high-affinity critical ligand associated with hamster aphrodisin that is being transferred to the ECAPH from vaginal discharge extracts. Purification and identification of the compound(s) transferred to the ECAPH and retained by the bacterial protein after adsorption chromatography may enable us to assess whether a specific ligand is independently electrogenic and/or functions to alter the conformation of the binding protein so that the protein interacts more effectively with its own receptors. A complementary possibility is that the N-acetylglucosamine residues of the hamster protein and/or the amino-terminal pyroglutamate normally plays an important role in the interaction with membrane receptors. It is conceivable, for example, that the pheromone initially evolved as a protein-ligand complex that is adequately simulated by the addition of a hamster-derived ligand to the nonglycosylated bacterial protein but that posttranslational modifications resulting in the presence of covalently bonded N-acetylglucosamine residues, which make presumably exposed asparagine side chains more polar, normally serve to maintain the protein in a pheromonally active state.

In conclusion, we are still unable to identify with certainty the mechanisms of pheromonal action by which aphrodisin produces sexual arousal in hamsters, but we appear to be closing in on some hard answers to our questions. As is common in basic research, our quest has led us down roads we had not expected to travel. When we first contemplated this research we expected to be seeking odorous volatile compounds that might serve as experimentally advantageous probes for studying central physiological and neurochemical mechanisms by which socially meaningful sensory stimuli influence affective functions (Macrides, 1976). At present we are pondering the evolutionary and functional relationships between binding proteins and chemosensory communication and how our probes might help elucidate mechanisms of stimulus transport and transduction in peripheral chemosensory neurons. When we return to the central nervous system, we are still hoping to be armed with a battery of both structurally and functionally well-defined "odorants" but are also hoping to be armed with an advantageous battery of nonvolatile products from genetic engineering.

ACKNOWLEDGMENTS

We thank Andrew N. Clancy, Steven Hecht, Taichang Jang, and Mary Murtland for their help in preparing this manuscript. Our research discussed here was supported in part by National Institutes of Health Grants NS12344 (to Macrides) and HD19764 (to Singer).

REFERENCES

Ali, S., and Clark, A. J. (1988). Characterization of the gene encoding ovine β-lactoglobulin. Similarity to the genes for retinol-binding protein and other secretory proteins. *J. Mol. Biol.* **199**:415-416.

Cagan, R. H. (1973). Chemostimulatory protein: A new type of taste stimulus. *Science* **181**:32-35.

Cavaggioni, A., Sorbi, R. T., Keen, J. N., Pappin, D. J. C., and Findlay, J. B. C. (1987). Homology between the pyrazine-binding protein from nasal mucosa and major urinary proteins. *FEBS Lett.* **212**:225-228.

Chou, P. Y., and Fasman, G. D. (1978). Prediction of the secondary structure of proteins from their amino acid sequence. *Adv. Enzymol.* **47**:45-148.

Clancy, A. N., Coquelin, A., Macrides, F., Gorski, R. A., and Noble, E. P. (1984a). Sexual behavior and aggression in male mice: Involvement of the vomeronasal system. *J. Neurosci.* **4**:2222-2229.

Clancy, A. N., Macrides, F., Singer, A. G., and Agosta, W. C. (1984b). Male hamster copulatory responses to a high molecular weight fraction of vaginal discharge: Effects of vomeronasal organ removal. *Physiol. Behav.* **33**:653-660.

Clancy, A. N., Singer, A. G., Macrides, F., Bronson, F. H., and Agosta, W. C. (1988). Experiential and endocrine dependence of gonadotropin responses in male mice to conspecific urine. *Biol. Reprod.* **38**:183-191.

Cogan, U., Kopelman, M., Mokady, S., and Shinitzky, M. (1976). Binding affinities of retinol and related compounds to retinol binding proteins. *Eur. J. Biochem.* **65**:71-78.

Coquelin, A., Clancy, A. N., Macrides, F., Noble, E. P., and Gorski, R. A. (1984). Pheromonally induced release of luteinizing hormone in male mice: Involvement of the vomeronasal system. *J. Neurosci.* **4**:2230-2236.

Finlayson, J. S., Potter, M., and Runner, C. C. (1963). Electrophoretic variation and sex dimorphism of the major urinary protein complex in inbred mice: A new genetic marker. *JNCI* **31**:91-107.

Finlayson, J. S., Asofsky, R., Potter, M., and Runner, C. C. (1965). Major urinary protein complex of normal mice: Origin. *Science* **149**:981-982.

Garnier, J., Osguthorpe, D. J., and Robson, B. (1978). Analysis of the accuracy and implications of simple methods for predicting the secondary structure of globular proteins. *J. Mol. Biol.* **120**:97-120.

Henzel, W. J., Rodriquez, H., Singer, A. G., Stults, J. T., Macrides, F., Agosta, W. C., and Niall, H. (1988). The primary structure of aphrodisin. *J. Biol. Chem.* **263**:16682-16687.

Kuhn, N. J., Woodworth-Gutai, M., Gross, K. W., and Held, W. A. (1984). Subfamilies of the mouse major urinary protein (MUP) multi-gene family: Sequence analysis of cDNA clones and differential regulation in the liver. *Nucleic Acids Res.* **12**:6073-6090.

Macrides, F. (1976). Olfactory influences on neuroendocrine function in mammals. In *Mammalian Olfaction, Reproductive Processes, and Behavior,* R. L. Doty (Ed.). Academic Press, New York, pp. 29-65.

Macrides, F., Johnson, P. A., and Schneider, S. P. (1977). Responses of the male golden hamster to vaginal secretion and dimethyl disulfide: Attraction versus sexual behavior. *Behav. Biol.* **20**:377-386.

Macrides, F., Clancy, A. N., Singer, A. G., and Agosta, W. C. (1984a). Male hamster investigatory and copulatory responses to vaginal discharge: An attempt to impart sexual significance to an arbitrary chemosensory stimulus. *Physiol. Behav.* **33**:627-632.

Macrides, F., Singer, A. G., Clancy, A. N., Goldman, B. D., and Agosta, W. C. (1984b). Male hamster investigatory and copulatory responses to vaginal discharge: Relationship to the endocrine status of females. *Physiol. Behav.* **33**:633-637.

Maraini, G., and Gozzoli, F. (1975). Binding of retinol to isolated retinal pigment epithelium in the presence and absence of retinol-binding protein. *Invest. Ophthalmol.* **14**:785-787.

Meredith, M., and O'Connell, R. J. (1979). Efferent control of stimulus access to the hamster vomeronasal organ. *J. Physiol. (Lond.)* **286**:301-316.

Pervaiz, S., and Brew, K. (1985). Homology of β-lactoglobulin, serum retinol-binding protein, and protein HC. *Science* **228**:335-337.

Pervaiz, S., and Brew, K. (1987). Homology and structure-function correlations between α_1-acid glycoprotein and serum retinol-binding protein and its relatives. *FASEB J.* **1**:209-214.

Pevsner, J., Trifiletti, R. R., Strittmatter, S. M., and Snyder, S. H. (1985). Isolation and characterization of an olfactory receptor protein for odorant pyrazines. *Proc. Natl. Acad. Sci. USA* **82**:3050-3054.

Pevsner, J., Sklar, P. B., and Snyder, S. H. (1986). Odorant-binding protein: Localization to nasal glands and secretions. *Proc. Natl. Acad. Sci. USA* **83**:4942-4946.

Pevsner, J., Reed, R. R., Feinstein, P. G., and Snyder, S. H. (1988). Molecular cloning of odorant-binding protein: Member of a ligand carrier family. *Science* **241**: 336-339.

Rask, L., and Peterson, P. A. (1976). In vitro uptake of vitamin A from the retinol-binding plasma protein to mucosal epithelial cells from the monkey's small intestine. *J. Biol. Chem.* **251**:6360-6366.

Sawyer, L. (1987). One fold among many. *Nature* **327**:659.

Singer, A. G., Agosta, W. C., O'Connell, R. J., Pfaffmann, C., Bowen, D. V., and Field, F. (1976). Dimethyl disulfide: An attractant pheromone in hamster vaginal secretion. *Science* **191**:948-950.

Singer, A. G., Macrides, F., and Agosta, W. C. (1980). Chemical studies of hamster reproductive pheromones. In *Chemical Signals*, D. Müller-Schwarze and R. M. Silverstein (Eds.). Plenum, New York, pp. 365-375.

Singer, A. G., Clancy, A. N., Macrides, F., and Agosta, W. C. (1984a). Chemical studies of hamster vaginal discharge: Effects of endocrine ablation and protein digestion on behaviorally active macromolecular fractions. *Physiol. Behav.* **33**: 639-643.

Singer, A. G., Clancy, A. N., Macrides, F., and Agosta, W. C. (1984b). Chemical studies of hamster vaginal discharge: Male behavioral responses to a high molecular weight fraction require physical contact. *Physiol. Behav.* **33**:645-651.

Singer, A. G., Macrides, F., Clancy, A. N., and Agosta, W. C. (1986). Purification and analysis of a proteinaceous aphrodisiac pheromone from hamster vaginal discharge. *J. Biol. Chem.* **261**:13323-13326.

Singer, A. G., Clancy, A. N., Macrides, F., Agosta, W. C., and Bronson, F. H. (1988). Chemical properties of a female mouse pheromone that stimulates gonadotropin secretion in males. *Biol. Reprod.* **38**:193-199.

Singer, A. G., Clancy, A. N., and Macrides, F. (1989). Conspecific and heterospecific proteins related to aphrodisin lack aphrodisiac activity in male hamsters. *Chem. Senses* **14**:565-575.

Thiessen, D. D., Clancy, A., and Goodwin, M. (1976). Harderian gland pheromone in the Mongolian gerbil *Meriones unguiculatus*. *J. Chem. Ecol.* **2**:231-238.

Unterman, R. D., Lynch, K. R., Nakhasi, H. L., Dolan, K. P., Hamilton, J. W., Cohn, D. V., and Feigelson, P. (1982). Cloning and sequence of several alpha-2u-globulin cDNAs. *Proc. Natl. Acad. Sci. USA* **78**:3478-3482.

14
Excretion of Transplantation Antigens as Signals of Genetic Individuality

Bruce Roser

Quadrant Research Foundation, Cambridge, England

Richard E. Brown

Dalhousie University, Halifax, Nova Scotia, Canada

Prim B. Singh

Institute of Animal Physiology, Cambridge, England

I. INTRODUCTION

Transplantation antigens, as the name implies, were originally identified as the entities that prevent the free exchange of grafted organs or tissues between individuals of a species. Using classic serologic techniques, Gorer (1937) showed that transplantation antigens could be divided into two broad classes: the major histocompability antigens and the minor histocompatibility antigens. The former have been shown to co-segregate in breeding experiments and to occupy a restricted contiguous stretch of DNA on a single chromosome. This area of DNA is known as the major histocompatibility complex (MHC). The genes that occupy the MHC code for molecules that have been shown by structural studies to fall into two classes.

The class I molecule consists of a heavy chain of about 45,000 molecular weight (MW) associated noncovalently with a 13,000 light-chain β_2-microglobulin. The class II antigens consist of two similarly sized chains of about 30,000, covalently linked by a disulfide bond. The class I molecules themselves can be divided again into two types, the so-called classic and nonclassic antigens. The classic class I molecules are extremely polymorphic within a species, there being 100 alleles at each of three genetic loci in humans and similar allele frequencies in other species (Klein, 1986). These molecules are expressed on all or nearly all cells in the body. The nonclassic class I antigens, although they have an overall structure similar to that of the classic molecules, are oligomorphic (for some of these antigens there is only a single allele within a species), occupy many more loci over a much larger stretch of DNA, and have a restricted tissue distribution. The class II molecules are also very highly polymorphic within a species but, unlike the classic class I molecules, are not expressed on all cells but are mostly confined to cells of the lymphomyeloid system.

The large number of classic class I alleles at each of three loci on each chromosome means that, within an outbreeding population, such as human or mouse, the class I MHC phenotype of virtually every individual in the population can be unique. It is this that makes tissue matching for transplantation purposes intractably difficult. These molecules are not restricted in their location to cell membranes but are also found naturally in the body fluids of normal animals (Callahan et al., 1975; Van Rood et al., 1970) and are constituitively excreted in the urine (Singh et al., 1987, 1988). It is known that the urine of individual animals within a species has a unique odor specific to that individual and that urinary odors have a powerful influence on social interactions among members of a species (Brown, 1979). We have therefore studied the physiology of soluble classic class I molecules, especially the role of excreted class I molecules in determining the unique odor phenotype of the urine of the individual.

An indirect mechanism for producing MHC-related odors by the action of immune responses, restricted by MHC class I gene products, to bacterial populations in the commensal flora has also been investigated. A possible interaction between excreted class I molecules and odorants derived from bacterial flora is discussed.

II. MATERIALS AND METHODS

A. Animals

Rats of the DA(RTI^{avl}), PVG(RTI^c), PVG.R1(RTI^{rl}), PVG.RTI^u, and F_1 hybrid strains were bred in conventional animal houses or a specific pathogen-free (SPF) unit or were raised germ free by the MRC Experimental Embryology and Teratology Unit (Carshalton, Surrey).

For behavioral experiments, rats were housed in a conventional animal house in pairs in opaque plastic cages (NKP, Ltd.) on a reversed 12:12 light-dark cycle with lights off at 06.00 h at a temperature of 20 ± 2°C. Food (CRM NUTS, Labsure irradiated diets) and water were freely available. At the beginning of the experiments, the subjects had a mean weight of 255 ± 20 g.

B. Enzyme Immunoassays (EIA)

For classic class I molecules in solution, EIA were carried out in rigid, non-sterile, flat-bottomed, 96-well polystyrene plates (Nunc Immunoplates II, GIBCO, Ltd., Middlesex, UK), as previously described (Singh et al., 1988).

C. Affinity Chromatography

Class I molecules were extracted from liver cell membranes, serum, urine, and lymph by affinity chromatography with specific monoclonal anti-class I antibodies covalently coupled to cyanogen bromide-activated Sepharose 4B (Pharmacia, Uppsala, Sweden), as previously described (Singh et al., 1988).

D. Sodium Dodecyl Sulphate-Polyacrylamide Gel Electrophoresis (SDS-PAGE)

SDS-PAGE Analysis of proteins purified by affinity chromatography was undertaken according to Laemmli (1970), and bands were stained with Coomassie blue.

E. Radiation Chimeras

SPF rats (3-5 months old) received 10 Gy (1000 rad) whole-body radiation from a ^{137}Cs source (Gammacell 40, Atomic Energy of Canada, Ltd.) at about 1 Gy/minute. On the same day they were given 10^8 donor bone marrow cells intravenously. In the parent into F_1 chimeras, graft-versus-host disease was avoided by using bone marrow cells from DA rats neonatally tolerant to (PVG × DA)F_1. Serum from all chimeras was assayed for the unshared soluble class I molecules by EIA. Staining of peripheral blood leukocytes for class I molecules by fluorescence using monoclonal antibodies and FACS 420 analysis showed the hematopoietic system of these chimeras to be >98% donor type at 9 months after reconstitution.

Scatchard and saturation analyses were as previously described (Singh et al., 1988).

F. Tissue Culture Methods

Cells were cultured at 10^6 per ml in tissue culture medium in 24-well tissue culture plates (Nunc, GIBCO, Middlesex, UK) in RPMI-1640 containing 25 mM HEPES, 5×10^{-5} M 2-mercaptoethanol, 2 mM glutamine, 0.06 mg/ml of benzylpencillin, 0.1 u/ml of streptomycin sulphate and 5% v/v FCS (Sera-Lab, Crawly Down, Sussex, England).

G. Half-life Measurement

Four DA rats rendered neonatally tolerant of PVG antigens were bled of 3 ml blood from the tail vein. They were then immediately injected intravenously (IV) with 5 ml PVG serum. After 10 minutes the first blood sample was taken. A blood sample was then taken from one pair of rats 2 h later and from the other pair after another 2 h. The pairs were then bled alternately at 4 h intervals.

H. Neonatally Tolerant Rats

Rats were tolerized according to a previously published protocol (Roser and Dorsch, 1982). Briefly, DA or PVG neonates were injected intravascularly with 75 μl of a 50% v/v suspension of (PVG $\times$ DA)F_1 hybrid bone marrow cells within 24 h of birth. As adults, these rats failed to reject skin grafts from donors homozygous for the unshared antigens of the F_1 for more than 50 days. Their lymph node cells were shown to be unreactive against donor antigens in popliteal lymph node graft-versus-host assays before they were used in experiments.

I. High-performance Liquid Chromatography (HPLC)

Under nondissociating conditions, 200 μl PVG serum and 200 μl purified RT1.A^c molecules from DA urine were fractionated on a LKB TSK G3000SW size exclusion column (LKB Produckter, Bromma, Sweden) running at 0.5 ml/minute in phosphate-buffered saline (PBS) and azide.

J. Bilateral Nephrectomy

Both kidneys were exposed through a midline incision and the vascular pedicle and ureters exposed by blunt dissection. The pedicle and ureters were ligated en bloc with a single silk ligature, and the kidneys were removed distal to the tie. The abdominal wound was closed with two layers of continuous silk suture.

K. Habituation-Dishabituation Tests

Subjects were habituated to the test arena (an opaque plastic cage measuring 29.6 $\times$ 23.6 $\times$ 14.6 cm with a top of 1 cm square wire mesh that rose another 5.6 cm above the cage rim) for 15 minutes and were then tested once per week in a habituation-dishabituation test in which they experienced nine separate sequential 2 minute odor presentations. Test liquid (0.1 ml) was placed on a 7 cm diameter disk of Whatman No. 1 filter paper taped to the mesh top of the cage, with the center of the disk 13.5 cm from the floor of the arena. The odor source was changed every 2 minutes by removing the top of the cage and replacing it with another top carrying a new disk. Tops and arenas were washed in 70% ethanol after each test, and filter papers were used only once.

On the first three 2 minute tests, water was placed on the filter paper. The first urine sample was presented for the next three 2 minute trials. On trial 7, the second urine sample was presented for three trials. Donor urine was collected from individual PVG and PVG.R1 males in Urimax metabolic cages. Urine from germ-free or SPF donors was collected using autoclaved, sterile equipment within the plastic film isolator and frozen in aliquots until used.

The time spent rearing on the hindlegs and sniffing, with the nose within 1 cm of the disk throughout each 2 minute period, was recorded by an observer using a stopwatch. Each of the rats was tested in three habituation-dishabituation tests. The observer was "blind" to the order of testing odorant samples, which were prepared by an assistant according to a random list.

L. Statistical Methods

Analysis of the behavioral data was done using separate randomized block (repeated-measures) analyses of variance for each experiment. Post hoc analyses were done using Newman-Keuls tests.

III. RESULTS AND DISCUSSION

A. Classic, Polymorphic Class I Antigens in Solution in the Body Fluids

Soluble class I antigens were first detected in the blood of normal rats by their ability to inhibit target cell lysis by a cytotoxic antiserum (Kamada et al., 1981). To show that they were not trace contaminants in the serum released either from damaged liver tissue or from formed elements of the blood during clotting, soluble class I molecules were detected in fresh serum, plasma, urine, and lymph (Figure 1A) using EIA. Analysis of the MHC molecules, recovered from lymph by affinity chromatography, by SDS-PAGE revealed a heterodimeric molecule with a heavy chain M_r 39,000 and a light chain M_r 13,000, typical of β_2-microglobulin (Figure 1B). Body fluids of the rat therefore contain authentic classic polymorphic class I antigens in solution, and fractionation of serum by gel filtration high-performance liquid chromatography under nondissociating conditions (Figure 1C) shows these molecules to be true monomers, not molecular aggregates.

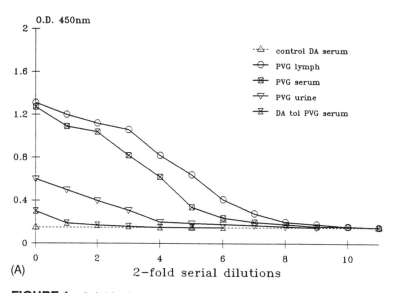

FIGURE 1 Soluble class I molecules in body fluids. (A) A two-site enzyme immunoassay of soluble class I antigens. Assay for the Ac molecule of the PVG (RT1Ac) strain using a sandwich of two noncompetitive monoclonal antibodies (Singh et al., 1987). Soluble molecules are found in serum (◧) and at about twice that concentration in central lymph from the thoracic duct (○). Smaller amounts are found in urine (▽). In DA strain animals made neonatally tolerant with (PVG × DA)F₁ bone marrow cells, very small quantities of donor antigen can be detected (⊠) because of the very low reproducible background with control sera (△). (B) SDS-PAGE analysis of membrane and soluble class I molecules. Membrane molecules solubilized in *n*-octyl glucoside or lymph molecules without detergent, purified by affinity chromatography, and run on SDS-PAGE show the same 13,000 β_2-microglobulin light chain, but the lymph heavy chain at 39,000 is considerably smaller than the membrane heavy chain at 47,000. (C) HPLC gel filtration of serum class I molecules. Serum (200 μl) fractionated into the classic peaks on gel filtration (light line). The class I molecule detected by sandwich EIA (heavy line) runs as a sharp peak at the trailing edge of the serum albumin peak with a MW between 45,000 (ovalbumen) and 68,000 (bovine serum albumin, BSA), showing that the molecule is a monomer in solution.

B. Quantitation and Turnover of Serum Class I Molecules

As both erythrocytes and leukocytes in the blood of normal rats also express membrane-bound class I molecules, the concentration of molecules in the cellular and liquid phases of the blood was measured. The number of molecules on the red blood cell (RBC) surface was measured by Scatchard analysis at 4,500 sites per PVG RBC and 10,800 sites per DA RBC. These

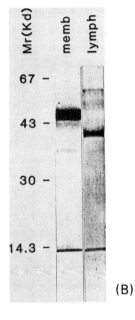

(B)

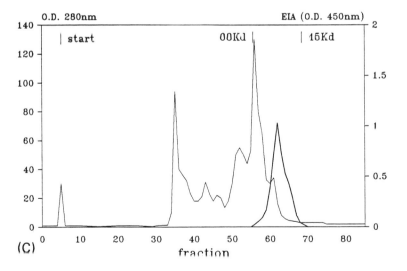

(C)

erythrocytes were then used as a calibrated source of antigen in competition experiments to measure the concentration of molecules in the serum by saturation analysis. The results showed that the concentration of class I molecules is 380 ng/ml in PVG serum and 350 ng/ml in DA serum (Singh et al., 1988).

The half-life of the serum molecule was measured by capitalizing on the fact that neonatally tolerant rats contain only trace amounts of donor antigen in their circulation (Figure 1A) and exhibit no reactivity against donor MHC molecules (Roser and Dorsch, 1982). Tolerant rats were injected with donor serum, and the level of donor antigens in their blood was measured by calibrated EIA at invervals. These experiments showed that the decay of donor molecule was exponential, with a half-life of 2.7 h (Figure 2).

Since neonatally tolerant rats are chimeras containing about 2-5% of donor hematopoietic cells, the origin of the small amount of soluble donor antigen in the serum of tolerant rats from the hematopoietic system seemed likely.

C. Cellular Origin of Soluble Class I Molecules

The origin of soluble molecules from bone marrow-derived cells was studied in radiation chimeras. In F_1 into parent chimeras, levels of serum class I

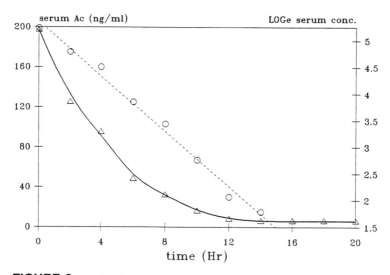

FIGURE 2 Half-life of soluble class I molecule in serum. Serum (5 ml) was injected into DA rats tolerant of (PVG × DA)F_1. The rate of loss of the A^c molecule was followed by EIA of sequential bleeds. The serum concentration shows an exponential fall (△). Log$_e$ transformation (○) produces a linear (----) regression curve. From classic pharmacokinetics the half-life is 2.66 h.

molecules derived from the donor reached maximal values at 17 days after reconstitution and remained high (Figure 3), showing that cells of the hematopoietic system were the source of at least half the serum molecules.

In the parent into F_1 chimeras the loss of soluble molecules of recipient type was slow but inexorable (Figure 3), suggesting that they were produced by cells with a long tissue residence time, which were slowly lost from the body after irradiation and replaced by donor-derived cells.

Since the level of host-type molecules remains high for >40 days after irradiation, we can exclude radiosensitive cells, such as lymphocytes, as the source of these molecules. A requirement for T lymphocytes in soluble antigen production was also formally excluded by showing that the levels of A^c molecules in the circulation of mutant athymic nude rats of the PVG-*rnu/rnu* strain, which lack peripheral T cells, were the same as the levels in normal euthymic PVG animals, that is, 380 ng/ml.

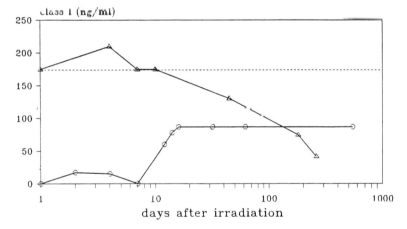

FIGURE 3 EIA of class I molecules from the F_1 in radiation chimeras. In chimeras of the parental DA (p) to (PVG × DA)F_1 type (△), the levels of A^c antigen, after a brief rise following irradiation and injection of bone marrow cells, falls progressively over the succeeding months, suggesting that the cell type secreting the molecules is radioresistant and only slowly replaced by donor DA cells. In chimeras of the (PVG × DA)F_1 → PVG type (O) the levels of A^{avl} antigen, after a brief rise following irradiation and injection of bone marrow cells, begin to rise permanently after day 7 and reach maximal levels of 50% of normal F_1 levels (----) at day 17 and remain there for at least 500 days. Since all chimeras were >95% reconstituted by donor cells, as shown by FACS analysis of PBL, this result suggests that about 50% of the secreted molecules in the blood of normal rats are derived from a long-lived relatively radioresistant cell type of hematopoietic origin, for example, a macrophage or a dendritic cell.

In vitro cultures of lymph node lymphocytes, spleen cells, and peritoneal cells were set up to ascertain whether release of soluble class I molecules could be detected in vitro. Peritoneal cells produced about 100 ng class I molecules per 10^6 cells per 24 h for the first 3 days of culture. Spleen cells produced about half this amount and lymph node cells much smaller amounts (Figure 4). No antigen was released by erythrocytes (data not shown). The most likely source of a major portion of soluble class I molecules in vivo therefore appears to be a cell of the macrophage or dendritic cell lineage. This is consistent with the observation that lymph draining from lymphoid tissues rich in cells of this type contains high levels of soluble antigen (Figure 1A).

D. Urinary Excretion of Soluble Class I Molecules

The presence of soluble class I molecules in the urine suggested that their short half-life in serum could be due to excretion. Calibration of the EIA with the serum molecule showed that the concentration of class I molecules in the urine was quite variable and ranged between 40 and 190 ng/ml. Proof that the urine molecules were derived from the blood was obtained by infusing

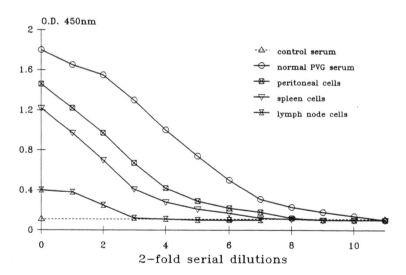

FIGURE 4 Secretion of class I molecules in vitro. Peritoneal cells (⊠) secreted about 100 ng/ml of A^{avl} molecules but spleen cells (▽) secreted less and lymph node cells (⊼) secreted very little, 48 h after establishing cultures of 10^6 DA cells per ml in RPMI-1640 medium with 10% fetal calf serum. No soluble antigen was produced by erythrocytes (data not shown). The amount of antigen produced was roughly proportional to the concentration of macrophages in the cultures.

known amounts of PVG donor antigen into the blood of DA rats tolerant of PVG and detecting its excretion in the urine. We could account for only 6.5% of the injected antigen in the urine collected over the subsequent 48 h (1.9 μg injected, 124 ng recovered).

A normal 300 g PVG rat loses from the blood about 18 μg class I molecules each 24 h (whole-blood concentration, 190 ng/ml; blood volume 7% of body weight = 21 ml; half-life 2.7 h). With an average 24 h urine volume of 14 ml, this gives an estimated urine concentration of 83 ng/ml, a figure in reasonable agreement with the measured urine concentrations of between 40 and 190 ng/ml. Thus, only a fraction of the class I molecules normally lost from the blood are detected by EIA in the urine. The class I molecules purified from the urine by affinity chromatography on monoclonal antibody columns not only have the typical 39,000 heavy chain and 13,000 light chain associated with soluble class I molecules but also show a major protein band at 30,000 (Figure 5A), suggesting fragmentation of the class I molecules is occurring in the urine.

HPLC analysis of class I molecules recovered from the urine by affinity chromatography showed that antigenic material emerges from the column as a broad band from about 50,000 down to 15,000 M_r (Figure 5B). This is in sharp contrast to the class I molecules in serum, which emerge from the HPLC column as a well-defined narrow peak between 69,000 and 45,000 (Figure 1C). Since only those fragments carrying two spatially separate epitopes are detected by our two-site EIA, it is probable that a major population of smaller degraded fragments is also present and escaped detection. This could explain the low yield of foreign class I molecules detected in the urine after their intravenous injection.

If excretion via the kidneys is the major pathway by which class I molecules are lost from the blood, interruption of renal excretion should cause an increase in the blood level of class I molecules in proportion to the fraction of molecules removed from the blood by the kidneys. Bilateral nephrectomy of four PVG rats followed by repeated blood sampling showed that, within 6 h of operation, the mean serum level of class I A^c molecules rose to 866 ng/ml by calibrated EIA. The calculated serum level, assuming that *all* the serum molecules are normally excreted in the urine, is 806 ng/ml. Renal excretion is therefore the major if not the only removal pathway for the class I molecules in solution in the blood. Once in the urine the majority of these molecules undergo rapid degradation to small fragments (Figure 5B).

In radiation chimeras of the F_1 into parent type, soluble class I molecules derived from the unshared antigens of the F_1, which were present in high concentrations in the blood of the donor (Figure 3), were also rapidly detectable in the urine (data not shown).

The excretion of class I antigens in the urine has also been reported in renal transplant patients. In these patients the class I molecules were shown

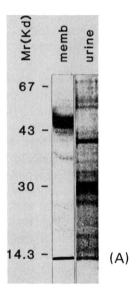

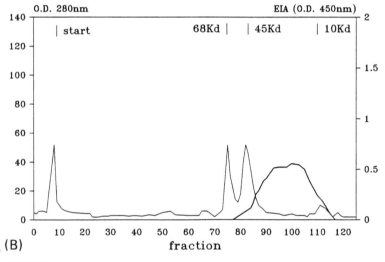

FIGURE 5 Fragmentation of class I molecules in urine. (A) SDS-PAGE of A^{av1} molecules affinity purified from urine. A heavy chain band at 39,000 and a β_2-microglobulin band at 13,000, together with a heavy band at around 30,000, suggested the presence of a two-domain fragment of the heavy chain. (B) HPLC gel filtration profile of this affinity-purified material. No E_{280} absorbing peaks were detected. The MW standards are shown as a light line. Antigenic material detected by EIA eluted as a broad MW peak from above 45,000 to below 10,000, indicating extensive fragmentation of the molecule.

to be of recipient type and did not carry the allelic specificities of the kidney donor (Robert et al., 1974).

The excretion of classic class I molecules in the urine of normal animals seemed to us important since olfactory discrimination in both mice and rats has been shown to include the ability to identify urine samples from congenic animals that differ only at the MHC (Beauchamp et al., 1985; Yamazaki et al., 1976). Also, up to 90% of pregnant female mice abort their preimplantation embryos when exposed to the odor of urine from a foreign male (Dominic, 1966) even when the male is from an inbred strain congenic with that of the original stud male and differs only at the MHC (Yamazaki et al., 1983) or is a mutant (*bml*) at one class I MHC locus (Yamazaki et al., 1986). The only known differences between *bml* and the C57BL/6 wild type are point mutations in the class I gene resulting in MHC molecules that differ by three amino acids. This implies that the olfactory cues in the urine are derived, not from the products of loci closely linked to class I but either directly or indirectly from the product of the class I genes themselves.

E. Olfactory Responses to MHC-Associated Urinary Odors

The presence of classic class I transplantation antigens of the A^{av1} and A^c type and their degradation products in the urine of normal rats prompted us to see whether rats could also detect urinary odors associated with these molecules. In these experiments we used the habituation-dishabituation method (Sundberg et al., 1982) in which untrained detector animals were habituated to the odor of urine samples from donors of one strain and then tested for dishabituation when exposed to urine from donors of the other strain. To exclude a role for any variable other than the MHC class I molecules in these experiments we used, throughout, male rats from the PVG congenic series (Butcher and Howard, 1986). Figure 6 shows that the PVG-$RT1^u$ rats could readily discriminate between the odor of PVG urine containing the A^c molecule and PVG.R1 urine containing the A^{av1} molecule. There were significant differences in time spent rearing and investigating odors over the nine tests for both groups (PVG then PVG.R1, $F = 14.89$, degrees of freedom = 8, 72, $p < 0.001$. PVG.R1 then PVG, $F = 16.24$, degrees of freedom = 8, 72, $p < 0.001$). Post hoc Newmann-Keuls tests indicated that for both groups of subjects, more time was spent investigating odors on trials 4 and 7 than on any other odor presentation ($p < 0.01$).

The time spent investigating urine from two individuals on the same strain showed that the odors from two different PVG males were not discriminated nor were two separate urine samples from the same individual (Singh et al., 1988).

The results obtained from these behavioral experiments indicate that the urine odors of normal male rats of the PVG and PVG.R1 strains are equally

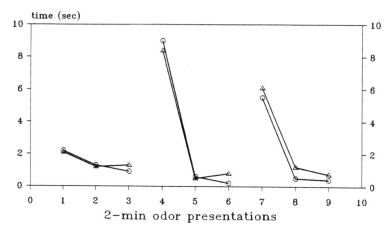

FIGURE 6 Detection of individuality odor in class I recombinant urine. The PVG and PVG.R1 strains differ only at the classic class I locus RT1A, where PVG is A^c and PVG.R1 is A^{av1}. All other genetic loci are identical. PVG-RT1^u detector animals were tested for their investigative behavior on samples of urine presented in both orders: PVG before PVG.R1 and PVG after PVG.R1. When exposed to water (trials 1-3) they showed little interest and within 2 minutes their investigative behavior was less than 2 s per 2 minute trial. When urine 1 was presented on trial 4, they showed a positive investigative response that habituated rapidly so that the same urine sample presented on trials 5 and 6 elicited very little investigation. Substitution of urine of the other strain on trial 7 resulted in dishabituation and elicited a striking investigative response that also habituated rapidly so that no further interest was shown on trials 8 and 9. Thus the detector animals could discriminate the smell of PVG and PVG.R1 urine irrespective of whether the samples were presented in the order PVG.R1 then PVG (△) or PVG then PVG.R1 (○). The detector animals could not discriminate between two separate samples from individual members of the same strain or two different urine samples from the same individual donors (data not shown). (From Singh et al., 1987, 1988.)

attractive as they are both investigated to the same degree on trial 4. PVG and PVG.R1 individuals are readily discriminated, as evidenced by the dishabituation on trial 7 (Figure 6).

F. Nature of Odorants

The most direct mechanism to explain the MHC discrimination by the detector animals is that the class I molecules themselves were the odorants. Preliminary evidence that intact class I molecules themselves were not discriminable came from habituation-dishabituation trials in which serum rather than urine was the odorant sample. Sera from PVG and PVG.R1 donors

EXCRETION OF TRANSPLANTATION ANTIGENS

were very attractive to rats, as shown by vigorous investigative responses in trial 4; however, rats completely failed to discriminate between PVG and PVG.R1 sera (Brown et al., 1987).

Furthermore, class I molecules purified from PVG (RT1.A^c) and PVG.R1 (RT1.A^{av1}) urine by monoclonal antibody affinity chromatography could not be used to discriminate between the two strains. The response to the class I molecule on trials 4 and 7 was no greater than that to saline or water. The urine from which the class I antigens had been removed could be used to discriminate between the two strains (Figure 7). There was a significant difference in sniffing time over the nine odor presentations ($F = 27.24$, degrees of freedom $= 8, 88, p < 0.001$), with more sniffing on trials 4 and 7 than on any other trial ($p < 0.01$).

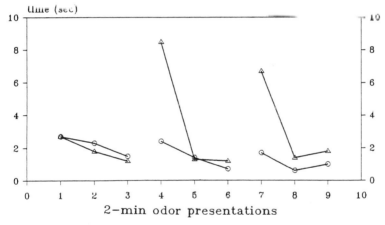

1-3 WATER 4-6 URINE 1 7-9 URINE 2

FIGURE 7 Intact class I molecules are not the individuality odorants in urine. Samples of PVG and PVG.R1 urine were passed over affinity columns to remove the class I molecules as confined by EIA. The class I molecules adsorbed to the columns were eluted with high-pH buffers and dialyzed against PBS. PVG-RT1^u detector animals were then tested for their ability to discriminate the samples. Urine from which the class I molecules were removed (remainder) could be detected (△), as well as normal urine (Figure 6), but the purified class I molecules themselves (O) elicited no response on trial 4: that is, they were not attractive to the rats and naturally, therefore, no dishabituation was seen on trial 7. Thus class I molecules or their fragments that bound to a monoclonal antibody against allelic determinants were not the individuality odorants.

We also showed formally that the molecules that confer the MHC-related odor are volatile. Figure 8 shows that nitrogen-purged urine cannot be used by PVG-RTI^u detector animals to discriminate between the PVG and PVG.R1 strains.

There were significant differences in sniffing times over the nine odor presentations for all 8 subjects ($F = 29.15$, degrees of freedom $= 7, 56, p < 0.001$). The urine remained attractive since more time was spent sniffing odors on trial 4 than any other trial ($p < 0.01$). For the untreated control urine there were significant differences in sniffing times over the nine presentations ($F = 12.145$, degrees of freedom $= 7, 56, p < 0.001$). More time was spent sniffing odors on trials 4 and 7 than on any other trial ($p < 0.01$).

However, dialysis experiments have shown that the volatile odorants in the urine are associated with a high-molecular-weight fraction. Using dialysis tubing of 12,000 MW cut off, 5 ml urine was dialyzed against 5 liters of PBS for 48 h, and the retentate was shown, using the habituation-dishabituation protocol, to enable detector animals to discriminate between PVG and

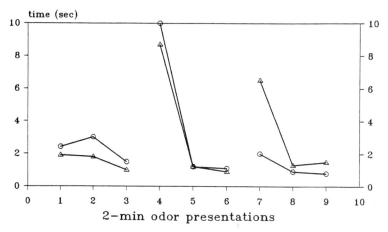

1-3 WATER 4-6 URINE 1 7-9 URINE 2

FIGURE 8 Individuality signals in urine are highly volatile. Purging samples of PVG and R1 urine with bubbled N_2 for 48 h removes the individuality signal, as shown by the failure of detector rats to dishabituate to such urine on trial 7 (O); the same urine sample before purging gave a strong dishabituation response (Δ). This treatment did not remove all volatile signals from the urine, as shown by a strong response on trial 4, indicating that both purged and nonpurged urine were equally attractive to the rats.

EXCRETION OF TRANSPLANTATION ANTIGENS

PVG.R1 rat strains. There were significant differences over all nine odor presentations for all eight subjects ($F = 7.934$, degrees of freedom = 7, 56, $p < 0.001$), and more time was spent sniffing odors 4 and 7 than any other odors ($p < 0.01$), although the dishabituation responses were diminished to 4.2 s. The dialysate could not be used to discriminate between PVG and PVG.R1 urine. The response to the dialysates was no different than that from PBS alone (Figure 9).

G. Role of Commensal Bacteria in Urine Odors

Behavioral studies like the habituation-dishabituation method used here, which revealed an intimate linkage between MHC class I genotype and urinary odor, have previously been interpreted as reflecting the action of immune response gene effects of the class I genetic locus (Howard, 1977), possibly

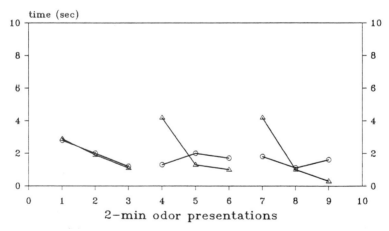

1-3 WATER 4-6 URINE 1 7-9 URINE 2

FIGURE 9 Individuality odors of urine associate with a high-MW component. When urine was dialyzed against PBS to equilibrium, low-MW components were diluted 1000-fold in the dialysate and the retentate. Under these circumstances the dialysate (O) was not attractive to rats (no investigative response on trial 4), nor was there any dishabituation on change of urine donor at trial 7. The retentate, however (△), showed a modest response on both trials 4 and 7, showing that it was attractive to the rats and they could discriminate between PVG and PVG.R1 retentate. Normal urine diluted 1:1000 failed to cause any response (data not shown). Thus the volatile odorants are associated with a component > 12,000 from which they dissociate only slowly.

controlling the host immune responses against commensal bacterial flora of the skin, urinary tract, or gut. Individual MHC haplotypes would then be associated with unique flora. The volatile odorants in the excretions were thought to be secondary metabolites derived from these organisms. However, it has been shown that in radiation chimeras of the F_1 into parent type, urinary odors are of F_1 type (Yamazaki et al., 1985). Since it is known that class I-restricted immune responsiveness is heavily biased toward the parental MHC (Bevan and Fink, 1978; Zinkernagel and Doherty, 1979), it is expected that the class I-associated immune response phenotype would remain parental in these animals. Urinary odors therefore do not show correlation with putative class I immune response gene effects but do show close correlation with the excreted soluble class I molecules themselves. In addition, Yamazaki et al. (1982) have shown that urine collected from mice that are congenic with respect to the nonclassic Qa/T1a class I loci can be distinguished by odor alone. These molecules have no known Ir gene function. They are, however, found in the circulation (Maloy et al., 1984; Kress et al., 1983).

Since bacteria, especially anaerobes, are well-known sources of biologic odors, the role of commensal flora as a possible source of the individuality signals was studied by using urine from germ-free donors in the habituation-dishabituation test.

Surprisingly, although the urine of germ-free rats was attractive to the detector animals as shown by marked dishabituation on trial 4, they could not discriminate between the urines of germ-free PVG and PVG.R1 animals (Figure 10). Within 1 week of re-exposure of germ-free rats to environmental organisms in a conventional animal house, the individuality signals returned to their urine (Figure 11).

Specific pathogen-free animals are derived from germ-free stock by inoculation of a cocktail of enteric organisms. This cocktail consists of *Streptococcus homionous, Escherischia coli, Streptococcus bovis,* and *Lactobacillus acidophilus.*

The urine of SPF animals appeared on first testing to contain some individuality signal, but on repeated study with larger panels of donors this was found not to be significant (Figure 12). When moved to a conventional animal house the urine of SPF animals again rapidly acquired individuality signals. Thus the flora that confer unique odors to the urine is made up of species found in conventional animals but not in SPF rats. The germ-free or SPF rat will thus be an ideal subject in which to identify, in future experiments, the precise bacterial flora that provides the source of the individuality odorant molecules, which can themselves be identified by qualitative and quantitative difference in urinary volatile chromatograms by gas chromatography-mass spectrometry (Holland et al., 1983).

Of course the unique odor profile may be a product of the complex, mutually interacting populations of bacterial species in the commensal ecosystem.

EXCRETION OF TRANSPLANTATION ANTIGENS

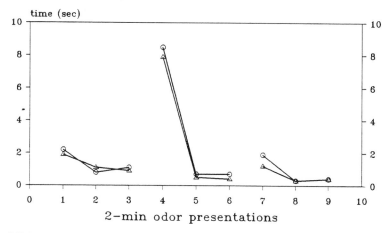

1-3 WATER 4-6 URINE 1 7-9 URINE 2

FIGURE 10 Failure of urine of germ-free rats to carry individuality signals. Urine of germ-free PVG and PVG.R1 rats is as attractive to PVG-RT1^u detector animals as the urine of normal rats as shown by marked investigative responses on trial 4. However, both urines clearly smell the same, as there is no dishabituation when the donor is switched on trial 7 whether the urines are presented in the order PVG then PVG.R1 (○) or PVG.R1 then PVG (△).

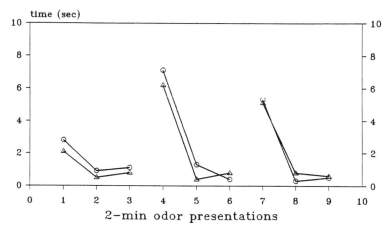

1-3 WATER 4-6 URINE 1 7-9 URINE 2

FIGURE 11 Conventionalized germ-free rats regain individuality signals. Within 7 days of transfer to a conventional animal facility, the urine of previously germ-free rats regained its individuality signal as shown by marked dishabituation responses on trial 7 when the urine donors were switched.

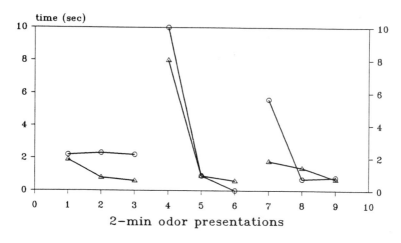

1-3 WATER 4-6 URINE 1 7-9 URINE 2

FIGURE 12 SPF rats regain individuality signals when conventionalized. Specific pathogen-free rats excrete urine (△) that, although attractive to detector animals (trial 4), is not discriminable between strains when the donor is switched (trial 7). Within 7 days of transfer to a conventionalized animal house (O), the urines regained their individuality signals.

This population may be precisely determined by the host immune system as controlled by the immune response gene profile of the individual (Howard, 1977), which maps into the MHC region on chromosome 14 in the rat (Oikawa et al., 1985). We, however, currently favor the idea that bacteria are a ubiquitous source of a common pool of odorant molecules from which the individual's own MHC molecules select and transport to the urine a unique mixture. This unique odorant cocktail is selected by specific association of odorants with the allelic determinants of the MHC molecules. Thus the commensal flora itself need not be unique to each individual in a species because the excreted class I molecules act as a selection mechanism, extracting from a common body pool of odorants a quantitatively or qualitatively specific cocktail for transport to the urine.

Experiments are in progress to test urine from normal animals that have been infused continuously with foreign MHC molecules. These are excreted in the urine along with the host animal's own class I molecules.

This experiment is a stringent test of current hypotheses. If (and only if) the infused MHC molecules have selected odorants for transport via the kidneys will the individuality signals present in the urine of these rats be a mixture of host and donor in type. The urine of such rats should smell identical to a mixture of normal urines of host and donor types or possibly to F_1 hybrid urine. If no donor type odor signal is detected in the urine then the simplest hypothesis, namely that associations between class I molecules and volatile molecules determine the excretion of a class I-specific odorant profile, will be disproved. An alternative possibility, that is, the formation of such associations at the cell membrane by class I molecules before their release into the circulation, will be harder to study. Finally, failure to alter the individuality signals in urine by manipulating only the class I molecules will be strong evidence that the uniqueness of urinary odor reflects an underlying uniqueness in the bacterial flora. The results of these experiments will be reported subsequently.

ACKNOWLEDGMENTS

This research was supported by the Natural Sciences and Engineering Research Council of Canada, Grant A7441 (Brown), the Medical Research Council of Great Britain (Roser), and a NATO Collaborative Research Grant. Singh is working under an AFRC New Initiatives Grant. We thank Terry Pendry, Lindsey Arnott, and David K. Tucker from the MRC Experimental Embryology and Teratology Unit, Woodmanstone Road, Carshalton, Surrey, SM5 4EF, England, for helping to rear the bacteria-free rats and collecting urine from them. We are grateful to Shev Sen and Rachael Pratt for helping with the EIA and Lynda Wisker for helping with the data analysis.

REFERENCES

Beauchamp, G. K., Yamazaki, K., and Boyse, E. A. (1985). The chemosensory recognition of genetic individuality. *Sci. Am.* **253**:86.
Bevan, M. J., and Fink, P. J. (1978). The influence of thymus H-2 antigens on the specificity of maturing killer and helper cells. *Immunol. Rev.* **42**:3.
Brown, R. E. (1979). Mammalian social odours: A critical review. *Adv. Study Behav.* **10**:103.
Brown, R. E., Singh, P. B., and Roser, B. (1987). The major histocompatibility complex and the chemosensory recognition of individuality in rats. *Physiol. Behav.* **40**:65.
Butcher, G. W., and Howard, J. C. (1986). The MHC of the laboratory rat. *Rattus norvegicus*. In *Handbook of Experimental Immunology*, Vol. 3, *Genetics and Molecular Immunology*. D. M. Weir (Ed.). Blackwell Scientific, Oxford.
Callahan, G. N., Ferrone, S., Poulik, M. D., Reisfeld, R. A., and Klein, J. (1976). Characterisation of Ia antigens in mouse serum. *J. Immunol.* **117**:1351.
Dominic, C. J. (1966). Observations on the reproductive pheromones of mice. I. Source. *J. Reprod. Fertil.* **11**:407.

Gorer, P. A. (1937). The genetic and antigenic basis of tumour transplantation. *J. Pathol. Bacteriol.* **44**:697.

Holland, M., Rhodes, G., DalleAve, M., Wiesler, D., and Novotny, M. (1983). Urinary profiles of volatile and acid metabolites in germfree and conventional rats. *Life Sci.* **32**:787.

Howard, J. C. (1977). H-2 and mating preferences. *Nature* **266**:406.

Kamada, N., Davies, H.ff.S., and Roser, B. (1981). Reversal of transplantation immunity by liver grafting. *Nature* **292**:840.

Klein, J. (1986). *The Natural History of the Major Histocompatibility Complex.* Wiley and Sons, New York.

Kress, M., Cosman, D., Koury, A., and Jay, G. (1983). Secretion of a transplantation-related antigen. *Cell* **34**:189.

Laemmli, U. K. (1970). Cleavage of structural proteins during the assembly of the head bacteriophage T4. *Nature* **227**:680.

Maloy, W. L., Coligan, J. E., Barra, Y., and Jay, G. (1984). Detection of a secreted form of the murine H-2 class I antigen with an antibody against its predicted carboxyl terminus. *Proc. Natl. Acad. Sci. USA* **81**:1216.

Oikawa, T., Yoshida, M. C., Satoh, H., Yamashina, K., Sasaki, M., and Kobayashi, H. (1984). Assignment of the rat major histocompatibility complex to chromosome 14. *Cytogenet. Cell Genet.* **37**:558.

Robert, M., Vincent, C., and Revillard, J. P. (1974). Presence of HL-A antigens and β_2-microglobulin in tubular proteinuria. *Transplantation* **18**:89.

Roser, B. J., and Dorsch, S. E. (1982). The cellular basis of transplantation tolerance in the rat. *Immunol. Rev.* **46**:55.

Singh, P. B., Brown, R. E., and Roser, B. (1987). MHC antigens in urine as olfactory recognition cues. *Nature* **327**:161.

Singh, P. B., Brown, R. E., and Roser, B. (1988). Class I transplantation antigens in solution in body fluids and in the urine. *J. Exp. Med.* **168**:195.

Sundberg, H., Doving, K., Novikov, S., and Ursin, H. (1982). A method for studying responses and habituation to odors in rats. *Behav. Neural Biol.* **34**:113.

Van Rood, J. J., Van Leeuwen, A., and Van Santen, M. C. T. (1970). Anti HL-A2 inhibitor in normal human serum. *Nature* **226**:366.

Yamazaki, K., Boyse, E. A., Mike, V., Thaler, H. T., Mathieson, B. J., Abbott, J., Boyse, J., Zayas, Z. A., and Thomas, L. (1976). Control of mating preferences in mice by genes in the major histocompatibility complex. *J. Exp. Med.* **144**:1324.

Yamazaki, K., Beauchamp, G. K., Bard, J., Thomas, L., and Boyse, E. A. (1982). Chemosensory recognition of phenotypes determined by the *Tla* and *H-2K* regions of chromosome 17 of the mouse. *Proc. Natl. Acad. Sci. USA* **79**:7828.

Yamazaki, K., Beachamp, G. K., Wysocki, C. J., Bard, J., Thomas, L., and Boyse, E. A. (1983). Recognition of H-2 types in relation to the blocking of pregnancy in mice. *Science* **221**:186.

Yamazaki, K., Beauchamp, G. K., Thomas, L., and Boyse, E. A. (1985). The hematopietic system is a source of odorants that distinguish major histocompatibility types. *J. Exp. Med.* **162**:1377.

Yamazaki, K., Beauchamp, G. K., Matsuzaki, O., Kupriewski, D., Bard, J., Thomas, L., and Boyse, E. A. (1986). Influence of a genetic difference confined to mutation of H-2K on the incidence of pregnancy block in mice. *Proc. Natl. Acad. Sci. USA* **83**:740.

Zinkernagal, R. M., and Doherty, P. C. (1979). MHC-restricted cytotoxic T cells: Studies on the biological role of polymorphic major transplantation antigens determining T-cell restriction—specificity, function and responsiveness. *Adv. Immunol.* **27**:51.;

15
Chemosensory Identity and Immune Function in Mice

Kunio Yamazaki and Gary K. Beauchamp
Monell Chemical Senses Center, Philadelphia, Pennsylvania

Judith Bard and Edward A. Boyse
University of Arizona, Tucson, Arizona

Lewis Thomas
New York Hospital, New York, New York

I. INTRODUCTION

The major histocompatibility complex (MHC) of genes, which is critical for immune response, is also the source of chemical sensory information that enables mice to identify one another as individuals. As such, a link between immunology and olfactory communication has been established. Any sensory communication system that reflects the genetic composition of individual members of a species and influences their social and reproductive behavior must be a vital factor in the evolution and biology of that species.

II. THE MHC OF THE MOUSE

The mouse has 20 pairs of chromosomes, with the MHC occupying a segment of chromosome 17. The importance of this group of linked genes can be gauged from the fact that a similar set of genes probably exists in all vertebrates (see Klein, 1986). The MHC of the mouse, called H-2, comprises many linked genes and is divided into regions, the main ones being H-2K (K), H-2D (D), and T1a. The mouse's "MHC type" or "H-2 type" is the total set of variable alleles of all genes in the MHC region. The set of MHC

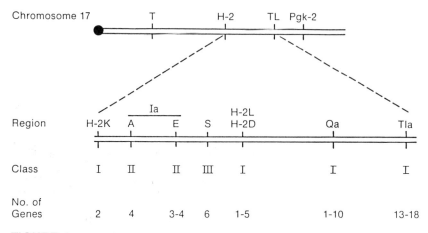

FIGURE 1 Genetic map of the mouse major histocompatibility complex. Solid circle at top left represents centromere. The mouse MHC is 2 centimorgans in length and encompasses the region from H-2K through T1a, as shown in more detail at the bottom. The class I genes are located at the H-2K, H-2D, Qa, and T1a loci. The class II genes are in the I region at the I-A and I-E loci. The S region includes class III genes.

alleles on a given chromosome 17 is called a haplotype, and a vast number of haplotypes is possible. The haplotype is denoted by a superscript letter or letters, such as $H-2^b$ (abbreviated b) or $H-2^k$ (abbreviated k).

The MHC is best known from studies on tissue transplantation because incompatibility of MHC types causes rapid rejection of grafts. The fate of organ transplants depends mainly on MHC compatibility. Throughout the MHC region there are also genes that determine the degree of response to particular antigens and other genes expressed selectively in lymphocytes. Thus the MHC is concerned in many aspects of how immune cells, lymphocytes equipped with specific receptors for antigen, handle chemical information from the environment.

A schematic genetic map of the H-2 region is shown in Figure 1. This region occupies 2 cM of a total haploid genome of 1600 cM and can accommodate perhaps 50 genes (Carroll et al., 1987). H-2 class I antigen is the product of at least four separate genetic loci, H-2K, H-2D, Qa, and T1a. The MHC includes a family of up to 30 "class I" genes, not all necessarily functional, that uniquely characterize the MHC throughout its length and encode transmembrane glycoproteins whose outer domains bear distinctive MHC cell surface antigens.

III. CONGENIC MICE

The use of inbred congenic mice has been essential in work on MHC-associated chemosensory communication. An H-2 congenic strain is produced by crossing two selected inbred strains of different H-2 types and by back-crossing many times to one of these strains, serologically selecting for the donor H-2 type in each generation (a simple outline is given in Beauchamp et al., 1985). Consequently, the final inbred congenic strain is genetically identical to the base strain, except for a segment of chromosome 17 bearing the H-2 haplotype, introduced from the donor strain. Any difference beteween the base strain and its congenic partner strain, if genetic, must be due to genes in the H-2 region, because this is the only genetic difference between the inbred base strain and its congenic partner (Boyse, 1977).

IV. INITIAL STUDIES

The original observation suggesting MHC odors were involved in chemical communication, made in the congenic mouse breeding rooms at Memorial Sloan-Kettering Cancer Center, was that mice of a particular inbred strain seemed socially more reactive to congenic mice of a different H-2 type than to members of their own strain. These chance observations appeared to agree with the suggestion by Thomas (1975) that histocompatibility genes might impart to each individual a characteristic scent. To substantiate this impression, a test system was developed in which an inbred male was presented simultaneously with two sexually receptive females, the two females being congenic for H-2 and therefore differing genetically only in their H-2 types. The trios were observed until the male had mated with one of the females as indicated by a vaginal plug.

An H-2-associated mating bias was demonstrated, commonly favoring the female whose H-2 type differed from that of the male (Yamazaki et al., 1976, 1978; Andrews and Boyse, 1978; Yamaguchi et al., 1978). This was evidently the first example of vertebrate reproduction behavior and selection that has been traced to variation of a particular gene or gene complex. These and subsequent studies of mating preference are discussed in more detail near the conclusion of this chapter.

V. OLFACTORY DISCRIMINATION

It was likely that mating preference was determined by chemical cues, but it was necessary to directly test for an H-2-associated communication system involving chemical sensation. To test this, a Y maze was used (Figure 2) in which air is drawn through two odor boxes containing urine of H-2 congenic mice. The air is then conducted to the left and right arms of the maze, which are thereby scented differentially by urine of mice whose only genetic difference is H-2. Some mice are trained to run toward the odor of one H-2 congenic type, whereas others are trained to run to the other. The incentive to run the maze is a drop of water, the test subject mice having been deprived of water for 23 h beforehand.

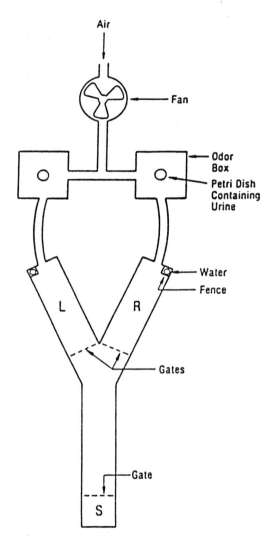

FIGURE 2 The Y maze. Air is conducted through two odor chambers, containing urine samples exposed in petri dishes, to the two arms of the maze. Gates are raised and lowered in timed sequence to permit the training or testing of each mouse in a series of up to 48 consecutive runs, the samples being changed for each run and left-right placement determined by random numbers. The reward is a drop of water, the mouse having been deprived of water for 23 h beforehand. The water dispenser in each arm of the maze is guarded by a fence, which is raised only if the mouse's choice is concordant with training (correct).

Mice could indeed be trained to distinguish between odor of urine of mice differing only at the MHC, which is proof that the MHC is involved in individual odor production (Yamazaki et al., 1979; Yamaguchi et al., 1981). The criterion employed in validating a distinction between two alternative odor sources includes a highly significant concordance score, generally 80% or more (chance would be 50%), observed not only in rewarded trials with the familiar odor sources but also in unrewarded blind trials of newly encountered odor sources that duplicate the genetic constitution of the familiar odor sources used in training. A demonstration of this discrimination can be seen in the learning curve (shown in Figure 3) for one mouse trained to go to the arm scented with $H-2^K$ urine in preference to $H-2^b$ urine.

Thus mice are able to distinguish individuals on the basis of genetic differences at the major histocompatibility complex of genes, and urine is a prime odor source (Yamaguchi et al., 1981). In addition to congenic strains differing throughout the entire H-2 complex, there are congenic mice that differ only at a part of this segment of chromosome 17. Testing of such fractional congenic strains shows that at least three subregions within the MHC can independently confer a distinctive odor (Yamazaki et al., 1982). Thus,

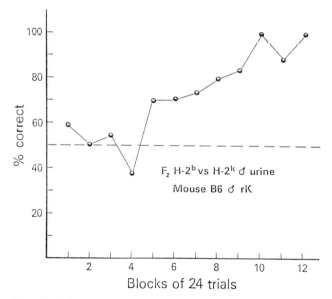

FIGURE 3 A learning curve for one trained mouse. Graphed is the percentage of correct responses as a function of blocks of successive 24 trials. Improvement is gradual, and by the last blocks almost perfect performance is evident.

each complete H-2-based odor reflects the action of several genes and is likely to comprise multiple elements.

VI. MHC MUTANTS

A central question is whether known MHC genes are involved in olfactory individuality. A test of this concerns the ability of mice to distinguish the odor of a known mutant strain from that of the otherwise genetically iden-

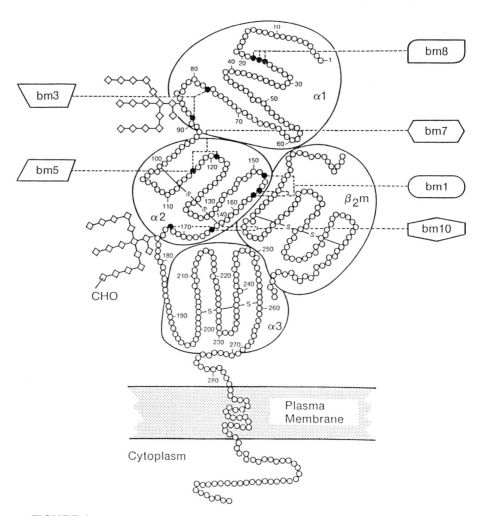

FIGURE 4 Location of amino acid substitutions caused by mutations in the H-2K^b molecule.

tical nonmutant strain. The first mutant chosen for tests of olfactory discrimination was a class I H-2K mutant, called bm1 (see Figure 4). Class I genes include the classic transplantation antigens that trigger rejection of organ grafts by the recipient's immune system. A schematic representation of the class I transplantation antigen, which is the product of the H-2K gene, is shown in Figure 4. In the case of bm1 mutant, three amino acids (152, 155, and 156), coded within a stretch of 13 nucleotides, comprise the only differences between the mutant protein and the nonmutant protein (Zeff et al., 1985). Therefore, any functional differences between mutant and nonmutant must depend only on the cluster of three amino acid differences in a glycoprotein containing 348 amino acids. At all other points throughout the entire genome, mutant and nonmutant mice are identical.

Mutant and nonmutant mice were successfully distinguished (Yamazaki et al., 1983b) based on odor alone (Figure 5). Additional class I mutants have also been studied. These are labeled bm5, bm7, and bm8, all of which can be discriminated from the inbred strains (Yamazaki et al., 1989). Surprisingly, there is no evidence of discrimination between mutant versus mutant, even though it is known that these mutants differ at a few amino acids.

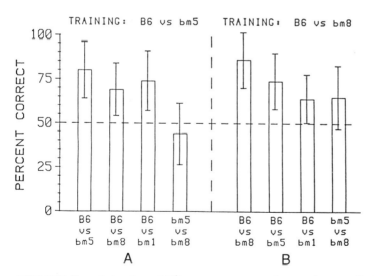

FIGURE 5 Distinction of K^b mutant mouse urines and generalization to other mutants. Generalization (transfer) trials with three mutant strains, bm1, bm5, and bm8. Trained mice discriminated B6 from bm1, bm5, and bm8 as shown in the first three bars of A and B. However, there was no evidence of discrimination between bm5 versus bm8 as shown in the last bar of each panel. Mean percentage correct trials ($\pm 95\%$ confidence interval).

The nonmutant may have a particular odor not shared by H-2 mutants as a group.

The class II genes occupy the I region between H-2K and H-2D and code a dimeric molecule of two roughly equally sized peptide chains expressed predominantly on the surface of certain cells of the immune system (see Figure 1). We used the class II I-A mutant B6.C-H-2^{bm12} mice. The A^{bm12} allele differs from the standard A^b allele in three nucleotides. Successful olfactory discrimination of B6 versus bm12 by trained mice has been accomplished, indicating that a class II genetic difference is sufficient to confer olfactory individuality (Yamazaki et al., 1989).

In summary, mutant and nonmutant mice were successfully distinguished based on odor alone. This is the strongest present indication that known MHC genes can influence individual odors because if class I and class II mutations change the odor, then H-2 genes themselves contribute to the odor.

VII. RADIATION CHIMERA MICE

The nature of the MHC-determined odorants perceived by the responding mouse remains a mystery. The use of cell chimeras permits an approach to

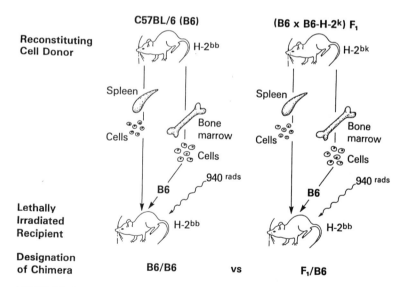

FIGURE 6 Constitution of radiation chimeras. Inbred homozygous mice (H-2 type bb) were given 940 rad gamma radiation, destroying the hematopoietic (blood-forming) system. Shortly thereafter, the mice were intravenously injected with 45-75 × 10^6 bone marrow and spleen cells from either identical donors (left) or heterozygous donors (H-2 type bk). Injected cells re-established the hematopoietic system, as shown by cytotoxicity tests conducted 5-11 weeks later.

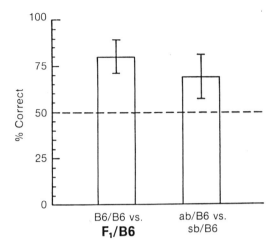

FIGURE 7 Distinction of urine of F_1-to-parent strain hematopoietic chimeras. Blind transfer (generalization) trials. Mean percentage correct trials ($\pm 95\%$ confidence interval).

the question of whether cells of particular lineages contribute to chemosensory identity. Radiation chimeras are especially favorable because virtually the entire hematopoietic system, but no other, is replaced by donor cells. The question raised was whether trained mice would respond to urine odors of chimera mice in a manner similar to their response to urine odors of strains of mice that provided the donor cells to the chimeras. Figure 6 is a schematic representation of how radiation chimera mice were made. Two types of chimera mice were produced, B6/B6, that is, irradiated B6 mice injected with cells from other B6 mice, and F_1/B6, which were irradiated B6 mice injected with F_1 cells. Results from Y maze tests clearly showed that the individual genetically determined olfactory identity of the mouse was changed by the procedure, as indicated by the greater than chance performance in discrimination (Figure 7). The urine of these radiation chimeras acquired a scent typical of the reconstituting donor's H-2 type (Yamazaki et al., 1985). Thus, cells of the hematopoietic system are responsible, at least in part, for the MHC-related odorant properties that enable mice to distinguish one another according to their H-2 types.

VIII. PREGNANCY BLOCK

Another system from which to view MHC-associated communication concerns the phenomenon known as pregnancy block, or the Bruce effect (Bruce, 1960). If a female mouse is separated from her mate shortly after mating and is then exposed to a male of a strain different from that of her first mate, or to the urine of such a male, there is an increased probability that pregnancy or pseudopregnancy will be terminated and she will return to normal estrous cycling.

Would sensory perception of MHC types play a role in this blocking of pregnancy caused by exposure of fertilized females to strange males? To test this, females of the inbred strain BALB, whose H-2 type is D, were mated with stud males who were either B6 (H-2^b) or B6-H-2^k. The test male, to which the fertilized female was exposed, was either the same stud male, a syngeneic male genetically identical to the stud male, or a congenic male: B6-H-2^b if the stud male was B6-H-2^k and B6-H-2^k if the stud male was B6-H-2^b.

Day 0 was the day of mating, when a vaginal plug was observed. On day 1, the female was isolated in a new cage. On days 2-4, she occupied one side of a divided cage. On the other side of a perforated screen was the test male. On days 5-7 she returned to the cage she occupied on day 1 and was examined for return to estrous until day 7, when the uterus was removed for inspection. Initiation of the estrous cycle during these 7 days indicates a blocked pregnancy or blocked pseudopregnancy. The incidence of pregnancy or pseudopregnancy block was substantially greater when the blocking male differed from the stud male at the MHC locus than when the blocking male had the same MHC type as the stud male (Table 1) (Yamazaki et al., 1983a). The effects of H-2 mutation in the circumstances of pregnancy block was next tested. The results showed the olfactory distinction of mice differing by a mutation of the H-2K gene can spontaneously influence neuroendocrine communication affecting reproduction (Yamazaki et al., 1986b).

Pregnancy block likely represents a neuroendocrine response to the genetic (individual) identity of the blocking male generally associated with a male signal. Thus, sensory recognition according to genotype may operate in the broader context of neuroendocrine responses linked to reproduction.

TABLE 1 Incidence of Blocking of Pregnancy or Pseudopregnancy in Isolated BALB Females Exposed to Males Whose H-2 Type Differed or Did Not Differ from That of the Stud Male

Test male	No. females returning to estrus (%)
Stud	8/73 (11)
Syngeneic	9/76 (12)[a]
Congenic	44/76 (58)[a]

[a]$X^2 = 33.49; p < 0.001$.

IX. DETERMINATION OF MATING PREFERENCE

In the standard test for H-2-associated mating preference, a male is presented with two females in estrus selected from two H-2 congenic mouse panels, B6 and B6-H-2^k for example, and watched until successful copulation is verified by presence of a vaginal plug. These studies revealed that B6 males, which are H-2^b, mated preferentially with congenic B6-H-2^k females, rather than with the alternative B6 females, and B6-H-2^k males mated preferentially with B6 females (Yamazaki et al., 1976). A large number of experiments have confirmed H-2-associated mating preference are generally characterized by preference for nonself H-2 type (Boyse et al., 1989b). Such preferences could act to promote valuable heterozygosity at H-2 as well as at other genetic loci (Dausset, 1981).

To determine whether this natural preference for the nonself H-2 haplotype is acquired during early life, the mating preferences of B6 males reared by B6-H-2^k foster parents and of B6-H-2^k males reared by B6 foster parents were studied (Yamazaki et al., 1988). This experimental design seemed most likely to reveal any influence that imprinting on parental H-2 types may have with respect to subsequent choice of a mate. Within 16 h of birth, entire litters were removed from their natural parents and transferred to foster parents whose own litters, born at approximately the same time, were simultaneously removed. At 21 days of age, the fostered mice were weaned and the

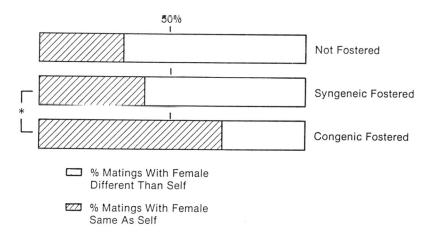

FIGURE 8 Mating preference of males (B6 and B6-H-2^k combined) from inbred nonfostered stock, males fostered to parents of the same (syngeneic) H-2 types as self, and males fostered to parents of a different (congenic) H-2 type than self. The first two groups do not differ, whereas the two fostered groups differ (*) signficantly, at $p < 0.01$.

males maintained in stock cages, containing only males of the same genotype and fostering history, until sexual maturity, when tests of mating preference began. The method of testing mating preference was as same as described earlier.

In previous mating tests, male mice showed a tendency to mate with females of an H-2 type different from their own. These preferences for the nonself H-2 haplotype were not significantly altered by fostering on syngeneic parents, which reproduces the genetic relations that obtain in the usual propagation of inbred strains. In contrast, when mice were fostered to parents of a different H-2 type, the usual mating preferences were reversed (Figure 8). Thus, B6 males fostered by B6-H-2^k parents mated preferentially with B6 females, and similarly, B6-H-2^k males fostered by B6 parents mated preferentially with B6-H-2^k females. Thus, H-2-selective mating preference is acquired by imprinting on familial H-2 types, and the prime basis of H-2-selective mating preference is clearly temporal. Whichever H-2 type is experienced during the rearing period of 3 weeks becomes the less favored H-2 type.

X. CONCLUSION

Genes known to be critically involved in regulating immune response also, in part, control the olfactory individuality of the mouse. Genes elsewhere in the genome, notably located on the X and Y chromosomes (Yamazaki et al., 1986a) and perhaps scattered throughout the remaining of the genome, also determine component of olfactory individuality (Beauchamp et al., 1989; Boyse et al., 1989a), but the MHC appears paramount (Yamazaki et al., 1986a). Similar results have recently been reported for the rat (Brown et al., 1987; Singh et al., 1987); presumably MHC-coded olfactory individuality is common in vertebrates and perhaps even in invertebrates (Grosberg and Quinn, 1986).

The nature of the odors produced remains a mystery. Genes in the MHC code for proteins that are unlikely to be volatile and hence odorous in the Y maze assay employed. Fragments of MHC molecules are found in urine, and they could conceivably act as odorants themselves (Singh et al., 1987) or bind other volatile substance (Beruter et al., 1973; Singer et al., 1988). Recent work implicates products of bacterial activity as a source of the volatile compounds (Roser et al., 1989). Regardless of the precise pathway between MHC genes and odor, it is likely that the odors themselves are made up of mixtures of volatile substances in particular proportions reflecting the MHC type of the individual mouse. These genetically determined patterns, or odor types (Boyse et al., 1987, 1989b), thus reflect individual identity and genetic relatedness and could serve as a basis for kin recognition and kin selection (Beauchamp et al., 1988; Boyse et al., 1989a).

ACKNOWLEDGMENTS

This work was supported in part by NIH Grant GMCA 32096. We thank D. Kupniewski, Y. Imai, Y. Habata, and H. Shirakawa for excellent technical assistance.

REFERENCES

Andrews, P. W., and Boyse, E. A. (1978). Mapping of an H-2-linked gene that influences mating preferences in mice. *Immunogenetics* **6**:265-268.

Beauchamp, G. K., Yamazaki, K., and Boyse, E. A. (1985). The chemosensory recognition of genetic individuality. *Sci. Am.* **253**:86-92.

Beauchamp, G. K., Yamazaki, K., Bard, J., and Boyse, E. A. (1988). Pre-weaning experience in the control of mating preferences by genes in the major histocompatibility complex of the mouse. *Behav. Genet.* **18**:537-547.

Beauchamp, G. K., Yamazaki, K., Duncan, H., Bard, J., and Boyse, E. A. (1989). Genetic determination of individual mouse odor. In *Chemical Signals in Vertebrates*, Vol. V. D. W. Macdonald et al. (Eds.). in press.

Beruter, J., Beauchamp, G. K., and Muetterties, E. L. (1973). Complexity of chemical communication in mammals: Urinary components mediating sex discrimination by male guinea pigs. *Biochem. Biophys. Res. Commun.* **53**:264-271.

Boyse, E. A. (1977). The increasing value of congenic mice in biomedical research. *Lab. Anim. Sci.* **27**:771-781.

Boyse, E. A., Beauchamp, G. K., and Yamazaki, K. (1987). The genetics of body scent. *Trends Genet.* **3**:97-102.

Boyse, E. A., Beauchamp, G. K., Yamazaki, K., and Bard, J. (1989a). Genetic components of kin recognition in mammals. In *Kin Recognition*, P. G. Hepper (Ed.). Cambridge University Press, London.

Boyse, E. A., Beauchamp, G. K., Bard, J., and Yamazaki, K. (1989b). Behavior and the major histocompatibility complex (MHC), H-2, of the mouse. In *Psychoneuroimmunology*, Vol. II, R. Ader, D. L. Felter, and N. Cohen (Eds.). Academic Press.

Brown, R., Singh, P. B., and Roser, B. (1987). The major histocompatibility complex and the chemosensory recognition of individuality in rats. *Physiol. Behav.* **40**:65-73.

Bruce, H. M. (1960). A block to pregnancy in the mouse caused by proximity of strange males. *J. Reprod. Fertil.* **1**:96-103.

Carroll, M. C., Katzman, P., Alicot, E. M., Koller, B. H., Geraghty, D. E., Orr, H. T., Strominger, J. L., and Spies, T. (1987). Linkage map of the human major histocompatibility complex including the tumor necrosis factor genes. *Proc. Natl. Acad. Sci. USA* **84**:8535-8539.

Dausset, J. (1981). The major histocompatibility complex in man. *Science* **213**:1469-1474.

Grosberg, R. K., and Quinn, J. F. (1986). The genetic control and consequences of kin recognition by the larvae of a colonial marine invertebrate. *Nature* **322**:456-459.

Klein, J. (1986). *Natural History of the Major Histocompatibility Complex.* Wiley & Sons, New York.

Roser, B., Brown, R., and Singh, P. (1990). Excretion of transplantation antigens into the environment as signals of genetic individuality. In *Chemical Senses*: Vol. 3 *Genetics of Perception and Communication*, C. J. Wysocki and M. Kare (Eds.). Marcel Dekker, New York.

Singer, A. G., Clancy, A. N., Macrides, F., Agosta, W. C., and Bronson, F. H. (1988). Chemical properties of female mouse pheromone that stimulates gonadotropin secretion in males. *Biol. Reprod.* **38**:193-199.

Singh, P. B., Brown, R. E., and Roser, B. (1987). MHC antigens in urine as olfactory recognition cues. *Nature* **327**:161-164.

Thomas, L. (1975). Symbiosis as an immunologic problem: The immunesystem and infectious diseases. In *Fourth International Congress of Immunology*, E. Neter and F. Milgrom (Eds.). S. Karger, Basel, p. 2.

Yamaguchi, M., Yamazaki, K., and Boyse, E. A. (1978). Mating preference tests with the recombinant congenic strain BALB.HTG. *Immunogenetics* **6**:261-264.

Yamaguchi, M., Yamazaki, K., Beauchamp, G. K., Bard, J., Thomas, L., and Boyse, E. A. (1981). Distinctive urinary odors governed by the major histocompatibility locus of the mouse. *Proc. Natl. Acad. Sci. USA* **78**:5817-5820.

Yamazaki, K., Boyse, E. A., Mike, V., Thaler, H. T., Mathieson, B. J., Abbott, J., Boyse, J., Zayas, Z. A., and Thomas, L. (1976). Control of mating preferences in mice by genes in the major histocompatibility complex. *J. Exp. Med.* **144**:1324-1335.

Yamazaki, K., Yamaguchi, M., Andrews, P. W., Peake, B., and Boyse, E. A. (1978). Mating preferences of F_2 segregants of crosses between MHC-congenic mouse strains. *Immunogenetics* **6**:253-259.

Yamazaki, K., Yamaguchi, M., Baranoski, L., Bard, J., Boyse, E. A., and Thomas, L. (1979). Recognition among mice: Evidence from the use of a Y-maze differentially scented by congenic mice of different major histocompatibility types. *J. Exp. Med.* **150**:755-760.

Yamazaki, K., Beauchamp, G. K., Bard, J., Thomas, L., and Boyse, E. A. (1982). Chemosensory recognition of phenotypes determined by the T1a and H-2K regions of chromosome 17 of the mouse. *Proc. Natl. Acad. Sci. USA* **79**:7828-7831.

Yamazaki, K., Beauchamp, G. K., Wysocki, C. J., Bard, J., Thomas, L., and Boyse, E. A. (1983a). Recognition of H-2 types in relation to the blocking of pregnancy in mice. *Science* **221**:186-188.

Yamazaki, K., Beauchamp, G. K., Egorov, I. K., Bard, J., Thomas, L., and Boyse, E. A. (1983b). Sensory distinction between $H-2^b$ and $H-2^{bm1}$ mutant mice. *Proc. Natl. Acad. Sci. USA* **80**:5685-5688.

Yamazaki, K., Beauchamp, G. K., Thomas, L., and Boyse, E. A. (1985). The hematopoietic system is a source of odorants that distinguish major histocompatibility types. *J. Exp. Med.* **162**:1377-1380.

Yamazaki, K., Beauchamp, G. K., Matsuzaki, O., Bard, J., Thomas, L., and Boyse, E. A. (1986a). Participation of the murine X and Y chromosomes in genetically determined chemosensory identity. *Proc. Natl. Acad. Sci. USA* **83**:4438-4440.

Yamazaki, K., Beauchamp, G. K., Matsuzaki, O., Kupniewski, D., Bard, J., Thomas, L., and Boyse, E. A. (1986b). Influence of a genetic difference confined to mutation of H-2K on the incidence of pregnancy block in mice. *Proc. Natl. Acad. Sci. USA* **83**:740-741.

Yamazaki, K., Beauchamp, G. K., Kupniewski, D., Bard, J., Thomas, L., and Boyse, E. A. (1988). Familial imprinting determines H-2 selective mating preferences. *Science* **240**:1331-1332.

Yamazaki, K., Beauchamp, G. K., Bard, J., and Boyse, E. A. (1989). Single MHC gene mutations alter urine odour constitution in mice. In *Chemical Signals in Vertebrates*, Vol. V, D. W. Macdonald et al. (Eds.). in press.

Zeff, R. A., Geier, S. S., Gopas, J., Geliebter, J., Schulze, D. H., Pease, L. R., Pfaffenbach, G. M., Pontarotti, P., Mashimo, H., McGovern, D. A., and Nathenson, S. G. (1985). Mutants of the murine major histocompatibility complex: Structural analysis of in vivo and in vitro H-2K^b variants. In: *Cell Biology of the Major Histocompatibility Complex*, B. Pernis and H. Vogel (Eds.). Academic Press, New York, pp. 41-49.

Part IV
Mouse Model Systems

16
The Genetics of Bitterness, Sweetness, and Saltiness in Strains of Mice

Ian E. Lush

University College London, London, England

I. INTRODUCTION

I began to work on the genetics of tasting in mice in January 1980 after reading the short paper by Warren and Lewis (1970) describing the effect of the *Soa* gene. They showed that sucrose octaacetate (SOA) is extremely bitter to strain CFW mice but completely tasteless to mice from several other strains. They also stated that the difference was due to one gene, with the tasting allele dominant. I was surprised to find that nothing had been published on this gene in the intervening 10 years. It was a simple matter to confirm and extend their results (Lush, 1981); indeed it was so simple that I decided to try some other bitter substances using the same technique.

II. TECHNIQUE

Perhaps I should add a few words here about the technique before describing my results. The mice in their experimental cages can drink from either of two metal spouts that come down through the cage roof. One is attached to a burette containing the tastant solution, and the other is attached to a burette containing the solvent, usually water. After day 1, the amount that has been drunk from each spout-burette unit is recorded and the two units of each cage are changed around with respect to their positions. After day 2 the consumption from each unit is measured again. The same procedure is carried out for days 3 and 4, but with the solutions reversed with respect to the units. The amount of the tastant drunk each day is expressed as a percentage of that day's total fluid intake, and the final figure for each cage is the mean of the results of the 4 days. This simple experimental procedure theoretically compensates for any preference the mice might have for drinking from one unit or from one of the two positions (see Lush, 1984). A result above 50% means that the tastant is preferred; a result below 50% means that the tastant is avoided. Even though I have used this technique for several years I still find it remarkable to see what smooth concentration-response curves can be produced by this simple procedure.

If the results are to be meaningful, any preference or aversion for the tastant should remain constant during the 4 days of the experiment. This is usually the case, but not always. With phenylthiourea (PTC) some strains, particularly albino strains, show a slow but steady increase in their aversion to

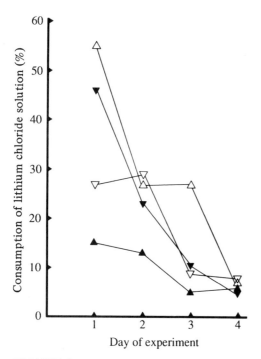

FIGURE 1 Consumption of a 75 mM solution of lithium chloride on each of the four consecutive days of a tasting experiment. The four strains are (△), A2G; (▼) 129/Sv; (▽) TO; (▲) C57BL/6Ty.

0.01% PTC over a period of 8-10 days (Lush, 1986b). A more extreme example is given by 75 mM lithium chloride, which is initially acceptable to some strains but at the end of 4 days is strongly rejected by all strains, as shown in Figure 1.

III. BITTERNESS

The only strains in my laboratory that are tasters of SOA are SWR and Schneider. The phenotypic difference between SOA tasters and non-tasters is extremely clear, as mentioned. This seemed to present an opportunity to discover what particular chemical structure confers the property of bitterness on the SOA molecule. The idea was to test the strains with a variety of chemicals similar to SOA and pick out those chemicals that are bitter to SWR and Schneider but tasteless to the other strains. This subset of bitter tastants must have something in common that is recognized by the SOA receptor. The same approach was used by Fox (1932) and, more recently, by Harris and Kalmus (1949) with the PTC tasting polymorphism in humans. The assumption that the (SOA) receptor is the site of the genetic variation in the mouse was shown to be sound by Shingai and Beidler (1985), who demonstrated by neurophysiologic methods that the glossopharyngeal and chorda tympani nerves of SWR mice responded to an SOA solution placed on the tongue, whereas the nerves of three nontaster strains did not.

In my own survey with acetylated sugars two acetylated disaccharides were tested, SOA and α,α-trehalose octaacetate, and they were both found to be equally bitter to SWR and Schneider mice. The α- and β-anomers of D-glucose pentaacetate, D-galactose pentaacetate, and D-mannose pentaacetate were all tested. The results showed that SWR and Schneider mice found the α-anomer to be slightly less bitter than the β-anomer in each case. The galactose acetates were slightly less bitter than the corresponding glucose acetates, and the mannose acetates were less bitter than the corresponding galactose acetates. In fact, the reaction of SWR and Schneider to α-D-mannose pentaacetate was not significantly different from the reaction of the SOA nontaster strains. This seems to show that a reduction in bitterness follows when any acetate group is moved from an equatorial to an axial valency of the same carbon atom in the ring. One curious finding was that L-glucose pentaacetate is considerably more bitter than the normal D-isomer. The oxygen atom in the ring is not essential for bitterness because 5-thio-glucose pentaacetate, in which the oxygen is replaced by a sulfur atom, is very bitter to the taster mice. Tri-O-acetyl glucal, in which carbons 1 and 2 are not acetylated but are joined by a double bond, is not bitter. Insofar as it is possible to draw any conclusion from the results with acetylated sugars, it is that the acetate attached to ring carbon 2 is of great importance in determining bitterness but changes elsewhere in the molecule can have an effect. It may be relevant to note here that humans find some sugars and glycosides bitter, and those with the β-conformation are more likely to be bitter than their α-counterparts (Birch and Lee, 1976; Birch et al., 1977).

The most unexpected finding was that SWR and Schneider mice are much more sensitive to the bitter taste of strychnine than are other strains. The sensitivity to strychnine and the ability to taste SOA could not be separated in the progeny of a backcross (Lush, 1982). This indicates that the *Soa* gene determines the ability to taste strychnine in addition to acetylated sugars. The molecular structure of strychnine bears no obvious resemblance to that of an acetylated sugar, but there must be some feature they both have that is recognized by the SOA receptor in taster mice. The acetylated sugars are not ideal material for identifying this common feature because they are very flexible molecules. Strychnine, on the other hand, is a polycyclic molecule and is extremely rigid. It would be helpful if one could find some other rigid molecules that taste bitter only to SWR and Schneider mice.

In 1981 I decided to test some other bitter substances, and I started with quinine. The procedure I then adopted, and have used ever since, is to choose about six strains for a preliminary experiment. The tastant concentration for this first experiment is usually about 1 mM, unless there is some reason for choosing a different value. If the result shows that none of the strains has reacted to the tastant, either positively or negatively, I double the concentration and try again. By raising and lowering the concentration it is usually possible to find a concentration range within which strain differences are detectable. Having found the concentration at which the strain differences are maximal, I then use this concentration to survey all the strains in my laboratory. As a result of this survey it is sometimes possible to pick out one or two strains that are different in some interesting way from the original strains. For example, they may be more extreme in their reaction to the tastant. These strains are also tested with a range of concentrations. At the end of this procedure one has concentration-response curves for six or eight strains and a survey of all the strains at one concentration. The survey may show that the strains tend to fall into two groups, tasters and nontasters, and this is good news because it is evidence that there is one gene that has a major effect. Even when such a division is not evident it may still be possible to detect a major gene by doing crosses and, in particular, by analyzing sets of recombinant inbred (RI) strains. It is important to realize that strain differences are often differences in *sensitivity* to a particular tastant. In other words, the concentration of a bitter substance can be lowered to a value at which no strain can taste it, and it can also be raised to a value at which all the strains can taste it and avoid drinking it. Somewhere between these two extremes the strain differences may be large enough to permit a genetic analysis. The choice of a suitable tastant concentration is therefore vital for success in this kind of work.

The importance of choosing the right concentration is illustrated by the quinine concentration-response curves shown in Figure 2A, where it can be seen that the strain differences are of a useful size only between 0.2 and 0.8 mM. Figure 2B illustrates the other important aid to success in this work, which is the use of RI lines. The theory and practice of these (see Chapter 1) was discussed by Dr. Taylor in this symposium. Figure 2B shows the quinine concentration-response curves of the seven CXB RI lines. These lines were derived by Dr. D. Bailey from a cross between BALB/c and C57BL/6. A comparison of the RI line curves with the curves in Figure 2A shows that four of the RI curves are similar to BALB/c and the other three are similar to C57BL/6. This is evidence that the phenotypic difference between BALB/c and C57BL/6 is due to a single gene, which has been given the symbol *Qui*. It is usually advisable to do conventional crosses to confirm the conclusions drawn from work with RI lines, although this is not always possible.

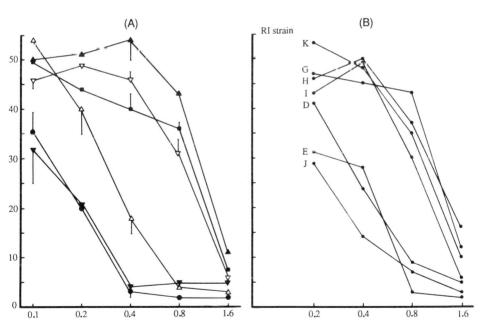

FIGURE 2 Concentration-response curves with quinine of (A) six strains. The strains are (▲) A2G; (▽) BALB/cBy; (■) DBA/2; (△) C57BL/6By; (▼) 129/Sv; (●) SWR. (B) Seven CXB RI lines. Vertical bars are SEMs.

During the work on acetylated sugars I noticed that raffinose acetate (RUA) is bitter to SWR and Schneider mice. This was not surprising since raffinose is a trisaccharide (O-α-D-galactopyranosyl-$(1\rightarrow 6)$-O-α-D-glucopyranosyl-$(1\rightarrow 2)$-β-D-fructopyranoside) and SWR and Schneider are the two strains that have the Soa^a allele and can therefore taste SOA and a number of other acetylated sugars, as explained earlier. The unexpected finding was that RUA is also bitter to three strains that have the Soa^b allele and are therefore nontasters of SOA and the other acetylated sugars. These three strains are DBA/2, C3H, and BALB/c, and their concentration-response curves with RUA are included in Figure 3. Further work (Lush, 1986a) showed that the ability to taste RUA is determined by a different gene, Rua. The Rua^a allele present in DBA/2, C3H, and BALB/c confers the ability to taste the RUA molecule without the ability to taste the constituent parts of the molecule, for example galactose acetate, glucose acetate, or SOA.

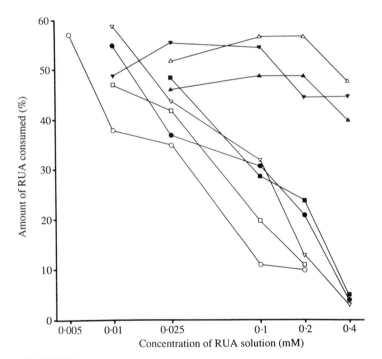

FIGURE 3 Concentration-response curves of eight strains tested with raffinose undecaacetate (RUA). The strains are ($\triangle$) C57BL/6By; ($\blacktriangledown$) 129/Sv; ($\blacktriangle$) A2G; ($\blacksquare$) DBA/2; ($\triangledown$) BALB/c; ($\square$) Schneider; ($\bullet$) C3H; ($\bigcirc$) SWR. Each point is the mean of between two and eight experiments.

The next bitter substance to engage my attention was cycloheximide. I chose it because Tobach et al. (1974) showed that cycloheximide is bitter to laboratory strains of rats at concentrations as low as 2 μM. They tested three strains, Wistar, Long-Evans, and fawn-hooded, and found that when the concentration was lowered to 1 μM the Wistar and Long-Evans rats still rejected it but the fawn-hooded rats accepted it, or possibly even preferred it to water. The testing procedure used by Tobach et al. was not entirely satisfactory, and one cannot be sure whether the fawn-hooded rats actually preferred 1 μM cycloheximide to water or merely could not taste its bitterness at that concentration, but in any case it seems clear that there is a difference between fawn-hooded and the other two strains in their sensitivity to the bitter taste of cycloheximide.

I found that mice are very similar to rats in their reaction to cycloheximide. There is strain variation in the degree to which the bitterness of a 1 μM solution can be tasted (Lush and Holland, 1988). A conventional genetic analysis was not done, but it was possible to show that the BXD RI lines fall into two groups, and this is evidence that one gene has a major effect on the phenotype. This gene was given the symbol *Cyx*.

IV. SWEETNESS

At this stage it was becoming clear that the genetics of bitterness tasting in mice involves many genes. Even with the limited number of tastants already tested, three new genes, *Qui, Rua,* and *Cyx,* had been identified in addition to *Soa*. Since there are hundreds of substances that are bitter to humans (and possibly also to mice) it seemed that I was in danger of spending my remaining years collecting more and more bitterness genes. I therefore decided to see if the same simple experimental technique that had been so productive with bitterness could also be applied to sweetness. Glycine was chosen as a suitable sweet substance because it is said to have a very pure sweet taste without bitter component (von Békésy, 1964). Eight strains were tested with a range of glycine concentrations (Lush and Holland, 1988), and the results are shown in Figure 4. At 0.1 mM none of the strains has a value significantly different from 50%, indicating that glycine is tasteless at that concentration. As the concentration is increased, however, some very large strain differences develop. Strain 129/Sv does not vary much from the 50% value, so for this strain glycine seems to be tasteless at any concentration. The two C57BL/6 substrains show a rapidly increasing preference for glycine when the concentration rises above 10 mM, but TO shows a rather more gradual rise. Presumably these strains are responding to the sweet taste of glycine at higher concentrations. The unexpected aspect of the results is that all the other strains begin to avoid the glycine solution as the concentration increases.

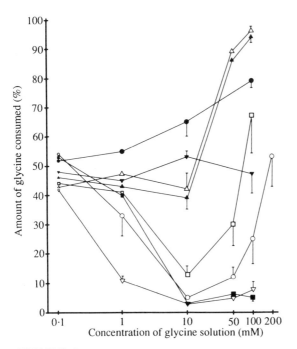

FIGURE 4 Concentration-response curves of eight strains tested with glycine. The strains are (△) C57BL/6By; (▲) C57BL/6Ty; (●) TO; (▼) 129/Sv; (□) Schneider; (○) SWR; (■) DBA/2Ty; (▽) BALB/cBy. The vertical bars are SEMs.

DBA/2 and BALB/c both fall to a very low level at 10 mM and remain low at the higher concentrations. Schneider and SWR also show an aversion at 10 mM, but they start to rise again as the concentration increases, as though the sweet taste begins to overcome their aversion.

The simplest explanation of these results is that glycine has both a sweet and a bitter taste and there is strain variation in the ability to detect each of these tastes. 129/Sv can detect neither the sweet nor the bitter taste. C57BL/6 can detect only the sweet taste. DBA/2 and Balb/c can detect only the bitter taste. SWR and Schneider can detect both tastes, but since the bitter taste is detectable at a lower concentration than the sweet taste the SWR and Schneider curves first fall and then rise again at the higher concentrations. When 27 strains were tested with 10 mM glycine they fell into two groups. One group of 15 strains had a mean consumption of 6.7%, and these were therefore clearly able to taste the glycine bitterness. The other 12 strains formed a group of nontasters, with a mean consumption of 45%. Further work with crosses and RI lines showed that the difference is due to a single gene, which was given the symbol *Glb* (glycine bitterness).

It could be said that, although glycine clearly has an unpleasant taste to some strains, one cannot assume that the unpleasant taste is bitterness. This is a valid objection, but there is circumstantial evidence that at least suggests that the unpleasant taste of glycine is similar to the unpleasant taste of cycloheximide, quinine, and raffinose acetate. This evidence comes from the linkage relationships of the four genes. The distribution patterns of the alleles of *Cyx, Qui, Rua,* and *Glb* in the BXD RI lines are shown in Table 1. That there are very few differences between them indicates that the four genes are tightly linked. They are probably arranged in the order shown in Table 1, since this is the order that minimizes the number of crossovers. The genes were found to have identical distribution patterns in the CXB RI lines, and taking all the RI data together, the genetic distances (in centimorgans) between the genes can be estimated to be as follows: *Cyx*-2.78-*Qui*-0.88-*Rua*-0.91-*Glb* (Silver, 1985). This little cluster of genes is located on chromosome 6 and is also very closely linked to a cluster of genes that specify the structure of some proline-rich salivary proteins (Azen et al., 1986; Azen, personal communication).

It seems probable that the four linked tasting genes have arisen from a single, original, tasting gene by a process of local duplication and all the present members of the cluster still have the same basic function. For this reason I believe that when mice show an aversion to cycloheximide, quinine, raffinose acetate, or glycine they do so because they experience a very similar unpleasant taste in each case. It seems reasonable to refer to this shared taste as bitterness since three of the four substances are bitter to humans.

TABLE 1 A Comparison of the BXD RI Distribution Patterns of the Four Bitterness Tasting Genes[a]

Gene	BXD RI strains																				
	1	2	5	6	8	9	11	12	15	16	18	19	22	24	25	27	28	29	30	31	32
Cyx	D	D X	B	B	B	D	D	B	B	B	B	D	B	D	B X	B	B	D	B	B	D
Qui	D	B	D	B	B	D	D	B	B X	B	B	D	B	D	D	B	B	D	B	B	D
Rua	D	B	D	B	B X	D	D	B	D	B	B	D	B	D	D	B	B	D	B	B	D
Glb	D	B	D	B	D	—	D	B	D	B	B	D	B	D	D	B	B	D	B	B	D

[a]X indicates crossover region. BXD 9 was not tested with glycine.

There are other examples, in addition to glycine, of substances that taste differently to mice and to humans. For example, 1 mM chloramphenicol is extremely bitter to humans but tasteless to mice (Lush, unpublished data). Denatonium benzoate (Bitrex), which is probably the most bitter substance known (Saroli, 1984), is only about as bitter as quinine to mice. These differences in tasting ability, both within and between species, should remind us that taste is not an inherent quality of a substance. Taste is an interaction between a substance and an organism and can be altered or abolished by a change in the organism.

It is not yet known if *Soa* is in the gene cluster discussed here, but it seems very probable that there are several more bitterness tasting genes in the cluster. On the assumption that the genes within the cluster are uniformly distributed and that the four known genes are a random sample, Dr. Michael Turelli (personal communication) has calculated that the probable total genetic length of the cluster is given by the expression $L[n + 1)/(n - 1)]$ centimorgans, where n is the number of known tasting genes in the cluster and L is the genetic distance they occupy. In the present case this gives 4.57 (5/3) = 7.6 cM. This is about 10% of the known genetic length of chromosome 6.

My attempt to study the sweetness of glycine had produced another bitterness gene. Because the bitterness of glycine interferes with the sweetness, it seemed to be a rather unsuitable substance with which to investigate the genetics of sweetness. I therefore decided to study some intensely sweet substances that are sweet at concentrations well below those at which any bitter component might be detectable. Saccharin was an obvious choice because Fuller (1974) had already reported that one gene, *Sac*, has a major effect on saccharin preference in DBA/2 and C57BL/6 mice. I also chose acesulfame and dulcin, and I included sucrose as a representative of the sugars.

The taste-testing procedure for sweetness was exactly the same as for bitterness, and it proved to be equally successful. The concentration-response curves for five strains tested with acesulfame are shown in Figure 5. It can be seen that DBA/2 and 129/Sv show little or no response, but the other strains have a greatly increased preference for acesulfame at the higher concentrations. The curves obtained with saccharin are almost the same as those in Figure 5 (Lush, 1989). Results from conventional crosses and RI lines confirmed that the difference between C57BL/6 and DBA/2 is largely due to one gene. All the mice in the crosses and the RI lines were tested both with saccharin and with acesulfame, and the results were the same for both tastants. This seems to mean that there is a common genetic determination of the abilities to taste saccharin and acesulfame, perhaps by affecting a common receptor. This idea was supported by the finding that 26 different inbred strains are highly correlated in their ability to taste saccharin and acesulfame. Similar

BITTERNESS, SWEETNESS, AND SALTINESS IN MICE 237

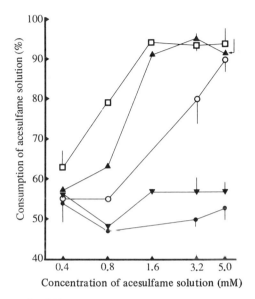

FIGURE 5 Concentration-response curves of five strains tested with acesulfame. The strains are (■) STS; (▲) C57BL/6Ty; (○) SWR; (▼) 129/Sv; (●) DBA/2Ty. Vertical bars are SEMs.

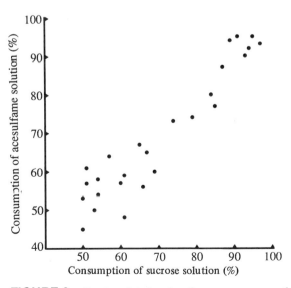

FIGURE 6 Scatter plot showing the mean consumption of 50 mM sucrose and 3.2 mM acesulfame by 26 strains. The coefficient of correlation $r = 0.946$.

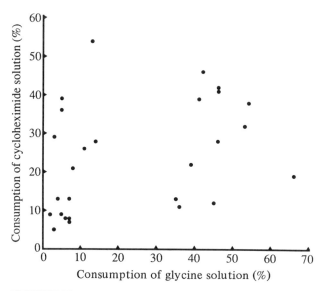

FIGURE 7 Scatter plot showing the mean consumption of 10 mM glycine and 1.0 μM cycloheximide by 27 strains.

correlations were found when the same strains were tested with dulcin and sucrose. As an example, the strain correlation between sucrose and acesulfame is shown in Figure 6. The conclusion seems to be that there is only one sweetness receptor. The sensitivity of this common receptor is determined by the gene *Sac* and perhaps by other genes as well (Lush, 1989). This is in complete contrast to the genetics of bitterness tasting in which many genes are involved but each gene determines a receptor that has a relatively narrow specificity. Because of the diverse origins of the inbred strains there are no correlations between tasting abilities with respect to bitter substances detected by different receptors. This is illustrated in Figure 7, where 27 strains are compared with respect to glycine and cycloheximide tasting (data from Lush, 1988).

V. SALTINESS

Bitterness and sweetness are associated with a great variety of complex organic molecules. Saltiness is caused by sodium chloride, although a few other salts are said to have a similar taste (Moncrieff, 1967). There is no confirmed report of genetic differences in behavioral response to saline solutions (Wolf and Lawrence, 1963), but in view of the large amount of variation found in bitterness and sweetness, it seemed worthwhile to have a fresh look at saltiness. The concentration-response curves for five strains are shown in Figure 8, where it can be seen that there are very large strain differences at concentrations between 30 and 300 mM. Genetic work on these differences is now in progress. It should be particularly interesting, for example, to discover why TO and 129/Sv like drinking 75 mM saline but C57BL/6 shows an equal dislike at the same concentation. Does saline taste different to these strains, or do the strains react differently to the same taste? Finding the answer to this sort of question will require the use of techniques from genetics, physiology, and animal psychology.

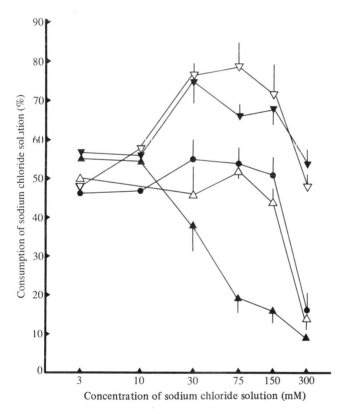

FIGURE 8 Concentration-response curves of five strains tested with sodium chloride. The strains are (▽) TO; (▲) 129/Sv; (●) DBA/2; (△) A2G; (▼) C57BL/6Ty. Vertical bars are SEMs.

VI. CONCLUSION

I hope that this brief account of the genetics of tasting in mice has impressed the reader with the extraordinarily large amount of variation to be found between different laboratory strains. Tasting ability seems to be an unusually variable characteristic in this species. Presumably this strain variation is due to variation present in the domesticated wild mice from which these strains descend. One would therefore expect wild mice to be highly polymorphic with respect to tasting ability. There are some hints that this is true (Warren and Lewis, 1970), but this question has not yet been systematically investigated. Finding the selective pressures (if any) that maintain the polymorphisms in wild populations would be the ultimate goal, but this is a notoriously difficult task and I am happy to leave it to the ecologic geneticists. From the evolutionary point of view it is also possible to see why the genetics of bitterness is more complex than the genetics of sweetness. Bitterness has presumably evolved as a warning system to ensure that harmful chemicals are avoided, and for that purpose a high degree of specificity and sensitivity would be advantageous. Sweetness does not play such a vital role in the life of a mouse and therefore does not need more than one type of receptor.

Perhaps the most fruitful application of the genetic approach will be in the study of the physiology of taste. Genetic variation of the peripheral receptors opens up the possibility of locating and isolating the relevant genes. From this will come more knowledge of the structure of the receptors. We shall then be in a better position to theorize about how the brain builds up complex tastes using the information coming from the tongue.

ACKNOWLEDGMENT

This work was supported by the Wellcome Trust.

REFERENCES

Azen, E. A., Lush, I. E., and Taylor, B. A. (1986). Close linkage of mouse genes for salivary proline-rich proteins (PRPs) and taste. *Trends Genet.* **2**:199-200.

Birch, G. G., and Lee, C. K. (1976). Structural features and taste in the sugar series: The structural basis of bitterness in sugars. *J. Food Sci.* **41**:1403-1407.

Birch, G. G., Lee, C. K., and Ray, A. (1977). The chemical basis of bitterness in sugar derivatives. In *Sensory Properties of Foods*, G. G. Birch, J. G. Brennan, and K. J. Parker (Eds.). Elsevier Applied Science, London, pp. 101-111.

Fox, A. L. (1932). The relationship between chemical constitution and taste. *Proc. Natl. Acad. Sci. USA* **18**:115-120.

Fuller, J. M. (1974). Single locus control of saccharin preference in mice. *J. Hered.* **65**:33-36.

Harris, H., and Kalmus, H. (1949). Chemical specificity in genetical differences of taste sensitivity. *Ann. Eugen.* **15**:32-45.

Lush, I. E. (1981). The genetics of tasting in mice I. Sucrose octaacetate. *Genet. Res.* **38**:93-95.

Lush, I. E. (1982). The genetics of tasting in mice II. Strychnine. *Chem. Senses* **7**: 93-98.

Lush, I. E. (1984). The genetics of tasting in mice III. Quinine. *Genet. Res.* **44**:151-160.

Lush, I. E. (1986a). The genetics of tasting in mice IV. The acetates of raffinose, galactose and β-lactose. *Genet. Res.* **47**:117-123.

Lush, I. E. (1986b). Differences between mouse strains in their consumption of phenylthiourea (PTC). *Heredity* **57**:319-323.

Lush, I. E. (1989). The genetics of tasting in mice VI. Saccharin, acesulfame, dulcin and sucrose. *Genet. Res.* **53**:95-99.

Lush, I. E., and Holland, G. (1988). The genetics of tasting in mice V. Glycine and cycloheximide. *Genet. Res.* **52**:207-212.

Moncrieff, R. W. (1967). *The Chemical Senses*, 3rd Ed. Leonard Hill, London.

Saroli, A. (1984). Structure-activity relationship of a bitter compound: denatonium chloride. *Naturwissenschaften* **71**:428-429.

Shingai, T., and Beidler, L. M. (1985). Interstrain differences in bitter taste responses in mice. *Chem. Senses* **10**:51-55.

Silver, J. (1985). Confidence limits for estimates of gene linkage based on analysis of recombinant inbred strains. *J. Hered.* **76**:436-440.

Tobach, E., Bellin, J. S., and Das, D. K. (1974). Differences in bitter taste perception in three strains of rats. *Behav. Genet.* **4**:405-410.

von Békésy, G. (1964). Sweetness produced electrically on the tongue and its relation to taste theories. *J. Appl. Physiol.* **19**:1105-1113.

Warren, R. P., and Lewis, R. C. (1970). Taste polymorphism in mice involving a bitter sugar derivative. *Nature (Lond.)* **227**:77-78.

Wolf, G., and Lawrence, G. H. (1963). Saline preference curve for mice: Lack of relationship to pigmentation. *Nature (Lond.)* **200**:1025-1026.

17
Congenic Lines Differing in Ability to Taste Sucrose Octaacetate

Glayde Whitney, David B. Harder, Kimberley S. Gannon, and John C. Maggio

Florida State University, Tallahassee, Florida

The key to analysis of many problems in experimental biology has been simplification through isolation of components. With mice, congenic strain development was rationally designed to facilitate isolating and studying the effects of individual loci on phenotypes of interest. Originating with the work of Snell in the 1940s (Morse, 1981; Snell, 1978; Snell et al., 1976), the congenic strain approach has contributed much to basic knowledge of mammalian genetics and to knowledge in a variety of phenotypic domains. A congenic line consists of mice that are virtually identical to those of another inbred line except for the substitution of genes that influence a specific trait. Such isolation of specific trait-influencing genes on an otherwise homogeneous genetic background provides an extremely simplified model system that, in theory, could facilitate analysis of any trait domain.

About a decade ago we set out to attempt an application of the congenic line approach to an analysis of chemosensory function. The plan was to employ a simple and efficient behavioral assay (two-bottle preference testing) to survey inbred mouse strains across a variety of chemicals (tastants). The search was for contrasting extreme strains different with regard to a specific tastant or class of tastants. If any were found the next step was to employ Mendelian cross-breeding procedures to see what might be the genetic architecture underlying a strain difference. For efficiency, the procedures envisioned to test for Mendelian segregation were primarily those breeding designs that are used to develop congenic strains. With regard to genetic architecture, the extreme possibilities are a polygenic, quantitative difference or a monogenic (single-locus) qualitative difference. Our notion was that polygenic systems had historically been rather complicated and recalcitrant to analysis of mechanisms, but monogenic systems have contributed immensely to ease of analysis of mechanisms. Thus, the plan was that if differences were found among strains that appeared to be quantitative or proved to be polygenic, the search would continue. If differences were found among strains that appeared to be monogenic, then congenic strain development would proceed.

I. SUCROSE OCTAACETATE

After investigation of a variety of tastants in which quantitative variation seemed predominant, the first tastant chosen for development of congenic lines was sucrose octaacetate (SOA). Individual differences among mice to this apparently nontoxic bitter substance were initially reported by Warren (1963). In 1970, Warren and Lewis reported results from two-bottle preference testing of SOA versus water for several inbred strains and a sample of wild mice. One strain (CFW/NIH) avoided SOA, as did some of the wild mice. Subsequent backcross results were consistent with the possibility that SOA detection involved a single autosomal locus with a dominant "taster" allele. Lush (1981a) surveyed 31 inbred strains of mice (including a CFW strain) for SOA avoidance. In individual tests none of the CFW mice avoided SOA. However, one other inbred strain, the SWR, displayed a profound aversion to SOA at a concentration of 10^{-4} M. Backcrosses involving SWRs yielded taster-nontaster ratios consistent with single-locus expectations. Lush (1981b) subsequently named two presumptive alleles: Soa^a (aversion-dominant); Soa^b (blind-recessive). We too surveyed inbred strains of mice with regard to SOA (Harder and Whitney, 1982; Harder et al., 1984). Consistent with Lush's (1981a) results, of 12 strains tested only the SWR/J reliably avoided 10^{-4} M SOA in unconditioned two-bottle preference tests of SOA in distilled water versus distilled water alone. However, 8 of 10 "nontasting" inbred strains display a profound avoidance of the near-saturation 10^{-3} M SOA when tested with a conditioned aversion procedure. The SWR strain appeared to be sensitive to SOA at all concentrations down to about 10^{-7} M, but both tested C57 strains (C57L/J and C57BL/6J) appeared to be nontasters of SOA even at 10^{-3} M, the strongest concentration tested (Harder et al., 1984).

Two-bottle preference tests are widely employed in chemosensory genetics, largely because they are amenable to testing the large number of subjects required by the probabilistic nature of Mendelian segregation. Unfortunately, a preference ratio score from an animal in a choice test is a long way from a chemosensory receptor; many variables, including centrally mediated motivation, in addition to or instead of sensory mechanisms can influence preference measures (Fuller, 1974; Harder et al., 1989; Kare and Ficken, 1963). Thus, although we planned to employ preference tests for genetic studies it was of interest to determine if the strain contrasts would be robust across assessment procedures in implicating sensory differences. To try a very different psychophysical procedure, Frye laboriously trained individual mice to lick water from a single drinking spout immediately upon presentation of the spout. Trained mice, which were naive with regard to SOA, were then presented with SOA solution in an apparatus that permitted computer recording of latencies for individual licks from the spout. If SWR mice avoided SOA in two-bottle preference tests because of an aversion to its taste, they were expected to quickly cease SOA licking in the single-spout test. In contrast, if C57 mice were indifferent to SOA in two-bottle tests because they could not taste it, they were expected to treat SOA like water in the single-spout test. As illustrated in Figure 1, SWR mice quickly ceased licking strong concentrations of SOA upon their first exposure (e.g., at 10^{-3} M SOA, after less than 1 s and about five licks), whereas C57 mice continued licking as if the SOA solutions were indistinguishable from water. Such results are consistent with taste mediation of SOA avoidance in two-bottle preference tests (Harder et al., 1984). Employing electrophysiologic recording from peripheral taste nerves (both chorda tympani and glossopharyngeal), Shingai and Beidler (1985) found relatively greater integrated responses to tongue application of SOA for the SWR taster strain than for a variety of inbred strains that were nontasters in two-bottle preference tests. These consistent strain differences across very disparate measures tend to validate the taster-nontaster inference for SOA from two-bottle preference tests.

II. SOA GENETICS

Further investigations of the genetics of SOA tasting have usually been consistent with the single major locus interpretation of Warren and Lewis (1970) and Lush (1981a). Taylor of the Jackson Laboratories produced a set of recombinant inbred (RI) strains from SWR/J (taster) and C57L/J (nontaster) progenitor strains. Six of the SWXL/Ty RI strains were tested for SOA preference (Whitney and Harder, 1986). The resulting dichotomous strain distribution pattern (SDP, Figure 2) was consistent with single-locus expectations. The specific SDP found for SOA among the six lines was identical to the SDP of three previously typed loci, two of which had been mapped to chromosome 1 and the third had been mapped to chromosome 6 (Whitney and Harder, 1986). Matching SDPs from such a small RI set could easily result from chance alone ($p = 0.016 \times$ the total number of phenotypes examined).

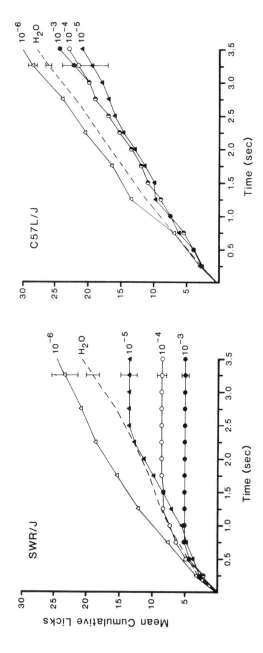

FIGURE 1 Mean cumulative licks by naive SWR/J (T) and C57L/J mice during the first 3.5 s of exposure to one of four SOA concentrations (M) in a single-bottle test, compared to strain water baseline means.

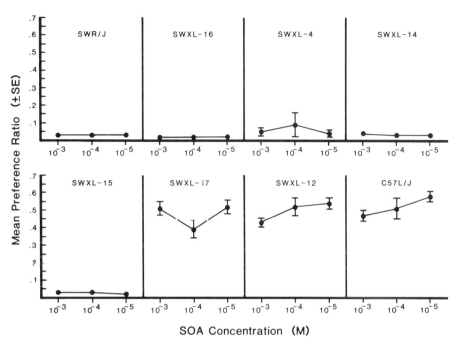

FIGURE 2 Mean preference ratios across three SOA concentrations for six SWXL/Ty recombinant inbred strains compared to means for the inbred SWR/J (T) and C57L/J (N) strains. (From Whitney and Harder, 1986.)

From the SWR (taster) and SJL (nontaster) inbred strains Beamer, also at the Jackson Laboratories, produced the SWXJ/Bm RI set, consisting of 14 lines. Upon taste testing with SOA (see Figure 3) a dichotomous SDP was once again obtained. However, only 3 of the 14 RI strains were nontasters, rather than the 7 of 14 that might be expected from a single-locus difference between the progenitor lines ($\chi^2_{(1)}$ = 4.57, p < 0.05). The results from the SWXJ/Bm RI set are compatible with a model invoking two loci differentiating the SWR and SJL lines with regard to SOA tasting (two loci (25:75), $\chi^2_{(1)}$ = 0.095, p > 0.30). The SDP for SOA among the SWXJ/Bm RI set did not perfectly match the SDPs for any previously typed loci, including markers on both chromosome 1 and 6 (Beamer, personal communications).

Mouse lines designated CFW were reported to be SOA tasters (Warren and Lewis, 1970) and SOA nontasters (Lush, 1981a). Such a discrepancy could be due to genetic differences among separate lines designated CFW resulting from heterozygosity in the original CFW stock (Staats, 1981; Lush, 1981a). Consistent with this possibility, animals from a currently available

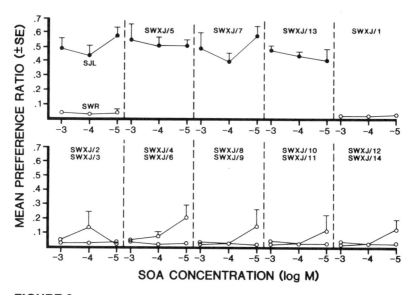

FIGURE 3 Mean preference ratios across three SOA concentrations for 14 SWXJ/Bm recombinant inbred strains, compared to means for the progenitor SWR/Bm (T) and SJL/Bm (N) strains.

outbred CFW stock were found to be dimorphic for SOA tasting (Whitney and Harder, 1986). Gannon and Whitney (1989) investigated SOA segregation among outbred CFW mice in an extensive series of crosses spanning three consecutive generations. Their results were consistent with segregation at a single *Soa* locus within the outbred CFW stock. Interestingly, both CFW mice and the SWR/J inbred strain are descended from nine albino mice received by Lynch from Switzerland in 1926 (Lynch, 1969; Whitney and Harder, 1986). It is quite possible therefore that the same allele producing SOA tasting by SWR inbred mice could be segregating within the present CFW outbred stock. Results from crossing various CFW individuals with taster (SWR/J) and nontaster (C57BL/6J) inbred lines were consistent with the possibility that an allele identical by descent to that in the SWR/J strain is segregating in the CFW stock (Gannon and Whitney, 1989).

III. CONGENIC DEVELOPMENT

SOA aversion (tasting) versus nonaversion (nontasting) as measured in two-bottle preference tests of 10^{-4} and 10^{-5} M SOA in distilled water, versus distilled water alone, seemed a very good candidate for attempting construction of congenic lines. Since the SWR/J (taster) versus C57L/J or C57BL/6J (nontaster) inbred strains provided a maximal phenotypic contrast across a wide range of SOA concentrations (Harder et al., 1984), they were chosen as progenitor inbred strains for congenic development. There were other considerations, including: the SWR/J strain (SW) was the only SOA taster inbred strain then available; the existence of the SWXL/Ty recombinant set favored use of the C57L/J strain (L) since there are advantages for further analysis of having both congenic and recombinent inbred lines from the same progenitors; and the C57BL/6J strain (B6) was favorable because of the large amount of information available and its vigor and fecundity.

A set of two congenic lines is developed through a successive series of backcrosses in which one strain provides a genetic background (called a first parent or *inbred partner*) and another strain provides a specific chromosome segment containing a differential locus (called a second parent or *donor strain*). When a second set of congenic lines is also developed with the inbred partner and donor strain reversed, the resulting set of four lines is called a congenic quartet. Congenic quartets are particularly useful in detecting any genetic interactions (genomic background effects) involving the phenotypic expression of either allele at the differential locus (Flaherty, 1981). The variant of the basic breeding design used for the development of congenic lines in each case depends on the mode of inheritance of the target phenotype, specifically on dominance (ability to identify heterozygotes). Since SOA tasting appeared fully dominant with the strains chosen, we used the backcross (NX) protocol for transfer of a dominant allele to the genetic background of a strain bearing the recessive allele. The cross-intercross (M) protocol was used for transfer of the recessive allele (Flaherty, 1981).

As diagrammed in Figure 4 for one replicate, the backcross (NX) procedure is as follows. First an SW (TT) donor animal is mated to a B6 (tt) inbred partner. All the F_1 progeny (F_1 is designated N1 in congenic nomenclature) are expected to be heterozygotes (Tt) at the *Soa* locus and are expected to be uniformly phenotypic SOA tasters. All offspring are phenotypically tested, and then one N1 donor (Tt, taster) is backcrossed to a B6 inbred partner mouse (tt, nontaster). The N2 progeny of this backcross are expected to be of two classes, both genotypically and phenotypically. One-half should be heterozygous, Tt, tasters and half homozygous, tt, nontasters. After phenotypic taste testing one taster (presumably Tt) donor is again backcrossed to a B6 inbred mouse. This cycle of behavioral taste testing and selective backcrossing continues across generations. With each backcross generation one-half of the remaining genetic material from the SW donor strain is expected to be lost, except for material closely linked on the chromosome bearing the differential *Soa* locus. By convention, after 10-12 generations of backcrossing, heterozygous taster siblings are interbred. These are the progenitors of the new SOA taster congenic strain. Henceforth the new strain (B6.SW-*Soa^a*)

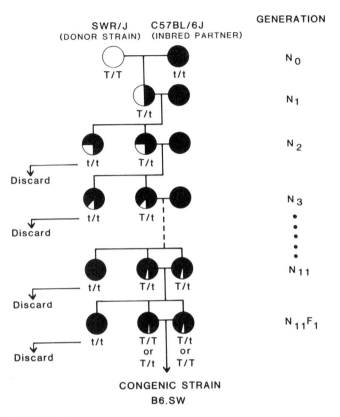

FIGURE 4 Backcross (NX) protocol for B6.SW congenic strain development involving transfer of the autosomal dominant SOA taster allele from the SW (T/T) donor strain onto the B6 (t/t) inbred partner strain genomic background. Only phenotypic taster (T/t) mice from generations N1-N10 are selected for backcrossing to the B6 mice. Tasters from generation N11 (and all subsequent generations) are sib intercrossed. (Modified from Flaherty, 1981.)

is maintained by full sib matings. Its congenic partner is the standard B6 inbred strain.

In practice any single replicate of the breeding protocol could become extinct at any generation. To safeguard against loss of the new congenic line during its development, the protocol just described and diagrammed in Figure 4 was established with 12 replicates. The maintenance of 12 replicates was not only to ensure against loss of incipient congenics but also to provide enough progeny for behavioral testing across generations so that the data set would contribute to segregation analysis (Whitney and Harder, 1986;

Whitney et al., 1989). Such a multiple-member congenic set is valuable because in any one congenic there is always a small possibility of a remaining unlinked "passenger gene" from the donor strain, and the exact size and contents of the chromosome segment transmitted in linkage with the differential locus is expected to vary among replicates. For investigating the pleiotropic effects of the target taste locus and looking at closely linked loci, a replicate congenic set derived from two already inbred strains combines some of the advantages of congenic strains and recombinant inbred strains (Flaherty, 1981; Bailey, 1981).

Sex has not been mentioned as a variable in SOA tasting because it has not had a measurable effect on SOA tasting in either our work or in the reports of others. However, to be able to evaluate sex effects in the context of segregation analyses, the replicates of the congenic breeding protocols were completely balanced with regard to reciprocal matings (Whitney and Harder, 1986).

For the second congenic pair of the complete quartet, the SW inbred strain (TT, taster) is the inbred partner and the B6 inbred strain (tt, nontaster) is the donor strain. In this case the transfer is of a recessive allele (t, nontaster) at the Soa locus onto the genomic background of an inbred partner that carries a dominant allele (T, taster) at the differential locus. Therefore, backcross progeny (TT and Tt) are expected to be phenotypically indistinguishable (all SOA tasters). For this reason the breeding protocol contains an extra generation for each cross to the inbred partner. As diagrammed in Figure 5, the cross-intercross or M protocol begins as before with the making of an F_1 generation. However, these F_1 animals are then mated inter se to produce an F_2 segregating generation. From behavioral taste testing among these segregants a phenotypic nontaster (presumably tt, expected to constitute 0.25 of the generation) is selected for the next cross to the inbred partner. Thus each cross is of a homozygous nontaster (tt) donor to a homozygous taster (TT) inbred partner, and each progeny set of tasters must then be intercrossed to obtain a nontaster for the next cycle. This cross-intercross protocol has the disadvantage of doubling the number of generations required to produce the new congenic line. However, it has the advantage of producing many animals across generations for segregation analysis. As before, the cross-intercross protocol was also established with 12 replicates balanced with regard to reciprocal matings. The resultant new congenic line is an SW genome bearing the Soa^b nontaster allele transferred from the B6 strain (SW.B6-Soa^b). Its congenic partner is the standard SW inbred strain.

Matings were established to begin replicated congenic lines from SW-B6 progenitor strains and from SW-L progenitor strains. From the first crosses (N1) there were phenotypic differences between the sets. In the NX protocol with SW-B6 crosses, F_1 and later heterozygotes were essentially uniformly

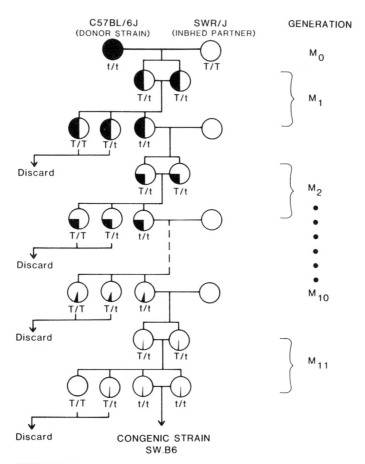

FIGURE 5 Cross-intercross (M) protocol for SW.B6 congenic strain development involving transfer of the autosomal recessive SOA nontaster allele from the B6 (t/t) donor strain onto the SW (T/T) inbred partner strain genomic background. Phenotypic taster (T/t) mice are intercrossed to produce nontaster (t/t) offspring, which are then selected for crossing to the SW mice. Nontasters from generation M11 (and all subsequent generations) are sib intercrossed. (Modified from Flaherty, 1981.)

phenotypic tasters at concentrations of both 10^{-4} and 10^{-5} M SOA (Whitney and Harder, 1986; Whitney et al., 1989). However, as illustrated in Figure 6, SW-L heterozygotes (N1-N4) were uniformly phenotypic tasters at 10^{-4} M SOA but quite variable at 10^{-5} M SOA. The phenotypic variability of SW-L crosses when tested with 10^{-5} M SOA was attributed to mildly incomplete dominance in heterozygotes because homozygous tasters from the SWXL/Ty

ABILITY TO TASTE SUCROSE OCTAACETATE

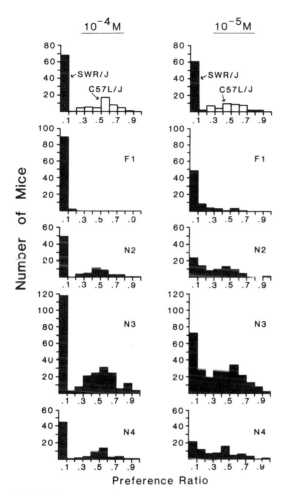

FIGURE 6 Individual preference ratio frequency distributions at 10^{-4} M SOA for SWR/J (T) and C57L/J (N) mice, their F_1 progeny, and three lineal backcross generations compared to distributions at 10^{-5} M SOA.

recombinant inbreds (Figure 2) were consistently tasters of 10^{-5} M SOA. When preference testing with 10^{-5} M SOA was extended across 8 days rather than our usual 2 day procedure, the variable expressivity was found to rather uniformly characterize the heterozygotes. That is, it was not the case that some SW-L F_1 individuals consistently strongly avoided 10^{-5} M SOA across days but other individuals were consistently mildly avoiding "semitasters";

rather, across days of repeated testing the same individual would accept 10^{-5} M SOA on some days and avoid it on other days.

The phenotypic difference between crosses of SW mice to Ls and SW to B6s could be because the L and B6 inbred strains carry different nontasting alleles at the *Soa* locus, or the difference could be because of background genomic (possibly polygenic) differences between the L and B6 lines. To attempt an investigation of these possibilities we defined three phenotypic categories: tasters (PR < 0.1 at both 10^{-4} and 10^{-5} M); semitasters (PR < 0.1 at 10^{-4} M but ≥ 0.1 at 10^{-5} M); and nontasters (PR ≥ 0.1 at both 10^{-4} and 10^{-5} M). We then tested subjects from additional crosses between the two C57 lines (L and B6), as well as three-way crosses of B6 mice with SW X L backcross progeny. We initially interpreted the results as consistent with a comprehensive model of one locus with three alleles and less consistent with an alternative polygenic (background genome influence) model (Harder and Whitney, 1985). However, later data have caused us to reconsider this interpretation. In the cross-intercross (M) protocol involving transfer of the B6 (nontaster)-derived allele onto the SW genomic background, we once again encountered semitaster phenotypes. For M2 (see Figure 5) cross progeny, 97% avoided 10^{-4} M SOA, but at the weaker 10^{-5} M SOA, fully 15%

TABLE 1 Observed Ratios of Taster (T) to Nontaster (N) Mice from the SW.B6 Cross-Intercross Congenic Strain Development Protocol in Preference Tests at Two SOA Concentrations[a]

Cycle	Generation	n Tested	% Expected T:N	% Observed 10^{-4} M SOA T:N	% Observed 10^{-5} M SOA T:N
M1	Cross (F$_1$)	46	100:0	100:0	98:2
	Intercross (F$_2$)	282	75:25	73:27	72:28
M2	Cross	199	100:0	98:2	92:8[b]
	Intercross	157	75:25	75:25	73:27
M3	Cross	151	100:0	99:1	91:8[b]
	Intercross	161	75:25	76:24	68:32
M4	Cross	142	100:0	100:0	95:5[c]
	Intercross	163	75:25	82:18	79:21
M5	Cross	151	100:0	100:0	84:16[c]
	Intercross	146	75:25	77:23	75:25

[a]These ratios are compared to ratios expected with single autosomal locus control of SOA detection, with complete dominance of the T phenotype. T = preference ratio < 0.15. N = preference ratio ≥ 0.15. A 100:0 Fisher's exact probability test; a 75:25 chi-square test.
[b]$p < 0.01$.
[c]$p < 0.001$.

of 236 progeny did not meet the taster criterion. Thus, 15% semitasters were observed from a cross that did not involve the L line. We chose at that time to discontinue the M protocol with SW-B6 mice. We began a new M protocol to see if the semitaster phenomenon would replicate. As enumerated in Table 1, when the B6 nontaster allele is placed on the SW background, a substantial proportion of animals (apparently heterozygotes) strongly avoid 10^{-4} M SOA but not 10^{-5} M SOA. This is similar to the result encountered with SW-L crosses in the NX protocol. However, when the SW allele is placed on the B6 background, almost equally strong avoidance is seen for both 10^{-4} and 10^{-5} M SOA (Table 2). Thus at present our interpretation favors background genomic influence: The taster allele, at least when heterozygous, is apparently expressed less on the SW and L backgrounds than on the genomic background provided by the B6 line.

After generation N4 we discontinued the lines with L progenitors because of generally poor breeding of both C57L inbreds and their derived hybrids.

TABLE 2 Observed Ratios of Taster (T) to Nontaster (N) Mice from the B6.SW Backcross Congenic Strain Development Protocol (Plus an F_2 Generation) in Preference Tests at Two SOA Concentrations[a]

Generation	n Tested		Observed (%)				Expected (%) T/N
	10^{-5} M	10^{-4} M	10^{-5} M T	N	10^{-4} M T	N	
SWR/J	405	387	98.0	2.0	100.0	0.0	
C57BL/6J	371	363	0.8	99.2	0.3	99.7	
F_1(N1)	141	126	97.9	2.1	100.0	0.0	
F_2	481	483	72.8	27.2	74.8	25.2	75:25
N2	84	—	58.3	41.7	—	—	50:50
N3	107	57	40.2	59.8	52.6	47.4	50:50
N4	227	227	55.1	44.9	55.5	44.5	50:50
N5	211	212	53.6	46.4	52.1	47.9	50:50
N6	187	186	55.6	44.4	55.4	44.6	50:50
N7	192	189	53.1	46.9	52.4	47.6	50:50
N8	229	229	42.4	57.6	43.9	56.1	50:50
N9	211	210	48.8	51.2	49.3	50.7	50:50
N10	304	306	48.7	51.3	49.8	50.2	50:50
N11	359	358	51.3	48.7	51.3	48.7	50:50
N11F_1	489	485	75.9	24.1	75.7	24.3	75:25

[a]For the 12 segregating generations (F_2-N11F_1), observed ratios are compared to ratios expected with single autosomal locus control of SOA detection, with complete dominance of the T phenotype (suggested by results from the progenitor SWR/J, C57BL/6J, and F_1 mice). T = preference ratio < 0.15; N = preference ratio ⩾ 0.15; all comparisons $p > 0.01$.

The second set of the M protocol established with B6 donor and SW as inbred partner is proceeding well. The 12 replicates are at approximately M6, and at 10^{-4} M SOA, the avoidance phenotype continues to segregate according to expectations for a single autosomal locus (Table 1). The B6.SW-Soa^a congenics from the NX protocol have finished 10 generations of backcrossing and are currently being sib mated as 12 lines (Whitney et al., 1989). Of the 12 lines now in existence, 7 have been separate since generation N2, 4 separate since N3, and 1 a separate lineage since N6. At generation N11, when sib mating began (Figure 4), conventional calculations (Flaherty, 1981) suggested that the donor chromosome segment retained by congenic taster mice would average 18.2 cM (centimorgans) in length. Unlinked SW genetic material totaling, on average, 1.5 cM was also expected to have been retained. This 19.7 cM of donor strain material represents about 1.2% of the ~1600 cM mouse haploid genome. In terms of loci, residual heterozygosity would be expected in N11 at ~0.1% of the unlinked loci for which the B6 and SW inbred strains carry different alleles.

IV. TASTE PROFILE OF CONGENIC MICE

The congenic mice are undoubtedly not coisogenic; however, the estimated 99.9% genetic identify for unlinked loci provides a near-homogeneous genetic background against which to investigate a major allele influence on taste. With regard to closely linked loci that may differentiate the B6.SW lines from their B6 congenic partner, the multiple-member set of congenic lines should aid eventual differentiation of linkage from pleiotropic manifestations of alleles at the Soa locus. Our initial endeavors with the B6.SW congenic mice have been to further characterize the influence of the Soa^a allele with regard to taste and to provide animals to another laboratory for characterization of mechanisms by molecular and physiologic approaches.

Figure 7 schematically illustrates some possible outcomes from the testing of other tastants. In the upper left panel B6 and B6.SW congenics are similar but the SW inbred strain is clearly different from each of them. The interpretation would be that the Soa^a allele does not influence the measured response to that tastant. In contrast, the upper right panel of Figure 7 illustrates the expected outcome for a tastant influenced only by the Soa^a allele: The B6.SW is similar to the SW donor strain and clearly different from its B6 congenic partner. The lower left panel illustrates a possible outcome if the mechanism influenced by the Soa^a allele mediates responsiveness to a tastant at one concentration and other mechanisms influence response to the same tastant at a second concentration. Finally, the lower right panel illustrates one of many possibilities in which the B6.SW are unlike either progenitor line. Such outcomes could occur if the Soa^a mechanism interacts with

ABILITY TO TASTE SUCROSE OCTAACETATE

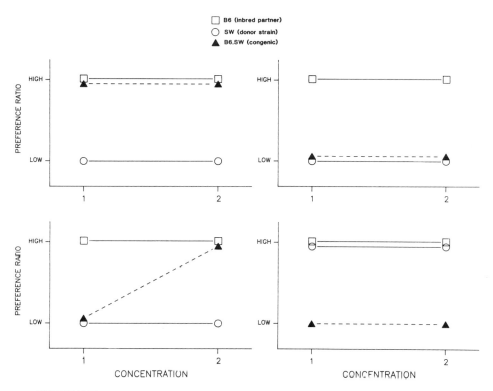

FIGURE 7 Four possible outcomes of testing Soa^a involvement in tastant preference. Top left, Soa^a not implicated; top right, Soa^a implicated; bottom left, Soa^a implicated with concentration specificity; bottom right, Soa^a possibly implicated with genic interaction.

other mechanisms influenced by other loci to determine responsiveness to a tastant. In the comparisons that follow, generally 10-15 subjects of each kind were tested with each tastant concentration. The B6.SW animals were from generations $N11F_0$ to $N11F_4$. Since SOA aversion is a dominant phenotype, heterozygotes as well as homozygotes at the Soa locus are represented among B6.SW mice in these early generations (see Figure 4).

In a sampling of sweet substances (sucrose, maltose, and saccharin), neither preference nor amount consumed was influenced by the Soa gene (Figure 8). As illustrated in Figure 9, the Soa gene apparently influences response to some but not all bitter tastants. Quinine sulfate is relatively uninformative since the B6 and SW progenitor strains did not differ in concentration-response profile. However, with quinine hydrochloride the progenitors are different and the Soa gene is apparently not involved since the B6 and B6.SW animals

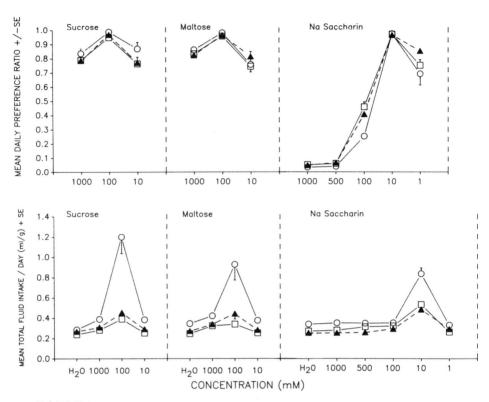

FIGURE 8 Responses of B6 (□), SW (○), and B6.SW congenic (▲) mice to three ("sweet") tastants. Top panels are preference ratios; bottom panels are total fluid intakes.

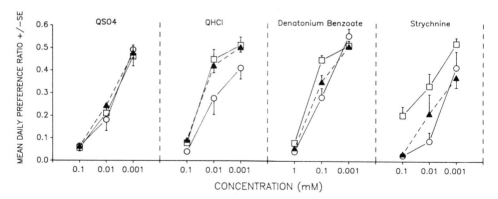

FIGURE 9 Preferences for four ("bitter") tastants by B6 (□), SW (○), and B6.SW congenic (▲) mice.

are similar. For denatonium benzoate the *Soa* gene is implicated because the B6 and B6.SW congenics were different ($p < 0.02$) when tested at the 0.1 mM concentration. As previously reported by Lush (1982), the *Soa* gene apparently influences avoidance of strychnine since the B6 differed from the B6.SW congenics at 0.1 mM ($p < 0.001$) and 0.001 mM ($p < 0.02$). This strychnine sensitivity is interesting in that it ties the SOA polymorphism to a naturally occurring bitter toxin as well as suggests a possibly similar polymorphism among humans. Barrows (1947) reported humans to be polymorphic in response to brucine, a compound very similar to strychnine, with a familial distribution consistent with the possibility that relative insensitivity (nontasting) was a Mendelian recessive trait.

For two salts (Figure 10) there was only one indication of a possible influence; the B6 and B6.SW differed ($p < 0.05$) when tested with 100 mM KCl. With acids, apparent *Soa* gene influences were substance and concentration specific. As illustrated in Figure 11, citric acid and HCl were not influenced by the *Soa* gene, but tannic acid may have been (B6 versus B6.SW, $p < 0.02$ at 0.05 mM). Acetic acid was strongly influenced at the strong 10 mM concentration (B6 versus B6.SW, $p < 0.01$) but was not detectably influenced at weaker concentrations (all comparisons $p < 0.05$ for acetic acid at 1.0 and 0.1 mM).

The *Soa* gene also apparently interacts with other mechanisms to influence the response to a variety of acetylated compounds. Four examples are illustrated in Figure 12. The effect for sucrose octaacetate in the upper left panel is, of course, the criterion phenotype used to develop the B6.SW congenic

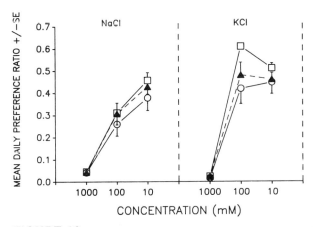

FIGURE 10 Preferences for NaCl and KCl salts: B6 (□), SW (○), and B6.SW (▲) congenics.

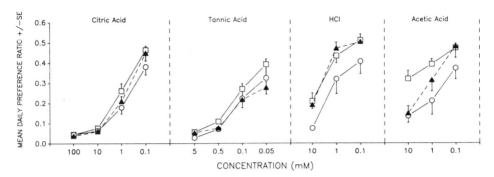

FIGURE 11 Preferences ratios of B6 (□), SW (○), and B6.SW congenic (▲) mice for four ("sour") acids.

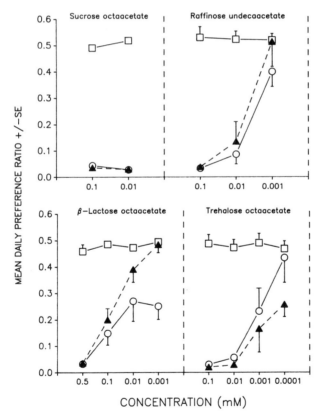

FIGURE 12 B6 (□), SW (○), and B6.SW congenic (▲) mouse preferences for three acetylated disaccharides and one acetylated trisaccharide (raffinose undecaacetate).

mice. The *Soa* gene influence generalizes to raffinose undecaacetate (also Lush, 1986). However with β-lactose octaacetate, an *Soa* allele effect is obvious at 0.5 mM but not at 0.001 mM. At the weaker concentration SW inbreds remain different from B6 inbreds ($p < 0.02$), but the B6 and B6.SW congenics are indistinguishable. In contrast, when tested with trehalose octaacetate the inbred B6 and SW strains are similar at the weak 0.0001 mM concentration, but the B6.SW differ from the congenic B6 ($p < 0.001$). The future availability of the complete congenic quartet should aid interpretation of such apparent interactions.

From these examples, as well as from much other data, it is abundantly clear that taste is a rather complex sensory domain. Because of the complexities of taste an isolation of components, as is possible with congenic strains, seems a useful approach. A congenic model system consisting of the presence and absence of a single major allele influence, expressed on otherwise near-homogeneous genetic backgrounds, may represent a simplification that will facilitate functional integration spanning levels from molecular mechanisms to behavioral phenomenology. Congenic taster mice could contribute as much to the future understanding of the chemical senses as congenic resistant mice have already contributed to the understanding of histocompatibility and immunology.

ACKNOWLEDGMENT

This research was supported in part by Grant NS 15560 from the NINCDS.

REFERENCES

Bailey, D. W. (1981). Recombinant inbred strains and bilineal congenic strains. In *The Mouse in Biomedical Research*, Vol. I, H. L. Foster, J. D. Smith, and J. G. Fox (Eds.). Academic Press, New York, pp. 223-239.

Barrows, S. L. (1947). The inheritance of the ability to taste brucine. Masters Thesis, Stanford University.

Flaherty, L. (1981). Congenic strains. In *The Mouse in Biomedical Research*, Vol. I, H. L. Foster, J. D. Smith, and J. G. Fox (Eds.). Academic Press, New York, pp. 215-222.

Fuller, J. L. (1974). Single-locus control of saccharin preference in mice. *J. Hered.* **65**:33-36.

Gannon, K. S., and Whitney, G. (1989). Sucrose octaacetate tasting in a heterogeneous population of CFW mice. *Behav. Genet.* **19**:417-431.

Harder, D. B., and Whitney, G. (1982). Taste psychophysics of sucrose octaacetate in mice. *Behav. Genet.* **12**:586.

Harder, D. B., and Whitney, G. (1985). Evidence for a third allele at the SOA locus controlling sucrose octaacetate tasting in mice. *Behav. Genet.* **15**:594.

Harder, D. B., Whitney, G., Frye, P., Smith, J. C., and Rashotte, M. E. (1984). Strain differences among mice in taste psychophysics of sucrose octaacetate. *Chem. Senses* **9**:311-323.

Harder, D. B., Maggio, J. C., and Whitney G. (1989). Assessing gustatory detection capabilities using preference procedures. *Chem. Senses* **14**:547-564.

Kare, M. R., and Ficken, M. S. (1963). Comparative studies on the sense of taste. In *Olfaction and Taste*, Vol. I. Zotterman, Y. (Ed.). Macmillan, New York, pp. 285-297.

Lush, I. E. (1981a). The genetics of tasting in mice. I. Sucrose octaacetate. *Genet. Res.* **38**:93-95.

Lush, I. E. (1981b). Mouse pharmacogenetics. *Symp. Zool. Soc. Lond.* **47**:517-546. [In *Biology of the House Mouse*, R. J. Berry (Ed.). Academic Press, New York.]

Lush, I. E. (1982). The genetics of tasting in mice. II. Strychnine. *Chem. Senses* **7**: 93-98.

Lush, I. E. (1986). The genetics of tasting in mice. IV. The acetates of raffinose, galactose, and B-lactose. *Genet. Res.* **47**:117-123.

Lynch, C. J. (1969). The so-called swiss mouse. *Lab. Anim. Care* **19**:214-220.

Morse. H. C., III (1981). The laboratory mouse—a historical perspective. In *The Mouse in Biomedical Research*, Vol. I, H. L. Foster, J. D. Smith, and J. G. Fox (Eds.). Academic Press, New York, pp. 1-16.

Shingai, T., and Beidler, L. M. (1985). Interstrain differences in bitter taste responses in mice. *Chem. Senses* **10**:51-55.

Snell, G. D. (1978). Congenic resistant strains of mice. In *Origins of Inbred Mice*, H. C. Morse, III (Ed.). Academic Press, New York, pp. 119-155.

Snell, G. D., Dausset, J., and Nathenson, S. (1976). *Histocompatibility*. Academic Press, New York.

Staats, J. (1981). Inbred and segregating inbred strains. In *The Mouse in Biomedical Research*, Vol. I, H. L. Foster, J. D. Small, and J. G. Fox (Eds.). Academic Press, New York, pp. 177-213.

Warren, R. P. (1963). Preference aversion in mice to bitter substance. *Science* **140**: 808-809.

Warren, R. P., and Lewis, R. C. (1970). Taste polymorphism in mice involving a bitter sugar derivative. *Nature* **227**:77-78.

Whitney, G., and Harder, D. B. (1986). Single-locus control of sucrose octaacetate tasting among mice. *Behav. Genet.* **16**:559-574.

Whitney G., Harder, D. B., and Gannon, K. S. (1989). The B6.SW bilineal congenic SOA-taster mice. *Behav. Genet.* **19**:409-416.

18
Taste Preference and Taste Bud Prevalence Among Inbred Mice

Inglis J. Miller, Jr.

*Wake Forest University,
Winston-Salem, North Carolina*

We have tested the hypothesis that mice that are "tasters" for sucrose octaacetate (SOA) possess more taste buds than "nontaster" mice (Miller and Whitney, 1988, 1989). The rationale for this hypothesis comes from several observations in humans and mice. Taste bud densities vary by 100-fold in human cadaver tongues (Miller, 1986, 1988), and human taste thresholds vary by 1-2 log arithmic units (Bartoshuk et al., 1986). Studies of human perception show that stimulation of increasing numbers of taste buds with the same concentration of stimulus produces a more intense taste sensation (Smith, 1971). Human subjects with higher taste bud densities in fungiform papillae rate taste perceptions as more intense for sucrose, NaCl and propyl thiouracil than subjects with lower taste bud densities (Miller and Reddy, 1990). The issue at hand is not whether or how genetic control is exacted over the relative number of taste buds in two strains of mice; rather, the issue is whether tasters have more taste buds than the nontasters. Two populations of mice that have the ability to taste or not to taste SOA are probably different for several attributes that may affect taste sensitivity to SOA. Since the glossopharyngeal nerve is more response to bitter stimuli than the chorda tympani and since the IXth nerve yields larger taste responses in tasters than in nontaster mice (Shingai and Beidler, 1985), we chose to count and compare taste buds in the vallate, foliate and fungiform papillae of two strains of mice: SWR/J, tasters for SOA, and C57BL/6J, nontasters for SOA.

Preference testing was performed on all mice to confirm that taster mice (SWR/J) prefer water to SOA in the concentration range of 10^{-3}-10^{-5} M; nontasters (C57BL/6J) are indifferent to water and SOA in the same concentration ranges. After testing, mice were sacrificed by an overdose of ether followed by decapitation. A region of the tongue containing the vallate papilla was removed and prepared by serial sections in paraffin with hematoxylin and eosin. Each section containing the vallate papilla was examined to count individual taste buds.

The distribution of taste buds in the vallate papillae of SWR/J (SOA tasters) and C57BL/6J (SOA nontasters) is shown in Figure 1. The subjects were ordered along the x axis by their rank in numbers of taste buds in the vallate papillae. SOA taster mice had a mean of 195.5 ± 39 (SD, $N = 13$) taste buds per vallate papilla, with a range from 142 to 255 tb/pap. The SOA nontaster mice had a mean of 144.5 ± 39 tb/pap (SD, $N = 10$), with a range from 102 to 179 tb/pap. Total average taste buds in the vallate papillae for the two strains are significantly different (t-test, $p < 0.005$, degrees of freedom $= 21$, $t = 3.719$). The difference between the two strains is most apparent at the high and low extremes. Half the animals of the taster strain had 200 tb/pap without equivalent among the nontasters. Of the nontasters, three (23%) had 105 tb/pap or fewer; the lower limit of the taster group was 142 tb/pap. Among these two strains, the mice that are genetic tasters for SOA (SWR/J) contain a higher mean number of taste buds per vallate papilla

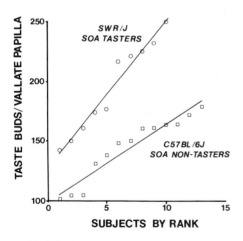

FIGURE 1 The number of taste buds in the vallate papilla of SOA taster (circles) and SOA nontaster (squares) mice. The number of taste buds in each papilla is plotted in increasing order by rank. (From Miller and Whitney, 1988b.)

than mice of the SOA nontaster (C57BL/6J) strain. The number of taste buds were also compared in foliate and fungiform papillae of the same two strains of mice in a preliminary report (Krimm and Miller, 1989). One of the bilateral foliate papillae was removed from each tongue of 9 SWR/J male mice and 13 C57BL/6J male mice. Taste buds were counted in serial paraffin sections as in vallate papillae. The SWR/J mice had an average of 81.1 ± 15.7 (SD, N = 9) taste buds/foliate papilla; while the C57BL/6J strain had an average of 61.4 ± 6.6 (SD, N = 13) tb/fol. pap. These observations are probably significantly different (t-test, $p < 0.01$, df = 20, t = 3.902). Foliate taste buds, which are innervated by the glossopharyngeal nerve like those in the vallate papillae, are more prevalent in the SOA taster strain than in the non-taster strain.

Bilateral fungiform papillae were counted by videomicroscopy, and the presence of taste buds was verified for a sample of tissue prepared by paraffin sections for light microscopy. In contrast to the vallate and foliate papillae, C57BL/6J mice had more fungiform taste buds than mice of the SWR/J strain. Mice of the C57BL/6J strain had an average of 98.1 ± 9.1 (SD, N = 13) fungiform taste buds, while SWR/J mice had an average of 86.2 ± 11.6 (SD, N = 13) fungiform tb. The total numbers of fungiform taste buds are probably significantly different between the two strains (t-test, $p < 0.05$, df = 24, t = 2.179).

Thus, the SWR/J mice, which are tasters for SOA, appear to have more vallate and foliate taste buds, but fewer fungiform taste buds, than the C57BL/6J (SOA non-taster) strain. The total number of lingual taste buds was estimated for individual animals by adding the total of fungiform and vallate papillae plus twice the total of foliate taste buds. Individual SWR/J mice had an average total of 429 ± 63.9 (SD, N = 7, range 371-535) lingual taste buds, and the C57BL/6J strain had an average total of 366 ± 20.3 (SD, N = 8, range 337-397) taste buds. The average total numbers of lingual taste buds are probably significantly different between the two strains (t-test, $p < .025$, df = 13, t = 2.64). While the SWR/J mice have an average of about 257 more lingual taste buds innervated by the glossopharyngeal nerve (vallate + 2 × foliate) than C57BL/6J mice, the latter strain have about 12% more fungiform taste buds than the SWR/J mice.

Differences in the prevalence of taste buds probably contribute to the sensory factors which influence the disparities in taste sensitivity and intake behaviors among inbred strains of mice.

ACKNOWLEDGMENTS

These experiments were conducted in collaboration with Glayde Whitney of Florida State University, who bred and tested the mice. Robin Krimm,

Mark Oliver and Paula Thomas made important contributions to this work. Support was provided by NIH Grants NS 20101 and NS 15560.

REFERENCES

Bartoshuk, L. M., Rifkin, B., Marks, L., and Bars, P. (1986). Taste and aging. *J. Gerontol.* **41**:51-57.

Krimm, R. F. and Miller, Jr., I. J. (1989). Taste buds of the foliate and fungiform papilla compared for two strains of mice. *Chem. Senses* **14**:719-720.

Miller, I. J., Jr. (1986). Variation in human fungiform taste bud densities among regions and subjects. *Anat. Rec.* **216**:474-482.

Miller, I. J., Jr. (1988). Human taste bud density across adult age groups. *J. Gerontol. Biol. Sci.* **43**:B26-30.

Miller, I. J., Jr., and Reedy, Jr., F. E. (1990). Variations in human taste bud density and taste intensity perception. *Physiol. & Behav.* **47**:1213-1219.

Miller, I. J., Jr., and Whitney, G. (1988). Genetic factors in taste bud density and taste preference. *Neurosci. Abstr.* **14**:1063.

Miller, I. J., Jr., and Whitney, G. (1989). SOA-taster mice have more vallate taste buds than non-tasters. *Neurosci. Lett.* **360**:271-275.

Shingai, T., and Beidler, L. M. (1985). Interstrain differences in bitter taste responses in mice. *Chem. Senses* **10**:51-55.

Smith, D. V. (1971). Taste intensity as a function of areas and concentration: Differentiation between compounds. *J. Exp. Psychol.* **87**:163-171.

19
Taste Receptor Mechanisms Influenced by a Gene on Chromosome 4 in Mice

Yuzo Ninomiya, Noritaka Sako, Hideo Katsukawa, and Masaya Funakoshi

Asahi University, Motosu, Japan

I. INTRODUCTION

Taste responses to sweet substances have been investigated in a variety of organisms with behavioral, physiological, and biochemical techniques by a number of researchers. However, the mechanism underlying sweet taste perception at the receptor level is not yet fully understood (Sato, 1985). In mammals, whether the sweetener receptor site is single or multiple is one major issue that is still controversial. This chapter briefly reviews backgrounds of the issue and describes our recent genetic studies on the mouse taste sensitivity to D-phenylalanine (D-phe), which provide direct evidence for the existence of multiple sweetener receptor sites in mammals.

II. SWEETENER RECEPTOR SITE OF MAMMALS

It has been proposed (Shallenberger and Acree, 1967) that, since all sweet substances have one common feature, an AH-B system, within their structure (where the A and B are the electronegative atoms separated by a distance of 0.25-0.4 nm and the H is a hydrogen atom attached to the electronegative atom by a covalent bond), only one type of receptor site possessing a complementary AH-B system would be responsible for the perception of sweetness in humans. This notion is supported by the evidence that the lingual treatment of gymnemic acid in humans (Diamant et al., 1965; Kurihara, 1969) and chimpanzees (Hellekant et al., 1985) and that of pronase E and semialkaline protease in rats and humans (Hiji, 1975; Hiji and Ito, 1977) more or less suppress taste responses to all sweet chemicals employed, but these drugs hardly suppress those to other salty, sour, and bitter substances.

On the other hand, electrophysiologic studies in dogs (Anderson et al., 1962), gerbils (Jakinovich, 1976), and rats (Yamada et al., 1982, 1983) suggested the possibility that the sugar receptor site is made up of a glucopyranose and a fructofuranose site. For example, Yamada et al. (1982, 1983) demonstrated that the three chemical modifiers, Woodward's reagent K, 1,4-dithiothreitol (DTT), and streptozotocin, caused a separate suppression between two pairs of sugars. The former two modifiers suppressed responses to sucrose and fructose, but not those to glucose and maltose; the remaining modifier depressed only responses to glucose and maltose. Zawalich (1973) reported that alloxan-depressed rat chorda tympani responses to sugars but not to saccharin Na, cyclamate Na, or glycine, suggesting the possibility that these nonsugar sweeteners activate receptor sites different from sugar receptor sites. This was further emphasized by electrophysiological and behavioral studies in gerbils (Jakinovich, 1981, 1982), in which two independent receptor sites, a sucrose and a saccharin site, were postulated. Other differences in sweetener receptor sites have been suggested between sugars and sweet-tasting amino acids (glycine and L-alanine) in rats and hamsters (Yamada et al., 1984) and mice (Iwasaki and Sato, 1986) by showing the separate suppression of chorda tympani responses to these sweeteners by the lingual application of extracts of leaves of the plant *Ziziphus jujuba* and Cu and Zn salts, respectively.

However, segregations of sweetener receptor sites employed in these studies are based on the differences in the relative magnitudes of responses among sweeteners or their suppression rates by drugs, which may vary with possible minor variations of microenvironments around the receptors. Furthermore, the neural response to a sweetener may not be simply quantified as a direct function of the amount of the stimulus-sweetener receptor complexes because the chemical may activate not only the sweetener receptor site but also receptor mechanisms for other tastes (Iwasaki and Sato, 1986; Funakoshi et al., 1987). Thus, although a considerable amount of experimental data have been accumulated, it is not fully established that multiple sweetener receptor sites exist in mammals.

A series of behavioral and electrophysiological studies by Hellekant and his colleagues (Brouwer et al., 1973; Hellekant et al., 1981, 1985; Hellekant, 1976; Glaser et al., 1978) shed light on the phylogenetic difference in taste sensitivity. They showed that aspartame, monellin, and thaumatin probably taste sweet only to members of infraorder Catarrhina, such as humans, chimpanzees, and Old World monkeys, but do not taste to other classes of mammals, such as New World monkeys, prosimians, and nonprimates. Furthermore, they showed that the lingual application of gymnemic acid suppressed sweet taste responses only in humans and chimpanzees, suggesting a fundamental difference in the gustatory effects of the sweetener receptors among mammals. This must have a genetic basis.

III. PREVIOUS GENETIC STUDIES ON TASTE SENSITIVITY

In mammals, the genetic approach to taste mechanism was first introduced by Fox in 1932, who found two kinds of humans who show markedly different taste thresholds for bitter substances containing an N-C=S group, such as phenylthiocarbamate (PTC) and several others (Kalmus, 1971). Genetic analyses soon indicated that the difference depends on a single autosomal gene (Snyder, 1932; Blakelee, 1932; Blakelee and Salmon, 1931, 1935). Similar polymorphism in PTC sensitivity was found among individuals of other mammalian species, such as anthropoid apes (Fisher et al., 1939), rats (Richter and Clisby, 1941), and mice (Hoshishima et al., 1962; Klein and DeFries, 1970). In mice, a single-locus control for PTC sensitivity was also suggested (Klein and DeFries, 1970).

The existence of dimorphism in taste sensitivity to another bitter substance, sucrose octaacetate (SOA), was subsequently found in mice. Warren and Lewis (1970) reported that one inbred strain (CFW/NIH) showed a strong aversion to SOA at concentrations of 10^{-3}-10^{-6} M to which several other inbred strains were indifferent. In addition, after examining further 31 inbred strains of mice, Lush (1981) found that only one strain (SWR/Lac) avoided 10^{-4} M SOA. This strain difference in SOA sensitivity was shown to depend on a single autosomal gene, *Soa* (Warren and Lewis, 1970). A recent electrophysiologic study (Shingai and Beidler, 1985) showed that SOA at concentrations of 10^{-3}-10^{-5} M produced marked responses in the glossopharyngeal and chorda tympani nerves of a taster strain of mice (SWR/J) (Harder et al., 1984), but not in those of several other nontaster strains. This indicates that the gene, *Soa*, probably affects the taste receptor mechanisms of SOA on the taste cell membrane. In other words, these genetic studies provided direct evidence for the existence of multiple receptor sites for bitter substances.

Single-locus controls for other bitter substances, *Qui* for quinine sulfate, *Rua* for raffinose acetate, and *Cyx* for cycloheximide, have also been reported (Lush, 1984, 1986). Azen et al. (1986) have found that these mouse tasting genes are closely linked to the genes for the salivary proline-rich proteins on chromosome 6.

With regard to sweet substances, Fuller (1974) reported that the preference of mice for a 0.1% solution of saccharin Na in a situation involving free choice is primarily regulated at a single locus for which the designation *Sac* is proposed. The allele present in the C57BL/6J strain of mice, Sac^b, is dominant over Sac^d, which was found in the DBA/2J strain. However, he suggested that the site of gene action is the central nervous system rather than the peripheral, because the mechanism of preference is probably related to the incentive value of the taste of saccharin rather than to the threshold for detection.

IV. GENETIC APPROACHES TO SWEETENER RECEPTOR MECHANISM

A. Strain Differences of Mice in Sweet Taste Response to D-phe

Genetic studies have provided a powerful tool for determining the existence of multiple receptor sites for bitter substances, as already mentioned. Therefore, genetically established variants with lack of the receptor site for a particular sweetener, if found, will give a direct evidence for the existence of at least two sweetener receptor sites: one is for the sweetener and the other is for other sweeteners. So, we first examined whether differences in sweet taste responses would be found among three inbred strains of mice by the use of both behavioral and electrophysiological techniques (Ninomiya et al., 1984a, b). Fortunately, we found that there are prominent strain differences in behavioral responses of mice to a sweet-tasting amino acid, D-phe. When the three inbred strains of mice, C57BL/6CrSlc, C3H/HeSlc, and BALB/cCrSlC, were conditioned to avoid D-phe, only C57BL mice showed a strong aversion (generalization) to sweet substances, sugars and saccharin Na, suggesting that D-phe presumably tastes sweet to C57BL mice (sweet taster) but not to C3H and Balb mice (nonsweet taster) (Ninomiya et al., 1984b). These strain differences corresponded quite well with those observed for the responses of sucrose-sensitive chorda tympani fibers to D-phe among the three strains of mice (Ninomiya et al., 1984a).

B. Single-Locus Control for Sweet Taste Response to D-phe

We next investigated the heredity of the ability to taste D-phe among the strains C57BL/6CrSlc (sweet taster) and BALB/cCrSlc (nonsweet taster) and their F_1 and F_2 hybrids, which were males and females of 8-16 weeks of age. We used a conditioned taste aversion paradigm (Ninomiya et al., 1984b) for the behavioral measure of taste similarities between D-phe and other test stimuli. We also used a two-bottle preference test for comparison. We quantified taste similarities by the strength of generalization from the conditioning stimulus to the test stimuli. Details of experimental procedures are described elsewhere (Ninomiya et al., 1984b). On the day for conditioning, each animal that had been trained for 5 days to drink distilled water within 30 minutes/day of the session time was given access to D-phe and then given an intraperitoneal injection of 230 mg/kg of LiCl to induce gastrointestinal malaise. The concentration of D-phe used for conditioning was 0.1 M, which is almost the maximum soluble concentration in distilled water at room temperature and gives a moderate sensation of sweetness to humans. Control mice drank only distilled water before the LiCl injection. On the test days, the number of licks for each of the test stimuli given by each animal was counted during the first 10 s after the animal's first lick. Test stimuli mainly used were 0.1 M D-phe, 0.1 M L-phenylalanine (L-phe), 0.1 M NaCl, 0.5 M sucrose, 0.001 M HCl, and 0.0001 M quinine HCl. When the mean number of licks of the stimulus per 10 s of three repeated trials during sessions with each experimental mouse was significantly smaller than that in control mice (t-test, $p < 0.05$), avoidance to that stimulus was considered to have occurred. The strength of generalization was described as a percentage suppression score. This was calculated according to the following equation: percentage suppression = [1 − (the number of licks of an experimental group to a test stimulus/the number of licks of the control group to that stimulus)] × 100.

All 20 C57BL mice conditioned with 0.1 M D-phe avoided D-phe and sucrose at concentrations greater than 0.01 M but did not avoid 0.1 M NaCl, 0.001 M HCl, and 0.1 M L-phe. Whether the animals avoided 0.0001 M quinine HCl varied considerably among individuals. The 20 Balb mice conditioned with 0.1 M D-phe avoided D-phe at concentrations of more than 0.001 M and avoided 0.001 M HCl, 0.0001 M quinine HCl, and 0.1 M L-phe but did not avoid sucrose even at concentrations of 0.5 M and 0.1 M NaCl. In two-bottle preference tests, C57BL mice showed a significant preference for D-phe at concentrations greater than 0.01 M, but BALB mice were indifferent to D-phe at concentrations from 0.001 to 0.1 M. The 42 F_1 hybrids conditioned with D-phe avoided D-phe and sucrose but did not avoid NaCl; this response was similar to that observed with C57BL mice, although they showed some individual differences in the responses to the other test stimuli. Genetically measurable differences among tasting abilities of mice to D-phe appeared generalized to sucrose, independently of the response to each of the other test stimuli. Figure 1 shows distributions of percentage suppression score for 0.5 M sucrose among two parent inbred strains and their F_1 and F_2 hybrids after conditioning with 0.1 M D-phe. Percentage suppression scores of 20 BALB mice ranged from 0 to 14%, whereas those of 20 C57BL mice ranged from 43 to 96%. There was no overlap between the two inbred

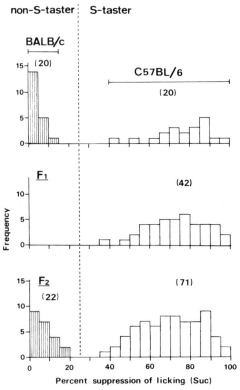

FIGURE 1 Distribution of percentage suppression of licking for 0.5 M sucrose after conditioning of aversion to 0.1 M D-phenylalanine for Balb and C57BL strains and their F_1 and F_2 hybrids. (n = 20 BALB, 20 C57BL, 42 F_1, and 93 F_2 subjects).

strains. The distribution of 42 F_1 hybrids, which came from two reciprocal classes of several single crossings between the parent inbred strains, is very similar to that of C57BL mice (sweet tasters). The 93 F_2 hybrids are clearly classified into two groups in the distribution; one shows less than 16% of suppression (similar to BALB mice: nonsweet tasters) and the other more than 38% of suppression (similar to C57BL and F_1 hybrids: sweet tasters). This classification in the percentage suppression score for sucrose corresponded to that in the occurrence of avoidance of the stimulus, which was statistically determined. The proportion of 71 sweet tasters to 22 nonsweet tasters in F_2 hybrids was statistically compatible with the expected 3:1 Mendelian ratio for the single-locus model with sweet taster determinant dominant (chi-square test). No sex differences were observed. We designated the possible single gene as *dpa*, which has a major effect on the tasting ability to D-phe. Linkage tests were performed between the gene *dpa* and the fur color genes *a* (nonagouti), *b* (brown), and *c* (albino). The observed numbers of the four phenotypic classes, sweet taster and nonbrown, sweet taster and brown, nonsweet taster and nonbrown, and nonsweet taster and brown, were 51, 3, 5, and 14, respectively, suggesting the presence of the genetic linkage between the gene *dpa* and the fur color gene *b* with statistical significance (chi-square test). This indicates that the gene *dpa* is located on chromosome 4.

C. Locus of the Gene *dpa* on Chromosome 4

To determine the locus of the gene *dpa*, we further examined recombination among *dpa* and other two maker genes, *b* and *Mup-1* (major urinary proteins) on chromosome 4 by using 104 nonalbino backcrosses that came from two reciprocal classes of several single crossings between F_1 hybrids and Balb mice. Segregation of *Mup-1* phenotypes, such as *Mup-1^a* or *Mup-1^{ab}*, was based on the pattern of the electrophoresis with urinary proteins (Hudson et al., 1967). Two classes of progeny in each phenotype were observed in backcrosses in ratios that did not significantly differ from the 1:1 ratio expected for a single gene (54 sweet versus 50 nonsweet tasters for *dpa*, 53 agouti versus 51 cinnamon mice for *b*, and 53 *Mup-1^{ab}* versus 51 *Mup-1^a* for *Mup-1*, chi-square test).

Recombination between *Mup-1* and *b* was 5.8 ± 2.3% (standard error of the mean), which is similar to the linkage value of 6.2 ± 1.8 previously reported (Nadeau and Eicher, 1982), and that between *dpa* and *Mup-1* was 8.7 ± 2.8%. The overall recombination between *dpa* and *b* was 14.4 ± 3.6%. This value is slightly higher but much more reliable than that obtained from F_2 hybrids (11.3 ± 11.5%) because of its lower standard error. Gene order among *dpa*, *Mup-1*, and *b* was determined on the basis that the recombinants between *Mup-1* and *b* were not recombinant between *dpa* and *Mup-1*. Thus, the following gene order and recombination percentages ± SEM were given: centromere-*dpa*-8.7 ± 2.8-*Mup-1*-5.8 ± 2.3-*b*.

D. Is the Site of Action of the *dpa* Gene on the Sweetener Receptor?

To determine whether the site of action of *dpa* is on the taste cell membrane, we examined the responses of single chorda tympani fibers to the four basic taste stimuli (0.5 M sucrose, 0.1 M NaCl, 0.01 M HCl, and 0.02 M quinine HCl), 0.1 M D-phe, and 0.1 M L-phe in the C57BL and BALB strains and their F_1 and F_2 hybrids. Fibers sampled were classified into two groups according to their responsiveness: one predominantly responded to sucrose among the four basic stimuli (S type) and the other predominantly responded to another stimulus (non-S type). The mean magnitude of response of S-type fibers to D-phe in C57BL F_1 or F_2 sweet tasters was 17.4 ($n = 11$), 13.8 ($n = 13$), or 14.3 impulses per 5 s ($n = 11$), respectively, which was significantly greater than that of non-S-type fibers in each group of mice (ranged between 3.0 and 4.1 impulses per 5 s, t-test, $p < 0.01$). In BALB and F_2 nonsweet tasters, the mean D-phe responses of S-type fibers (around 3.5 impulses per 5 s) were very small compared with those shown by sweet tasters and tended to be smaller than those of non-S-type fibers. A Pearson product moment correlation coefficient between numbers of responses to D-phe and sucrose, based on the data of 30, 37, or 30 chorda tympani fibers of C57BL F_1 or F_2 sweet tasters, was 0.749, 0.625, or 0.615, which was statistically highly significant ($p < 0.001$), whereas that based on the data of 30 or 32 fibers of BALB or F_2 nonsweet tasters was very small (-0.095 or 0.038). Instead, nonsweet tasters showed considerably higher correlations between D-phe and quinine HCl or HCl (around 0.4). These results clearly suggest that differences observed in behavioral responses to D-phe were derived from those in their peripheral neural responsiveness. Furthermore, we found that the lingual application of the proteolytic enzyme (2% pronase E) for 7 minutes reduced the D-phe responses of S-type fibers to about 25% of the control in sweet tasters but did not suppress those of non-S-type fibers in either sweet or nonsweet tasters. This indicates the protease-sensitive nature of the receptor for D-phe, which only sweet tasters possess on their taste cell membranes. Therefore, the site of action of the gene *dpa* must be on the taste cell membrane.

E. Selective Action of *dpa* on the Sweetener Receptor Site for D-phe

During the course of these studies, we have not extensively examined responses to D-amino acids. To determine whether the action of *dpa* is selective on the sweetener receptor site for D-phe, in this study we employed eight D-amino acids as taste stimuli and compared the behavioral and neural responses to the amino acids between C57BL and BALB mice.

In behavioral experiments, we examined the generalization patterns of each mouse from a conditioned aversion of one of the eight D-amino acids to the test stimuli, which were 0.1 M NaCl, 0.001 M HCl, 0.0001 M quinine HCl, 0.5 M sucrose, 0.5 M fructose, 0.5 M glucose, 0.5 M maltose, 0.1 M L-alanine, 0.1 M D-alanine, 0.1 M D-serine, 0.1 M D-valine, 0.1 M D-leucine, 0.1 M D-methionine, 0.1 M D-histidine, 0.1 M D-phe, 0.1 M L-phe, and 0.03 M D-tryptophan. Correlation coefficients between generalization patterns across 18 test stimuli shown by C57BL and BALB mice ranged from 0.471 to 0.904 among the seven D-amino acids other than D-phe, which was statistically significant ($p < 0.05$), suggesting significant similarities in behavioral responses to each of the seven D-amino acids between C57BL and BALB mice.

A correlation coefficient for D-phe was 0.198, which was not significant ($p > 0.05$). Therefore, prominent strain differences between C57BL and BALB mice exist in the behavioral response only to D-phe among the eight D-amino acids tested.

In electrophysiological experiments we examined the effect of pronase E on the integrated responses of the chorda tympani of the two strains to eight D-amino acids. Treatment with pronase E for 10 minutes reduced the D-phe response to about 50% of control in C57BL mice, but it hardly affected the D-phe response in BALB mice. However, the efficiency of the drug in responses to the other seven D-amino acids in BALB mice appeared to be parallel with that in C57BL mice. In both two inbred strains, a significant reduction (t-test, $p < 0.05$) following pronase treatment was observed in the responses to D-valine, D-leucine, D-methionine, D-histidine, D-tryptophan, and D-sucrose but was not observed in the responses to D-alanine or D-serine. Prominent strain differences between C57BL and BALB mice in this neural measure again appeared only in the D-phe response.

These results indicate that the mouse *dpa* gene selectively acts on a single sweetener receptor site for D-phe thereby controlling the sweet taste responses to this amino acid.

V. A SINGLE SWEETENER RECEPTOR SITE FOR D-phe IN MAMMALS

Our genetic studies provide direct evidence for the existence of a separate sweetener receptor site for D-phe in C57BL mice and that this site is different from the one responding to sucrose. This suggests that there are at least two different receptor site for sweeteners in C57BL mice. Further genetic studies in our laboratory (Ninomiya et al., in preparation) show that the sweet taste response to L-proline is also controlled by a single gene closely linked to but apparently different from *dpa* on chromosome 4. This suggests the possibility that there exists a multigene family controlling sweetener receptor sites for amino acids on chromosome 4. If this is the case, some parallelism in the mouse sweet taste sensitivities among several amino acids described in the following reports may be explained as a result of a close genetic linkage, even if the receptor molecule may vary with each amino acid. That is, Ninomiya et al. (1984a, b) reported that C57BL mice (sweet tasters) showed high similarities in behavioral and neural responses between D-phe and D-tryptophan and between L-proline and L-alanine. Iwasaki and Sato (1986) showed that in ddy mice Cu and Zn salts suppressed chorda tympani responses to sugars and saccharin Na to about 30% of their control responses and those to glycine, L-alanine, L-proline, and D-tryptophan to about 80% but that these salts hardly suppressed responses to L-serine and L-valine.

What is the physiologic significance for the existence of a specific receptor site for D-phe? It is generally believed that D-amino acids are not typical natural products and are rare in the metabolism of vertebrates. However, almost all vertebrates possess D-amino acid oxidase in their kidneys, livers, and brains (Krebs, 1935; Meister, 1965; Konno and Yasumura, 1981), although the physiologic role of this enzyme is not known. In another aspect, D-phe is known as a potent inhibitor of a proteolytic enzyme, carboxypeptidase A (Elkins-Kaufman and Neurath, 1949), which is contained in pancreatic juice in various mammals and is an exopeptidase that is specific for catalyzing the cleavage of certain carboxyl-terminal peptide bonds in peptides and proteins (Folk and Schirmer, 1963). L-phe is not a typical inhibitor of this enzyme (Elkins-Kaufman and Neurath, 1949). Several behavioral studies (Ehrenpreis et al., 1979; Cheng and Pomeranz, 1979) demonstrated that D-phe can induce a moderate analgesia in mice. This analgesia is prevented or reversed by specific opiate antagonists (naloxone or naltrexone).

This evidence concerning the possible function of D-phe may imply that D-phe is a substance by itself or a substitute of its analog that may transmit meaningful information for the animal's internal milieu through its specific receptor site.

REFERENCES

Anderson, H. T., Funakoshi, M., and Zotterman, Y. (1962). Electrophysiological investigation of the gustatory effect of various biological sugars. *Acta Physiol. Scand.* **56**:362-375.

Azen, E. A., Lush, I. E., and Taylor, B. A. (1986). Close linkage of mouse genes for salivary proline-rich proteins (PRPs) and taste. *Trends Genet.* **2**:199-200.

Blakelee, A. F. (1932). Genetics of sensory thresholds: Taste for phenylthiocarbamide. *Proc. Natl. Acad. Sci. USA* **18**:120-130.

Blakelee, A. F., and Salmon, M. R. (1931). Odor and taste blindness. *Eugen. News* **16**:105-109.

Blakelee, A. F., and Salmon, M. R. (1935). Genetics of sensory thresholds: Individual taste reactions for different substances. *Proc. Natl. Acad. Sci. USA* **21**:84-90.

Brouwer, J. N., Hellekant, G., Kasahara, Y., Van Del Wel, H., and Zotterman, Y. (1973). Electrophysiological study of the gustatory effects of the sweet proteins monnelin and thaumatin in monkey, guinea pig and rat. *Acta Physiol. Scand.* **89**:550-557.

Cheng, R. S., and Pomeranz, B. (1979). Correlation of genetic differences in endorphin systems with analgesic effects of D-amino acids in mice. *Brain Res.* **177**:583-587.

Diamant, H., Oakley, B., Strom, L., Wells, C., and Zotterman, Y. (1965). Comparison of neural and psychophysical response to taste stimuli in man. *Acta Physiol. Scand.* **64**:67-74.

Ehrenpreis, S., Balagot, R. C., Comaty, J. E., and Myles, S. B. (1979). Naloxone reversible analgesia in mice produced by D-phenylalanine and hydrocinnamic acid, inhibitors of carboxypeptidase A. In *Advances in Pain Research and Therapy*, J. J. Bonica (Ed.). Raven Press, New York, pp. 479-488.

Elkins-Kaufman, E., and Neurath, H. (1949). Structural requirements for specific inhibitors of carboxypeptidase. *J. Biol. Chem.* **178**:645-654.

Fisher, R. A., Ford, E. B., and Huxley, J. (1939). Taste-testing the anthropoid apes. *Nature (Lond.)* **144**:750.

Folk, J. E., and Schirmer, E. W. (1963). The porcine pancreatic carboxypeptidase A system. *J. Biol. Chem.* **238**:3884-3898.

Fuller, J. L. (1974). Single-locus control of saccharin preference in mice. *J. Hered.* **65**:33-36.

Funakoshi, M., Tanimura, T., and Ninomiya, Y. (1987). Genetic approaches to the taste receptor mechanisms. *Chem. Senses* **12**:285-294.

Glaser, D., Hellekant, G., Brouwer, J. N., and VanDel Wel, H. (1978). The taste responses in primates to the protein thaumatin and monnelin and their phylogenetic implications. *Folia Primatol.* **29**:56-63.

Harder, D. B., Whitney, G., Frye, P., Smith, J. C., and Rashotte, M. E. (1984). Strain differences among mice in taste psychophysics of sucrose octaacetate. *Chem. Senses* **9**:311-323.

Hellekant, G. (1976). On the gustatory effects of monnelin and thaumatin in dog, hamster, pig and rabbit. *Chem. Senses Flav.* **2**:97-105.

Hellekant, G., Glaser, D., Brouwer, J., and Van Del Wel, H. (1981). Gustatory responses in three prosimian and two simian primate species to six sweeteners and miraculin and their phylogenetic implications. *Chem. Senses* **6**:165-172.

Hellekant, G., Hard Af Segerstad, C., Roberts, T., Van Del Wel, H., Brouwer, J. N., Glaser, D., Haynes, R., and Eichberg, J. W. (1985). Effects of gymnemic acid on the chorda tympani proper nerve responses to sweet, sour, salty and bitter taste stimuli in the chimpanzee. *Acta Physiol. Scand.* **124**:399-408.

Hiji, Y. (1975). Selective elimination of taste responses to sugars by proteolytic enzymes. *Nature* **256**:427-429.

Hiji, Y., and Ito, H. (1977). Removal of sweetness by proteases and its recovery mechanism in rat taste cells. *Comp. Biochem. Physiol.* **58**:109-113.

Hoshishima, K., Yokoyama, S., and Seto, K. (1967). Taste sensitivity in various strains of mice. *Am. J. Physiol.* **202**:1200-1204.

Hudson, D. M., Finayson, J. S., and Potter, M. (1967). Linkage of one component of the major urinary protein complex of mice to the brown coat color locus. *Genet. Res.* **10**:195-198.

Iwasaki, K., and Sato, M. (1986). Inhibition of taste nerve responses to sugars and amino acids by cupric and zinc ions in mice. *Chem. Senses* **11**:79-88.

Jakinovich, W., Jr. (1976). Stimulation of the gerbil's gustatory receptors by disaccharides. *Brain Res.* **110**:481-490.

Jakinovich, W., Jr. (1981). Stimulation of the gerbil's gustatory receptors by artificial sweeteners. *Brain Res.* **210**:69-81.

Jakinovich, W., Jr. (1982). Taste aversion to sugars by the gerbil. *Physiol. Behav.* **28**:1065-1071.

Kalmus, H. (1971). Genetics of taste. In *Handbook of Sensory Physiology*, L. M. Beidler (Ed.). Springer-Verlag, Berlin, pp. 165-179.
Klein, T. W., and DeFries, J. C. (1970). Similar polymorphism of taste sensitivity to PTC in mice and men. *Nature (Lond.)* **227**:77-78.
Konno, R., and Yasumura, Y. (1981). Activity and substrate specificity of D-amino acid oxidase in kidneys of various animals. *Zool. Mag.* **90**:368-373.
Krebs, H. A. (1935). Metabolism of amino acids. III. Deamination of amino acids. *Biochem. J.* **29**:1620-1644.
Kurihara, Y. (1969). Antisweet activity of gymnemic acid A1 and its derivatives. *Life Sci.* **8**:537-543.
Lush, I. (1981). The genetics of tasting in mice. I. Sucrose octaacetate. *Genet. Res.* **38**:93-95.
Lush, I. (1984). The genetics of tasting in mice. III. Quinine. *Genet. Res.* **44**:151-160.
Lush, I. (1986). The genetics of tasting in mice. IV. The acetates of raffinose, galactose and β-lactose. *Genet. Res.* **47**:117-123.
Meister, A. (1965). *Biochemistry of the Amino Acids*. Academic Press, New York, pp. 297-304.
Nadeau, J. H., and Eicher, E. M. (1982). Conserved linkage of soluble aconitase and galactose-1-phospate uridyl transferase in mouse and man:assignment of these genes to mouse chromosome 4. *Cytogenet. Cell Genet.* **34**:271-281.
Ninomiya, Y., Mizukoshi, T., Higashi, T., Katsukawa, H., and Funakoshi, M. (1984a). Gustatory neural responses in three different strains of mice. *Brain Res.* **302**:305-314.
Ninomiya, Y., Higashi, T., Katsukawa, H., Mizukoshi, T., and Funakoshi, M. (1984b). Qualitative discrimination of gustatory stimuli in three different strains of mice. *Brain Res.* **322**:83-92.
Richter, C. P., and Clisby, K. H. (1941). Phenylthiocarbamide taste thresholds of rats and human beings. *Am. J. Physiol.* **134**:157-164.
Sato, M. (1985). Sweet taste receptor mechanisms. *Jpn. J. Physiol.* **35**:875-885.
Shallenberger, R. S., and Acree, T. E. (1967). Molecular theory of sweet taste. *Nature (Lond.)* **216**:480-482.
Shingai, T., and Beidler, L. M. (1985). Interstrain differences in bitter taste responses in mice. *Chem. Senses* **10**:51-55.
Snyder, L. H. (1932). Inherited taste deficiency. *Science* **74**:151-152.
Warren, R. P. and Lewis, R. C. (1970). Taste polymorphism in mice involving a bitter sugar derivative. *Nature* **227**:77-78.
Yamada, Y., Imoto, T., and Hiji, Y. (1982). Effect of chemical modification of the rat tongue on the responses to various sugars. *Proc. Jpn. Symp. Taste Smell* **16**:29-32.
Yamada, H., Imoto, T., and Hiji, Y. (1983). Discrimination of molecular structure of various sugars in the sweet receptor sites of the rat tongue. *Proc. Jpn. Symp. Taste Smell* **17**:33-36.
Yamada, H., Imoto, T., and Yoshikawa, S. (1984). Suppression of sweet sensitivity by extracts of *Zyzyphus jujuba* leaves in the rat and hamster. *Proc. Jpn. Symp. Taste Smell* **18**:69-72.
Zawalich, W. S. (1973). Depression of gustatory sweet response by alloxan. *Comp. Biochem. Physiol.* **44**:903-909.

20
Linkage Studies of Genes for Salivary Proline-Rich Proteins and Bitter Taste in Mouse and Human

Edwin A. Azen

University of Wisconsin-Madison, Madison, Wisconsin

I. INTRODUCTION

A comparative genetic approach in elucidating the genomic localization of specific taste genes in mouse and human can give useful biologic information, such as the finding of homologous genes in mouse and human, evidence for more than one gene subserving a specific taste function, and as molecular markers for eventually cloning the taste genes. We were led into these investigations several years ago by the serendipitous finding that certain genes for bitter taste are closely linked to those for salivary proline-rich proteins (PRPs) in the mouse (Azen et al., 1986). Since closely linked genes tend to be conserved between mouse and human (Searle et al., 1987), we wondered whether homologous *PRP* and bitter taste genes would be linked in the human. Also, since initial genetic studies in recombinant inbred (RI) mice showed no recombinants between *Prp* and bitter taste genes, we considered the possibility that *Prp* and bitter taste genes may in fact be the same. Subsequently in human studies, we looked for a linkage between genes for tasting the bitter substance PTC (phenylthiourea) and genes for PRPs (O'Hanlon et al., 1989). Since we did not find a linkage, it seems likely that the gene for PTC taste may not be the same as those for certain bitter substances tested in the mouse, including quinine (*Qui*), raffinose acetate (*Rua*), cycloheximide (*Cyx*) and glycine (*Glb*), whose genetic determinants may be closely linked (Lush and Holland, 1988).

In parallel studies of *Prp* gene localization in the mouse, the *Prp/taste* gene cluster has now been localized to chromosome 6 (Azen et al., 1989). Since the *PRP* gene family is localized to human 12p13.2 (Mamula et al., 1985), there is the definite possibility that homologs of some genes for bitter taste in the mouse are also localized to chromosome 12p in the human.

II. HUMAN PRP GENES AND PROTEINS

Salivary PRPs constitute about 70% of the proteins of human saliva, and proline alone accounts for 21-42% of amino acids of PRPs. Proline, glycine, and glutamine (glutamic acid) together constitute 70-88% of amino acids of PRPs. PRPs can be roughly subdivided into three groups: glycosylated, basic, and acidic account for approximately 17, 23, and 30%, respectively, of the total proteins of parotid saliva (Bennick, 1982).

Some probable tooth-related functions (reviewed by Bennick, 1982) have been assigned to acidic PRPs in saliva; these include binding to calcium and maintaining it in the supersaturated state (thus avoiding calcium precipitation in salivary ducts or in the oral cavity), forming part of the dental pellicle, and binding to hydroxyapatite. Since PRPs are found in serous cells of the submucosal glands of the respiratory tract, they may subserve other functions, such as effects on viscosity (Warner and Azen, 1984; Ito et al., 1985). The functions for the basic and glycosylated PRPs are largely unknown. In the rat and mouse, the production of PRPs is stimulated by feeding sorghums containing high levels of tannins (reviewed by Mehansho et al., 1987) or by injecting the β-adrenergic agonist, isoproterenol (reviewed by Carlson et al., 1986). Tannins are bitter substances known to be toxic and carcinogenic and to bind strongly to salivary PRPs. Therefore, it was proposed that salivary PRPs may bind tightly to tannins and thus protect dietary proteins, digestive enzymes, and the gastrointestinal tract from detrimental effects.

The many PRPs show complex electrophoretic patterns and marked polymorphic variations in different individuals, and these are very useful for genetic linkage studies in families (reviewed by Azen and Maeda, 1988; Azen, 1989). A total of 13 genetic protein polymorphisms have been determined among salivary PRPs. Although it was originally thought that the 13 polymorphisms arose from individual genes, it is now known that they are all coded by the six *PRP* genes. The larger number of PRPs in saliva originating from a smaller number of genes can be explained by differential RNA processing, proteolytic cleavage of large precursor proteins, and the recognition that some acidic PRP phenotypes represent allelic products of one gene, rather than each being the product of a separate gene. Most of the PRP polymorphisms have been specifically assigned to the six *PRP* genes (Lyons et al., 1988). The six *PRP* genes have been localized on chromosome 12p13.2

(Azen et al., 1985; Mamula et al., 1985) and are clustered over an approximately 600 kbp (kilobasepairs) of DNA (Kim et al., 1990).

The *PRP* genes contain multiple tandem repeats that vary slightly and thus are the basis for subclassification of the genes into two groups, termed *Hae*III and *Bst*NI types (Maeda, 1985), since the repetitive region in exon 3 is cut frequently with either *Hae*III or *Bst*NI. The two *Hae*III genes code for acid PRPs, and the four *Bst*NI genes code for basic and glycosylated PRPs. The two *Hae*III-type genes (Kim and Maeda, 1986) and the four *Bst*NI-type genes (Kim et al., in preparation) have been completely sequenced. The six closely related *PRP* genes probably arose in evolution by a process of gene duplication, facilitated perhaps by the prominent tandem repeats in exon 3.

III. MOUSE PRP GENES AND PROTEINS

Many PRPs, homologous to those in humans, are synthesized in vivo in rat and mouse salivary glands after induction with tannins or isoproterenol (reviewed by Carlson et al., 1986). It is believed (Ann, personal communication) that there are two closely linked productive *Prp* genes in the mouse termed *MP2* and *M14*. Both genes have been completely sequenced (Ann and Carlson, 1985; Ann et al., 1988). The *Prp* gene family in the mouse is localized on chromosome 6 (Azen et al., 1989), and these studies are described in more detail subsequently.

IV. GENETIC LINKAGE STUDIES ASSIGNING PRP GENES TO MOUSE CHROMOSOME 6

DNAs from 16 inbred strains (including progenitors of RI strains) were digested with the restriction enzyme *Hind*III, Southern blotted, and hybridized to the rat PRP33 cDNA probe (Ziemer et al., 1984). Approximately seven to eight bands are seen, and the strains can be classified as belonging to one of four groups (Figure 1). We can assign these groups haplotype designations to identify particular combinations of restriction variants within a cluster of *Prp* genes that constitute the murine *Prp* gene complex (see Figure 1). Since *Prp* was previously assigned to chromosome 8 based on somatic cell hybrid analysis (Azen et al., 1984), we anticipated detecting linkage with chromosome 8 markers, such as *Gr-1, Es-1,* and *Emv-2* marking the proximal, middle, and distal parts of the chromosome, respectively. However, no evidence supporting linkage was found. Additional small sets of backcrosses, providing linkage data with respect to chromosome 8 markers, including *Gr-1, Es-1, Got-2, Emv-2,* and a DNA restriction fragment detected with a metallothionine gene (Mt-1) probe, showed no linkage with *Prp*. We could

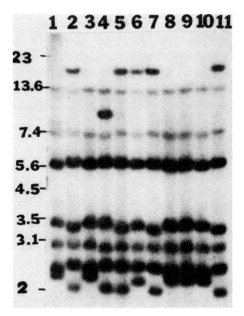

FIGURE 1 PRP *Hind*III fragment patterns in different mouse stains. Southern blots of *Hind*III-digested mouse DNAs were probed with radiolabeled rat PRP33 cDNA. Some representatives of four PRP haplotypes are shown. Lanes 1, 3, 8, 9, and 10 are PRP[a] haplotype (PL/J, NZB/B1NJ, AKR/J, C3H/HeJ, and DBA/2J, respectively). Lanes 2, 5, 7, and 11 are PRP[b] haplotypes (C57 BR/cdJ, SJL/J, C57L/J, and C57 BL/6J, respectively). Lane 6 is PRP[c] haplotype (SWR/J). Lane 4 is PRP[d] haplotype (129J). Size markers (kbp) are shown on the left margin. (The data are from Azen et al., 1989.)

also exclude linkage of *Prp* to either end of chromosome 8 by testing in backcrosses a Robertsonian fusion marker chromosome, Rb(3.8)2Rma, and also the E^{so} mutation. In neither case was there indication of chromosome 8 linkage. In the light of these negative data, as well as the subsequent positive data to be described, we assume that some unexpected rearrangement previously occurred in several of the somatic cell hybrids that led to spurious results.

Because of emerging evidence of extensive homology between the distal part of mouse chromosome 6 and human chromosome 12p (Searle et al., 1987), the previous localization of *PRP* genes on human 12p (Mamula et al., 1985; O'Connell et al., 1987), as well as our previously noted association between *Prp* and *Ldr-1* (a mouse chromosome 6 marker) in a small number of RIs, we began to suspect that *Prp* genes might be located on mouse chromosome 6. Therefore, we extended our typing of RI and backcross (or testcross) DNAs with respect to *Prp* genes and markers for the middle and distal parts of chromosome 6. The results are summarized in Table 1 (Azen et al., in press). Four markers, *Mtv-8, Ly-2, Igk,* and *Rn7s-6,* each map to a tight cluster near the middle of chromosome 6. Since the genes are located at some distance from the cluster, it is appropriate to pool data over the four loci. This pooling yields 30 recombinants among 75 RI strains for an estimated recombination frequency of 0.25 ± 0.09, a result not significantly different from random assortment. Combining data from two test crosses yields an estimated recombination frequency of 0.29 ± 0.07 for *Prp* with *Igk* (14:48, recombinants to total). The weighted average of the RI and test cross data is 0.280 ± 0.055. There are 16 recombinants among 43 RI strains between *Prp* and the *Kras-2* locus for an estimated recombination frequency of 0.21 ± 0.09, a result not significantly different from random assortment. However, combining data from three test crosses yields an estimated recombination frequency of 0.15 ± 0.04 for *Prp* with *Kras-2* (11:72, recombinants to total). The weighted average of the RI and test cross data is 0.168 ± 0.038. From analysis of the recombinants in two 3-point crosses and consideration of the previously determined locations of *Igk* and *Kras-2* on chromosome 6, we can order the genes thus: *Igk-Prp-Kras-2,* with *Igk* toward the centromere. *Igk* is represented by a homolog on human chromosome 2p, but *Kras-2* has a homolog on human chromosome 12p, as do other distal mouse chromosome 6 markers, including *Tpi-1, Ldh 2,* and *Gapd.* We also found close linkage between two markers (not previously assigned to a mouse chromosome) and *Prp* in several sets of RIs. For *Es-12* and *Prp,* the recombination frequency is 0.049 (R to total, 5:33) with 95% confidence limits of 0.014-0.15.

TABLE 1 *Prp* Genes Are Linked to Mouse Chromosome 6 Markers *Mtv-8, Ly-2, Igk, Kras-2,* and *Rn7s-6* and to Previously Unassigned Markers *Es-12* and *Ea-10*

	RI			Test cross		
Prp with	R to total	r^a	*Prp* with	R to total	r^a	Weighted average
Mtv-8, Ly-2, Igk, Rn7s-6	30/75	0.25 ± 0.09 (NS)[b]	*Igk*	14:48	0.29 ± 0.07	0.280 ± 0.055
Kras-2	16:43	0.21 ± 0.09 (NS)[b]	*Kras-2*	11:72	0.15 ± 0.04	0.168 ± 0.038
Es-12	5:33	0.049 (0.014, 0.15)				
Ea-10	8/56	0.045 (0.018, 0.11)				

[a]Recombination frequency ± standard error of the mean or 95% confidence limits.
[b]Not significantly different from 0.5 at the 0.05 significance level.
Source: From Azen et al., 1989.

For *Ea-10* and *Prp*, the recombination frequency is 0.045 (R to total, 8:56) with 95% confidence limits of 0.018-0.11. In summary, we can now confidently assign *Prp* genes and, as is discussed in more detail next, the closely linked genes for bitter taste to distal mouse chromosome 6.

V. LINKAGE OF PRP GENES AND TASTE GENES FOR BITTER SUBSTANCES IN THE MOUSE

In a surprising convergence of two lines of research on mice, one on the genetics of bitter tasting and the other on chromosomal localization of genes for PRPs (described earlier), it was found that the strain distribution pattern (SDP) for *Qui* (quinine), *Rua* (raffinose acetate), and *Cyx* (cycloheximide) was the same as that for *Prp* genes in 27 RI strains (Azen et al., 1986). Thus from these earlier experiments it appears that some bitter taste genes in mice are either closely linked to *Prp* genes (95% confidence limits for *Qui-Prp* gene distance = 0.0-3.9 cM) or they may be the same genes.

In Table 2 is shown a comparison of SDPs for the bitter tastants Qui and Rua with that for *Prp Hind*III genomic fragment patterns as reported previously (Azen et al., 1986) in 20 BXD and 7 CXB RIs. As previously described, the SDP patterns are identical. Subsequently, Lush and Holland (1988) have extended their studies of the genetics of bitter tasting in mice to include glycine (Glb), which tastes both bitter and sweet to mice. There are large differences between strains in their ability to detect each taste. This difference, as with Rua, Cyx, and Qui, is probably determined by a single gene. Thus, loci for Glb, Rua, Cyx, and Qui appear to be distinct but closely linked as judged by comparison of SDPs in the RIs. There are a few apparent recombinants between the taste and *Prp* genes when their SDPs are compared. Table 3 summarizes the recombinants observed in the BXD and CXB RI sets. There are no recombinants with the *Prp/Rua* comparison, but some are seen with the *Prp/Qui, Prp/Glb*, and *Prp/Cyx* comparisons. No recombinants are seen with the smaller number of CXB RI comparisons. Thus, these newer data may offer supportive evidence that *Prp* genes are not the same as those for some (and possibly all) genes for bitter tasting in the mouse, although these data need to be confirmed.

Shingai and Beidler (1985) found large interstrain differences in the response of the glossopharyngeal nerve after exposing taste buds on the tongue to the bitter substance *SOA* (sucrose octoaacetate). These differences in neural response correlate with the ability of these strains to taste *SOA*. They postulated that the difference between "tasters" and "nontasters" depends on differences in the receptor membranes that react with SOA, although a long term effect of saliva is possible. It was recently reported that PRP RNAs are found in von Ebner's gland of mice and macaques (Azen et al., in press).

TABLE 2 Strain Distribution Patterns for Taste Sensitivity and PRP Genomic Fragment Patterns in BXD and CXB RI Sets

Strain	Mean tastant consumed (% ± SEM)	Rua taste SDP	Mean tastant consumed (% ± SEM)	Qui taste SDP	PRP SDP
BXD					
1	19 ± 10	D	32 ± 2.2	D	D
2	42 ± 4.5	B	6 ± 2.4	B	B
5	5 ± 0.3	D	32 ± 6.4	D	D
6	40 ± 4.2	B	18 ± 8.5	B	B
8	50 ± 4.1	B	3 ± 1.1	B	B
9	7 ± 2.2	D	30 ± 6.4	D	D
11	26 ± 7.4	D	39 ± 5.0	D	D
12	44 ± 11.4	B	7 ± 1.6	B	B
16	43 ± 3.8	B	11 ± 5.5	B	B
18	43 ± 3.8	B	3 ± 0.8	B	B
19	5 ± 0.3	D	34 ± 10.2	D	D
22	49 ± 3.8	B	16 ± 8.2	B	B
24	24 ± 2.5	D	33 ± 1.5	D	D
25	17 ± 11	D	44 ± 1.5	D	D
27	48 ± 2.6	B	19 ± 5.1	B	B
28	49 ± 2.9	B	17 ± 4.7	B	B
29	16 ± 6.3	D	36 ± 4.3	D	D
30	44 ± 5.5	B	6 ± 1.0	B	B
31	37 ± 2.9	B	15 ± 4.0	B	B
32	6 ± 0.7	D	34 ± 4.6	D	D
CXB					
D	46 ± 0.9	B	9 ± 4.7	B	B
E	51 ± 2.0	B	3	B	B
G	16 ± 6.4	C	43	C	C
H	20 ± 7.2	C	37	C	C
I	14 ± 9.2	C	30	C	C
J	47 ± 2.4	B	7	B	B
K	13 ± 8.9	C	35	C	C

Source: From Azen et al., 1986; Lush, personal communication.

TABLE 3 Summary of Recombinants Between Bitter Taste and *Prp* Genes Based on Comparison of SDP Patterns in RI Strains

Genes	No. recombinants	Total comparisons
B × D		
Prp/Cyx	2	21
Prp/Qui	1	21
Prp/Glb	1	20
Prp/Rua	0	21
C × B		
Prp/Cyx	0	7
Prp/Qui	0	7
Prp/Glb	0	7
Prp/Rua	0	7

Source: From Lush and Holland, 1988; Lush, personal communication; and Azen et al., 1986.

Since von Ebner's gland is believed to be important in taste stimulation, it is possible that PRPs may play a role in bitter taste, at least in the mouse, although this needs to be directly tested.

VI. GENES FOR PRPs AND TASTE FOR PTC ARE NOT CLOSELY LINKED IN HUMANS

As previously noted, we wished to learn whether *PRP* genes might be linked to those for bitter taste in humans (O'Hanlon et al., 1989). We used PTC for taste tasting, since it provides a useful polymorphic marker for human family linkage studies (reviewed by Kalmus, 1971). However, other workers adduced no strong evidence for single gene determination of PTC tasting in mice (Lush, 1986; Whitney and Harder, 1986). As previously mentioned, PRPs are also very polymorphic, so it is likely that informative linkage data between *PRP* genes and *PTC* taste genes can be obtained after study of a relatively small number of families. The gene for PTC taste is reported to be linked to that for the Kell blood group (Farrer et al., 1987), although there is some evidence for heterogeneity of this linkage (Spence et al., 1984). Also the *PTC-KEL* linkage group has not been mapped to a human chromosome (Farrer et al., 1987).

Large families (generally more than four children) were ascertained via word of mouth with no selection criteria other than family size. We used the sorting method of PTC tasting as described by Harris and Kalmus (1949)

with minor modifications. Participants were designated taster (threshold at solutions 7-14), nontaster (threshold at solutions 0-3), or uncertain (threshold at solutions 4-6). Parotid salivary PRP polymorphisms were tested by electrophoretic techniques as summarized by Azen (1989). Linkage analysis was performed using the computer program LIPED (Ott, 1974). PTC was treated as a dichotomous trait with the nontasting phenotype inherited as an autosomal recessive with a gene frequency of 0.445. Allele frequencies for salivary PRP polymorphisms were derived from reports in the literature as summarized by Azen and Maeda (1988).

In brief, there was no evidence of linkage between the *PTC* locus and any of the PRP markers (Table 4). Linkage could be excluded up to a recombination of 5% for *PRH1* and *PRB2* and up to 1% recombination for *PRH2*.

Nadeau and Taylor (1984) estimated that the mean length of conserved autosomal segments in the mouse genome, since divergence of human and mouse, is about 8 cM (centimorgans). The lack of close linkage between genes for PRPs and PTC taste in the human suggests that the gene for PTC taste and those in the *Qui-Rua-Cyx Glb* gene complex in the mouse are not homologous, although it is possible that close linkage between *PTC* and the taste genes noted has been disrupted in the human (for discussion see Nadeau and Reiner, 1989). Indeed, the gene for PTC taste may be involved in the recognition of the specific N-C=S group, which is limited to only certain bitter substances (see Kalmus, 1971). In view of the previously described linkage between the taste genes noted here and those for PRPs in the mouse, it may have been more appropriate to have used quinine or one of the other bitter substances alluded to earlier rather than PTC. However, taste polymorphisms for these substances has not yet been described in humans. There is clearly a great need for further genetic studies of bitter tastants other than PTC in humans. Our results, however, leave open the interesting possibility that there are human taste genes, other than *PTC*, that are homologous to some bitter taste genes in mice and are closely linked to the human *PRP* genes.

TABLE 4 LOD Scores for PTC by Marker Crosses

Locus	*PTC* with marker	No. of families	LOD scores at $\theta =$			Maximum LOD score		
			0.001	0.05	0.1	Score	θ Male	θ Female
(*PRB3*)	G1	4	0.12	0.12	0.11	0.12	0.30	0.001
(*PRH2*)	Pr	5	−4.08	−0.84	−0.40	0.00	0.5	0.5
(*PRH1*)	Db, Pa, PIF	6	−7.14	−2.19	−1.42	0.12	0.5	0.01
(*PRB2*)	Ps	9	−7.44	−2.33	−1.47	0.05	0.5	0.01

Source: From O'Hanlon et al., 1989.

If this is true, our data, as well as the recent studies in mice, support the idea that there are multiple genes subserving the function of bitter taste, at least in the human and the mouse.

VII. CONCLUSIONS

Close linkage of *Prp* and some bitter taste genes has been established in the mouse. *Prp* (and bitter taste) genes are localized to the distal portion of mouse chromosome 6, and a linkage group has been established with homologs on human chromosome 12p. Although the gene for PTC taste is not closely linked with those for PRPs in humans, there may be homologs of *Qui, Rua, Cyx,* and *Glb* in the human that are closely linked to *PRP* genes. The negative *PTC/PRP* linkage result in humans might be interpreted as supporting the presence of more than one bitter taste gene in the human, as is probably true in mice. There is no direct evidence as yet that *PRP* genes are involved in bitter taste.

ACKNOWLEDGMENTS

This work was supported by the National Institutes of Health Grant DE03658-23 and Grant 88096 from the University of Wisconsin Graduate School. This is paper 3026 from the Laboratory of Genetics.

REFERENCES

Ann, D. K., and Carlson, D. M. (1985). The structure and organization of a proline-rich protein gene of a mouse multigene family. *J. Biol. Chem.* **29**:15863-15872.

Ann, D. K., Smith, M. K., and Carlson, D. M. (1988). Molecular evolution of the mouse proline-rich protein multigene family. *J. Biol. Chem.* **263**:10887-10893.

Azen, E. A. (1989). Genetic protein polymorphisms of human saliva. In *Human Saliva: Clinical Chemistry and Microbiology*, Vol. I, Chapter 7, J. O. Tenovuo (Ed.). CRC Press, Boca Raton, Florida, pp. 161-195.

Azen, E. A., and Maeda, N. (1988). Molecular genetics of human salivary proteins and their polymorphisms. *Adv. Hum. Genet.* **17**:141-199.

Azen, E. A., Carlson, D. M., Clements, S., Lalley, P. A., and Vanin, E. (1984). Salivary proline-rich protein genes on chromosome 8 of mouse. *Science* **226**:967-969.

Azen, E. A., Goodman, P. A., and Lalley, P. A. (1985). Human salivary proline-rich protein genes on chromosome 12. *Am. J. Hum. Genet.* **37**:418-424.

Azen, E. A., Lush, I. E., and Taylor, B. A. (1986). Close linkage of mouse genes for salivary proline-rich proteins and taste. *Trends Genet.* **2**:199-200.

Azen, E. A., Davisson, M. T., Cherry, M., and Taylor, B. A. (1989). *Prp* (proline-rich protein) genes linked to markers *Es-12* (esterase 12). *Ea-10* (erythrocyte alloantigen) and loci on distal mouse chromosome 6. *Genomics* **5**:415-422.

Azen, E. A., Hellekant, G., Sabatini, L. M., and Warner, T. F. (1990). mRNAs for PRPs, statherin and histatins in von Ebner's gland tissues. *J. Dent. Res.* **68**.
Bennick, A. (1982). Salivary proline-rich proteins. *Mol. Cell Biochem.* **45**:83-99.
Carlson, D. M., Ann, D. K., and Mehansho, H. (1986). Proline-rich proteins: Expression of salivary multigene families. In *Microbiology*, Leive, (Ed.). American Society for Microbiology, Washington, D.C., pp. 303-306.
Farrer, L. A., Spence, M. A., Bonné-Tanner, B., Bowcock, A. M., Cavalli-Sforza, L. L., Cortessis, V., Frydman, M., Herbert, J., Kidd, J. R., Kidd, K. K., Marazita, M. L., and Sparkes, R. S. (1987). Evidence for exclusion of the Kell blood group (KEL) from more than one-half of the human genetic linkage map. *Cytogenet. Cell Genet.* (*HGM9*) **46**:613.
Harris, H., and Kalmus, H. (1949). The measurement of taste sensitivity to phenylthiourea (PTC). *Ann. Eugen.* **15**:24-31.
Ito, S., Suzuki, T., Momotsu, T., Isemura, S., Saitoh, E., Sanada, K., and Shibata, A. (1985). Presence of salivary protein C and salivary peptide P-C like immunoreactivity in the laryngo-tracheo-bronchial glands. *Acta Endocrinol.* (*Copenh.*) **108**:130-134.
Kalmus, H. (1971). Genetics of taste. In *Handbook of Sensory Physiology*, Vol. IV, *Chemical Senses*, Part 2, L. M. Beidler (Ed.). Springer-Verlag, Berlin, pp. 165-179.
Kim, H.-S., and Maeda, N. (1986). Structure of two *Hae*III-type genes in the human salivary proline-rich protein multigene family. *J. Biol. Chem.* **261**:6712-6718.
Lush, I. E. (1986). Differences between mouse strains in their consumption of phenylthiourea. *Heredity* **57**:319-323.
Lush, I. E., and Holland, G. (In press). The genetics of tasting in mice. V. Glycine and cycloheximide. *Genet. Res.*
Lyons, K. M., Azen, E. A., Goodman, P. A., and Smithies, O. (1988). Many protein products from a few loci: Assignment of human salivary proline-rich proteins to specific loci. *Genetics* **120**:255-265.
Lyons, K. M., Kim, H.-S., Saitoh, E., and Smithies, O. (submitted). Nucleotide sequences and evolution of the PRB3 and PRB4 genes from the human proline-rich protein multigene family.
Maeda, N. (1985). Inheritance of human salivary proline-rich proteins: A reinterpretation in terms of six loci forming 2 subfamilies. *Biochem. Genet.* **23**:455-464.
Mamula, P. W., Heerema, N. A., Palmer, C. G., Lyons, K. M., and Karn, R. C. (1985). Localization of the human salivary proteins complex (SPC) to chromosome band 12p13.2. *Cytogenet. Cell Genet.* **39**:279-284.
Mehansho, H., Butler, L. G., and Carlson, D. M. (1987). Dietary tannins and salivary proline-rich proteins: Interactions, induction and defense mechanisms. *Ann. Rev. Nutr.* **7**:423-440.
Nadeau, J. H., and Taylor, B. A. (1984). Lengths of chromosomal segments conserved since divergence of man and mouse. *Proc. Natl. Acad. Sci. USA* **81**:814-818.
Nadeau, J. H., and Reiner, A. H. (1989). Linkage and synteny homologies in mouse and man. In *Genetic Variants and Strains of the Laboratory Mouse*, M. F. Lyon and A. Searle (Eds.). Oxford University Press, London.

O'Connell, P., Lathrop, G. M., Law, M., Leppert, M., Nakamura, Y., Hoff, M., Kumlin, E., Thomas, W., Elsner, T., Ballard, L., Goodman, P., Azen, E., Sadler, J. E., Cai, G. Y., Lalouel, J.-M., and White, R. (1987). A primary linkage map for human chromosome 12. *Genomics* **1**:93-102.

O'Hanlon, K., Weissbecker, K., Cortessis, V., Spence, M. A., and Azen, E. A. (1989). Genes for salivary proline-rich proteins and taste for phenylthiourea are not closely linked in humans. *Cytogenet. Cell Genet.* **49**:315-317.

Ott, J. (1974). Estimation of the recombination fraction in human pedigrees: Effective computation of the likelihood for linkage studies. *Am. J. Hum. Genet.* **26**: 588-597.

Searle, A. G., Peters, J., Lyon, M. F., Evans, E. P., Edwards, J. H., and Buckle, V. J. (1987). Chromosome maps of man and mouse, III. *Genomics* **1**:3-18.

Shingai, T., and Beidler, L. M. (1985). Interstrain differences in bitter taste responses in mice. *Chem. Senses* **10**:51-55.

Spence, M. A., Falk, C. T., Neiswanger, K., Field, L. L., Marazita, M. L., Allen, F. H., Jr., Siervogel, R. M., Roche, A. F., Crandall, B. F., and Sparkes, R. S. (1984). Estimating recombination frequency for the PTC-Kell linkage. *Hum. Genet.* **67**:183-186.

Warner, T. F., and Azen, E. A. (1984). Proline-rich proteins are present in serous cells of the submucosal glands of the respiratory tract. *Am. Rev. Respir. Dis.* **130**: 115-118.

Whitney, G., and Harder, D. B. (1986). Phenylthiocarbamide (PTC) preference among laboratory mice: Understanding of a previous "unreplicated" report. *Behav. Genet.* **16**:605-610.

Ziemer, M. A., Swain, W. F., Rutter, W. J., Clements, S., Ann, D. K., and Carlson, D. M. (1984). Nucleotide sequence analysis of a proline-rich protein cDNA and peptide homologies of rat and human proline-rich proteins. *J. Biol. Chem.* **259**:10475-10480.

21
Genetics and the Neurobiology of Olfactory Bulb Circuits

Charles A. Greer

Yale University, New Haven, Connecticut

I. INTRODUCTION

The use of neurologic mutations as tools for investigating the principles and mechanisms influencing the organization of the central nervous system has a long and productive history. Although more than 100 genetic mutations are known to affect neural development, the use of mutations as tools for unraveling the complexities of the nervous system has been largely evident in their application to the cortex, where seven single-locus mutations have proven particularly useful in unraveling the mechanisms of cortical organization. These mutations have been successfully utilized to study developmental neurobiologic determinants of neuronal position, geometry, interaction, and connectivity (Caviness and Rakic, 1978). Studies have been conducted in mice whose tissue and cells all share the mutant genotype as well as in genetic mosaics or chimeras in which mutant and normal tissue is mixed (Mullen and Herrup, 1979). A feature common to these cortical neurologic mutants is an affected cerebellum. Consequently, the identification of these mutations has been in part serendipidous as a result of the distinctive gait phenotype associated with the affected cerebellum. These gaits have led to a variety of colorful names, including staggerer, weaver, lurcher, leaner, and reeler. Analyses of the organization of the cortex in these animals has provided important information about interactions of cellular elements during development and principles upon which the nervous system is organized. These studies have generally not sought to suggest that the synaptic connectivity of cortex, or any region is hardwired or prespecified by a particular gene locus. Rather, the emphasis has been on genetic specification as a stochastic series of events, each of which can influence the final outcome, much as depicted by Waddington's epigenetic landscape. For example, the postnatal death of cerebellar granule cells in the mutant staggerer has been attributed not to a direct gene effect but rather as secondary to a primary gene effect on the Purkinje cells. Normal migration and development of the granule cells requires interaction with the Purkinje cells, which in the case of staggerer are abnormal in their distribution and dendritic morphology. As a consequence granule cells fail to establish their normal connections and die. Stent (1981) has identified this as the instrumental perspective in the genetic approach to neurobiology. The emphasis is placed not on how the single gene might code for neurons and circuits, as is true of the ideologic perspective, but rather on genetics as another tool for understanding differences between phenotypes that may be accounted for by developmental or organizational principles. The hope is that any results that may be obtained will be relevant not only to understanding the mutation itself, but will generalize to developmental events, principles, and mechanisms throughout the nervous system.

The instrumental genetic approach to studying the developmental neurobiology and organization of the olfactory system has not been widely employed. This may reflect, in part, that of the many genes known to affect the organization of the central nervous system only three genotypes have been reported with morphologic anomalies of the olfactory system, the neurologic mutants reeler and Purkinje cell degeneration and the inbred strain Balb/c. There have also been provocative reports of altered olfactory acuity or sensitivity between different inbred lines of mice, although the mechanisms underlying these differences have not yet been clearly eludicated (Wysocki et al., 1977; Price, 1977; Pourtier and Sicard, 1988; Sicard et al., 1988; Sicard and Pourtier, 1988). In a similar vein, Gozzo and Fulop (1984) reported that C57BL/6 and SEC/1ReJ mice differ in the response to olfactory nerve transection. C57 mice showed extensive transneuronal degeneration in the olfactory bulb but SEC mice did not. The mechanisms underlying this type of observation is also unknown, although it may be of interest to study in more detail the capacity for reinnervation of the olfactory bulb in C57 versus SEC mice.

II. SPECIFIC ANOSMIA

The early report by Wysocki et al. (1977) of selective anosmia in C57BL/6 mice for isovaleric acid was exciting because it provided a basic animal model for studying how olfactory information was encoded. Sicard and coworkers (Sicard et al., 1988; Sicard and Pourtier, 1988) recently reported on 2-deoxyglucose studies of sensitivity to isovaleric acid in C57BL/6j mice and AKR mice. Based upon earlier work showing that olfactory bulb glomerular 2-deoxyglucose patterns reflected input from the olfactory receptor axons (Greer et al., 1981), Sicard and associates hypothesized that strain differences in sensitivity to an odor due to peripheral receptor mechanisms could be visualized through differences in 2-deoxyglucose deposits. Their data demonstrated that both strains of mice responded to isovaleric acid with an increase in 2-deoxyglucose uptake, although the patterns were qualitatively different. The authors suggest that C57BL/6 mice may be missing a subset of molecular receptors. The anosmia originally reported by Wysocki et al. (1977) would be reflected in the absence of this subset; the ability to respond to higher concentrations would reflect the recruitment of other subsets less sensitive to isovaleric acid. These initial studies are very exciting since they point strongly to genetic determinants of subsets of olfactory receptor cell types or molecular receptor types. Further research distinguishing between these alternatives will be important. Resolution of this problem would have a profound effect on the field of olfaction because it could resolve a basic dilema of whether molecular odor receptors are distributed among receptor cells in a generally homogeneous or heterogenous fashion.

III. REELER

Of the known olfactory system mutants, the first to receive attention was the mutant reeler (Caviness and Sidman, 1972; Devor et al., 1975). As implied earlier, reeler was initially recognized for abnormalities of gait due to an affected cerebellum. Expression of the mutant phenotype is limited to the homozygous recessive genotype, with no apparent abnormalities in the heterozygote. Subsequent research demonstrated, in addition to the cerebellar deficit, a widespread inversion of cellular laminae in cortex. In piriform cortex, as well as the anterior olfactory nucleus and olfactory tubercle, the larger pyramidallike neurons are displaced, forming a distinct zone that is beneath a layer of small polymorphic cells. The number of cells, however, is not different from controls based on analyses of DNA and RNA content as well as protein profiles (Mikoshiba et al., 1980). The inversion of piriform cortex has been of interest because the interaction of pyramidal neurons with growing lateral olfactory tract (LOT) axons to produce the normal highly specific laminar organization is poorly understood. Of significance, in reeler the LOT was reported to establish a normal sublaminar pattern in piriform cortex despite the malalignment of the pyramidal neurons. This means that the LOT axons were not going deep into piriform cortex to contact displaced pyramidal neurons whose dendrites did not extend into the molecular layer. This was taken as evidence of the independent gene effects on pyramidal cell migration of axonal elongation (Devor et al., 1975). This can also be taken as evidence that the gross topologic sorting of LOT axons in piriform cortex can occur independently of a normal distribution of target dendrites. However, because at least some of the displaced pyramidal neurons extended their apical dendrites into the molecular layer, the question of a dynamic and sublaminar interaction between these dendrites and the LOT axons remains; it must yet be determined if the highly specific distribution of LOT synapses on the pyramidal dendritic spines is normal or if the malalignment of dendrities results in a disruption of the precise topologic sorting that normally directs synapse placement along the dendrites (Price, 1987).

Interestingly, the mutant allele in reeler exhibits only partial penetrance in the olfactory bulb. The early reports suggested that laminar organization in the bulb was normal; indeed, the suggestion was that the mechanism of cell migration and sorting may be different for the olfactory bulb than for other laminated structures (Caviness and Sidman, 1972). However, more recent reports demonstrate that although the olfactory bulb neurons exhibit a typical laminar segregation, within the laminae neurons are displaced in a nonuniform fashion (Wyss et al., 1980). Moreover, the volume of the bulb laminae is reduced from 26 to 46%. The significance of these alterations for processing odors has not been established. Also, the developmental events or mechanisms affected by the *rl* locus in the olfactory bulb or cortex are as yet unknown. The decreased volume of the bulb laminae does suggest the loss of neurons, perhaps subpopulations, that may provide further insights into patterns of connectivity and the organization of parallel circuits normally found in the olfactory bulb. In general, the mutant reeler continues to offer an attractive, although little employed, model for investigating developmental events in the bulb as well as piriform cortex.

IV. Balb/c

Considerable excitement was generated by the early reports by White (1972, 1973) of intraspecific variation in the synaptic organization of the mammalian olfactory bulb glomerulus. Briefly, in the mammalian olfactory glomerulus the olfactory receptor cell axons make excitatory synapses onto three postsynaptic targets: mitral cell dendrites, tufted cell dendrites, and periglomerular cell dendrites. However, White reported that in Balb/c mice the synapse from the olfactory receptor axon onto the periglomerular cell dendrite was absent. These results suggested that there could be great specificity in the formation of synaptic connections. This is particularly interesting in the olfactory system, where the continual turnover of olfactory receptor cells would require that the mechanisms influencing the formation of synapses must remain operative throughout life. There appear to be two possible explanations for the observation in Balb/c mice. First, during the course of development the trophic or tropic signals influencing synapse formation onto the periglomerular cells fail, either in the axon or in the target. The second alternative focuses on the receptor cell axons and consists of two parts. First, that the receptor cell axons fail to arborize within the glomerulus. Developmentally, the mitral and tufted cells extend dendrites into the glomerulus prior to the periglomerular cells. Consequently, the mitral and tufted cells may compete more effectively for the innervating receptor cell axons and "occupy" all presynaptic sites. The second alternative is that a specific subpopulation of receptor cells normally innervate the periglomerular neurons and that these receptor cells are absent in Balb/c mice. Unfortunately, none of these alternatives have been explored further in Balb/c mice; moreover, the initial studies by White have not been replicated. This seems a significant oversight because the Balb/c mutation could be useful in understanding the contribution of glomerular circuitry to odor processing, although they do appear comparable to several other strains at least in their ability to detect isovaleric acid (Wysocki et al., 1977). An interesting issue that might arise with Balb/c mice is whether the distribution of subpopulations of olfactory receptor cells, as identified with immunocytochemistry, is comparable that to other mammals. If indeed the absence of the glomerular synapse is indicative of missing subpopulations of receptor cells, this may be apparent through analysis with several of the antibodies to subpopulations of olfactory receptor neurons that are now available (Mori et al., 1985; Allen and Akeson, 1985; Hempstead and Morgan, 1985; Key and Giorgi, 1986; Schwob and Gottlieb, 1988; Onada, 1988a, b; Onada and Fujita, 1988).

V. PURKINJE CELL DEGENERATION

Of the identified mutants the one found most interesting by my laboratory is Purkinje cell degeneration (PCD). These were first reported by Mullen et al. (1976) for the early degeneration of cerebellar Purkinje cells and the later occurring retinal degeneration. Subsequently they were reported on for the postnatal loss of thalamic neurons (O'Gorman and Sidman, 1985a, b). We were drawn to the PCD mice by the initial report, which included a comment on degenerating fibers in the lateral olfactory tract at approximately 12 months postnatal. It seemed apparent that two alternatives could account for this observation, either the degeneration of olfactory bulb projection neurons, mitral and tufted cells, or the degeneration of centrifugal fibers. Either alternative was appealing since both would provide the opportunity to study organization within the olfactory system utilizing developmental neurobiologic techniques and a instrumental genetic perspective.

A. Mitral Cell Loss

Utilizing conventional light microscopy we examined the olfactory bulbs of homozygous recessive PCD mutants and matched heterozygous controls at 2, 4, and 8 months postnatal (see Figure 1) (Greer and Shepherd, 1982). At 2 months the number and distribution of mitral and tufted cells did not differ. By 4 months, however, the number of mitral cells in the homozygous recessive had been reduced by more than 80% but the number of tufted cells was unchanged. At 8 months virtually all the mitral cells were lost, although we conservatively suggested that 4% remained based on the appearance of larger somata close to the residual mitral cell layer. Recent data suggest that those remaining cell bodies are most likely deep tufted cells rather than mitral cells. In Figure 2 the dendritic composition of the external plexiform layer following mitral cell loss is shown. As can be seen, the larger pale dendritics, typical of the mitral cells, are absent. Quantitive planimetric assessment confirmed this observation and, moreover, revealed that the largest decrease in area occupied by dendrites characteristic of mitral cells occurred in the deeper portion of the external plexiform layer (see Figure 3). This provided an independent confirmation of sublaminar organization within the external plexiform layer, with the dendrites of mitral cells distributed predominantly within the deeper portion of the laminae and the dendrites of tufted cells distributing within the more superficial portion (Mori et al., 1983; Orona et al., 1984). The PCD mice are one of the few neurologic mutants that exhibit normal organization until they are virtually adults. The loss of the Purkinje cells in the cerebellum also occurs following neuronal migration and synaptogenesis but relative to the olfactory bulb is still comparatively early at 2 weeks postnatal. Both, however, are in contrast with reeler, in which the primary gene expression occurs during cell migration in the embryo. The late loss of the mitral cell suggests strongly that the PCD mutants can be effectively utilized as animal models of neurodegenerative diseases in adult humans.

Analysis with the 2-deoxyglucose technique demonstrated that despite the loss of mitral cells metabolic patterns in the olfactory bulb glomeruli were not different from those of control animals (see Figure 4). Moreover, the cross-laminar patterns of 2-deoxyglucose distribution were comparable in the homozygous recessive and heterozygous condition. The observations with 2-deoxyglucose are important in that they provide a basic insight into the relative contribution of olfactory bulb postsynaptic structures in the odor-stimulated uptake of 2-deoxyglucose. The observation was also consistent with previous suggestions that the majority of 2-deoxyglucose uptake in the glomeruli occurs in the olfactory receptor cell axon terminals (Greer et al., 1981). The 2-deoxyglucose data also suggest strongly that the topography of olfactory nerve innervation of the olfactory bulb is not seriously compromised in the PCD mice as a result of mitral cell loss. If the topography

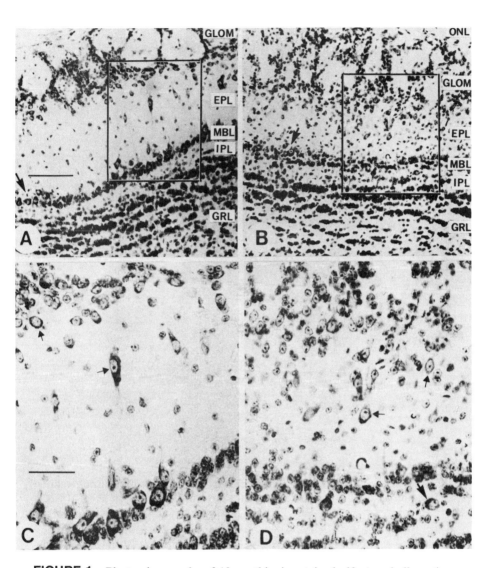

FIGURE 1 Photomicrographs of 10 μm thionine-stained olfactory bulb sections from affected (pcd/pcd) (B and D) and heterozygous control (pcd/++) (A and C), PCD mice. Boxes in A and B demaracate the areas shown at higher magnification in C and D, respectively. Arrows in A and B indicate clustering of mitral cell bodies. Small arrows in C and D indicate tufted cell somata. Large arrow in D indicates larger cell body in the internal plaxiform layer. ONL, olfactory nerve layer; GLOM, glomerular layer; EPL, external plexiform layer; MBL, mitral body layer; IPL, internal plexiform layer; GRL, granule cell layer. Calibration bars: A and B = 100 μm; C and D = 40 μm. (From Greer and Shepherd, 1982.)

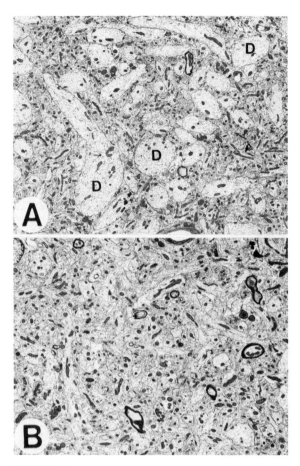

FIGURE 2 Low-magnification electron micrographs illustrating the dendritic composition of the external plexiform layer in control (A) and affected (B) PCD mice. Numerous large pale dendrites (D), characteristic of mitral cell dendrites, are seen in A, together with smaller flocculent profiles typical of granule cell spines. In B the larger pale profiles are absent, reflecting the loss of mitral cells. The remaining pale profiles are consistent with the features of tufted cell dendrites that are not lost. Magnification ×2400. (From Greer and Halasz, 1987.)

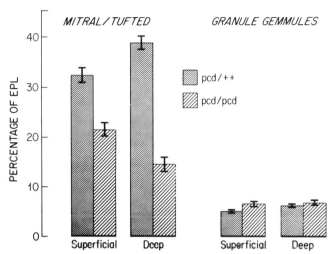

FIGURE 3 Mean (+ SEM) percentage of the superficial and deep external plexiform layer occupied by characteristic mitral and tufted cell dendrites and granule cell dendritic spines (gemmules) in affected homozygous recessive PCD mice (*pcd/pcd*) and heterozygous controls (*pcd/++*). (From Greer and Halasz, 1987.)

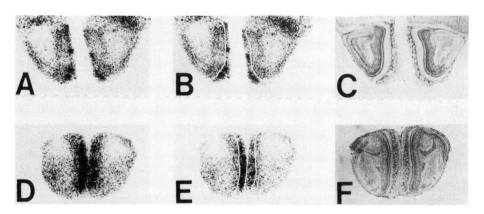

FIGURE 4 Transverse olfactory bulb sections showing the punctate accumulation of 2-deoxyglucose along the medial wall in control (A-C) and affected (D-E) PCD mice following odor stimulation. In B and E the inner border of the glomerular layer has been superimposed on the autoradiograph. In C and F the corresponding cresyl violet-stained sections are shown. Note in both genotypes that the uptake and distribution of odor-induced 2-deoxyglucose is similar. (From Greer and Shepherd, 1982.)

were significantly changed the topographic organization of glomerular domains exhibiting increased uptake of 2-deoxyglucose should have differed between the genotypes. This is a particularly interesting outcome because it suggests that during the continual turnover and reinnervation of the olfactory bulb by receptor cell axons the presence of mitral cells per se is not necessary for appropriate topography to be realized. It is widely recognized that olfactory nerve axons could form glomerularlike structures outside the olfactory bulb (Graziadei et al., 1978, 1979). Although these observations pointed to the "inductive" capacity of these axons, they did not address the issue of normal topography and the formation of odor domains within the olfactory epithelium and bulb. The PCD mice, however, suggest that olfactory bulb neurons, as targets for the growing axon, may contribute relatively little to the initial development and maintenance of epithelial bulb topography. The data of course do not account for the potential influence of either tufted or periglomerular cells, but nevertheless are provocative. Of equal importance was the observation that the PCD mice could perform simple behavioral tasks dependent upon olfactory function after mitral cell loss had occurred (Greer and Shepherd, unpublished observations). These data provided some of the first evidence that the two projection neurons, mitral and tufted cells, comprise parallel circuits in the olfactory system that differ in the subsets of genetic loci contributing to their final position, form, and connectivity.

B. Granule Cell Morphology

A significant secondary effect of mitral cell loss in the PCD mice is the denervation of a principal olfactory bulb interneuron, the granule cell. Typically, mitral cells and a subset of granule cells form local microcircuits, reciprocal dendrodendritic synapses, within the external plexiform layer. As a result of mitral cell loss that subset of granule cells is deprived not only of afferent input but also efferent targets. Unlike stagger in the cerebellum, the olfactory bulb interneurons, granule cells, are not lost as a result of mitral cell death. Although cell density in the granule cell layer is the same for both homozygous recessive and heterozygous PCD mice, it is apparent that the width of the external plexiform layer, where the granule cell apical dendrites arborize, is significantly reduced in width as a result of the loss of mitral cell dendritic arbors. In light of these changes it was of interest to establish the nature of any changes in the granule cell population that accompanied or followed mitral cell loss.

Utilizing Golgi impregnation procedures we described the organization of dendritic arbors and spine distribution in the external plexiform layer of normal heterozygous mice. The subpopulations of granule cells previously described for the rabbit (Mori et al., 1983; Mori, 1987) and the rat (Orona et al., 1983) were also generally applicable to the mouse (see Figure 5). Three primary subpopulations emerged from the analyses, one of which, type II, arborized exclusively within the deeper portion of the external plexiform layer where the dendrites of mitral cells are found. This normal organization suggests that it is the type II granule cell that is preferentially denervated with mitral cell loss. Following mitral cell loss it was no longer possible to identify subpopulations because the laminar segregation of granule cell dendrites was absent (see Figure 6). In contrast, dendrites of all granule cells distributed in a homogeneous fashion in the external plexiform layer. Consistent with the loss of a clear sublaminar organization of dendrites in PCD mice is a change in staining patterns with cytochrome oxidase. Mouradian

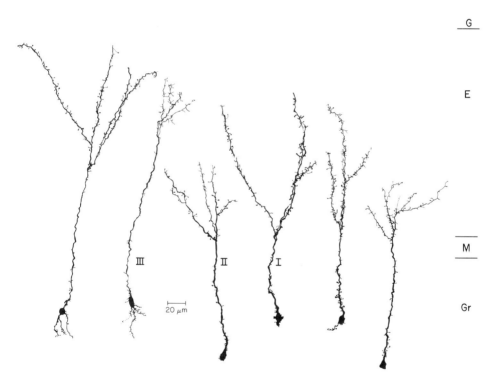

FIGURE 5 Montage of camera lucida reconstructions of typical granule cells from heterozygous PCD control mice. The three subpopulations of granule cells are identified (types I, II, and III). Note the radial orientation of apical dendrites into the external plexiform layer (E) and the sublaminar distribution of spinous processes. Glomerular layer (G); mitral cell layer (M); granule cell layer (Gr). (From Greer, 1987.)

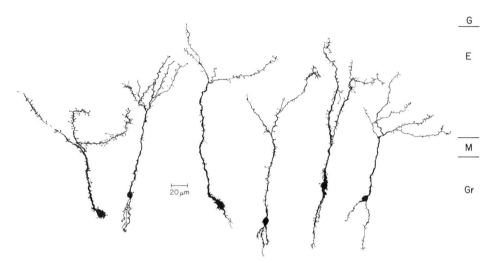

FIGURE 6 Montage of camera lucida reconstructions of typical granule cells from affected homozygous recessive PCD mice. Note that many of the apical dendrites exhibit a tendency toward an oblique or horizontal orientation within the external plexiform layer. There also appears a less distinct sublaminar distribution of spines than in control mice (see Figure 5). Abbreviations as in Figure 5. (From Greer, 1987.)

and Scott (1988) reported on a sublaminar organization in the external plexiform layer using cytochrome oxidase as a marker; a darkly stained intermediate zone is surrounded superficially and deeply by more lightly stained zones. A similar banding pattern is apparent in heterozygous PCD and other control mice, but the variation in cytochrome oxidase staining is absent in homozygous PCD mice.

Although the number of granule cells does not change in PCD mice and although subpopulations cannot be identified, the morphometric characteristics of the collective population do shift following mitral cell loss (see Table 1). Total cell length and the summed length of dendrities within the external plexiform layer are both decreased. The number of dendritic branches, however, is unchanged. This indicates that the terminal portions of dendrites are being lost rather than entire branches. In a similar vein, the sublaminar distribution of spines is unchanged. Moreover, the density of spines remains constant at 0.21 per 1 μm of dendrite although the total number of spines per cell decreases.

TABLE 1 Morphometric Characteristics of Control (*pcd*/++) and Mutant (*pcd/pcd*) PCD Mouse Granule Cells Following the Loss of Mitral Cells[a]

Attribute	*pcd*/++		*pcd/pcd*	
Total cell length, μm	203	(11.6)	148.4	(7.5)
Somata below MBL, μm	80.5	(6.9)	48.1	(6.7)
Extension above MBL, μm	121.7	(12.2)	100.3	(6.37)
Number of dendritic branches	4.5	(0.3)	4.4	(0.3)
Summed length of branches, μm	309.3	(29.0)	240.3	(18.4)
Number of spines per cell	64.9	(7.1)	52.3	(4.4)
Number of spines per 1 μm dendrite	0.21	(0.01)	0.21	(0.02)
Distance beteween soma and first bifurcation	107.5	(10.99)	94.1	(9.64)
Spine distribution in 20 μm EPL divisions				
0-20	5.93	(0.96)	7.29	(1.40)
20-40	7.90	(1.17)	9.80	(1.52)
40-60	10.11	(1.67)	11.41	(1.56)
60-80	11.46	(2.18)	11.11	(1.48)
80-100	12.90	(2.57)	14.67	(2.78)
100-120	9.94	(1.81)	6.67	(1.12)
120-140	10.50	(2.71)	6.00	(1.60)
140-160	11.90	(2.58)	7.00	(1.41)
160-180	11.60	(2.20)	—	
180-200	14.50	(1.97)	—	

[a]Values are presented as mean ($\pm$ SEM). Spine distribution within the external plexiform layer (EPL) was determined by plotting the density of spines along the apical dendrites as they extended into the EPL. At the 160-200 μm level in mutant PCD mice no values are presented because the EPL was no longer that thick. MBL, mitral body layer.

C. Synaptic Organization

PCD clearly offers a unique opportunity to study the effect of a highly specific pertubation on the organization of dendrodendritic microcircuits within the external plexiform layer. A principal question is the specificity of microcircuit formation between subsets of granule cells and mitral cells and the specificity of the parallel circuits between other subsets of granule cells and tufted cells. In addition, the special nature of the reciprocal dendrodendritic circuit provides the opportunity to study ultrastructurally the response of a neuron to the loss of both afferent input as well as efferent targets. An extensive body of literature attests to the capacity of the olfactory bulb for plasticity, particularly in the renewal of its afferent input (Graziadei and

Monti-Graziadei, 1978; Graziadei et al., 1978, 1979; Barber, 1981a, b; Costanzo and Graziadei, 1983), as well as the ability to respond to new odorants based upon early experience (Pederson and Blass, 1982) and respond to odors despite significant lesions of the olfactory bulb (Hudson and Distel, 1987; Slotnick et al., 1987). However, none of these experimental approaches provided insights into the potential of olfactory bulb local circuits for reorganization or plasticity. The emphasis in ultrastructural studies of PCD has been placed on the local circuits of the bulb external plexiform layer because it represents comparatively pure or well-defined genetic lesion. Although alterations of synaptic organization in the glomeruli of PCD mice is of great interest, analyses of that region are complicated by receptor cell axon turnover that may cloud alterations in local circuit plasticity.

Synaptic organization between the dendrites of mitral and tufted cells and the spiny dendritic appendages of their respective granule cell subpopulations in heterozygous PCD mice and homozygous PCD mice before mitral cell degeneration are consistent with earlier reports (Hirata, 1964; Reese and Brightman, 1965; Price and Powell, 1970a, b, c; Landis et al., 1974; Hinds and Hinds, 1976a). Mitral and/or tufted cell dendrites are easily recognized from their electron-lucent appearance, arrays of microtubules, comparatively large size, and the presence of small collections of spherical vesicles usually in close proximity to the membrane. In contrast, the granule cell spines, gemmules, are electron dense and lack microtubules but often have a membrane-bound cistern known as a spine apparatus and large collections of elongated or elliptical vesicles. In addition the granule cell spines often have mitochrondria, which sets them apart from the prototypical spines of cortex. Finally, clusters of ribosomes are often found in both the mitral and tufted cell dendrites as well as the granule cell spines (Shepherd and Greer, 1988). A classic example of a reciprocal pair of dendrodendritic synapses between a mitral and tufted cell dendrite and a granule cell spine from the external plexiform layer is shown in Figure 7. Each of the apposed profiles has two distinctive zones of membrane specialization. The zone of presynaptic specialization is seen in the mitral and tufted cell dendrite as an area of increased membrane density. A cluster of small spherical vesicles is seen closely apposed to the area of specialization. Among the vesicles can be seen an electron-dense granular substance that appears to extend from the area of membrane specialization. Postsynaptically, the granule cell spine has a zone of membrane specialization characterized by an increase in density and thickness in aldehyde-fixed tissue. The postsynaptic zone is thicker than the zone of presynaptic specialization, giving the appearance of asymmetry. Frequently, the cytoplasm adjacent to the postsynaptic specialization in the granule cell spine is free of vesicles and organelles. The zone of presynaptic specialization in the granule cell spine is indistinguishable from that of the mitral and

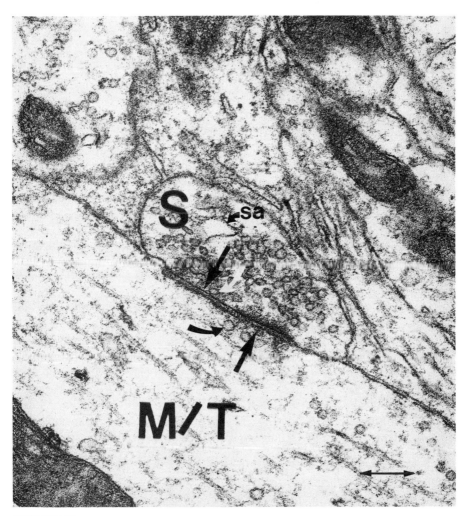

FIGURE 7 Electron micrograph of a typical reciprocal dendrodendritic synapse between a mitral and tufted cell dendrite (M/T) and a granule cell spine (S) in the external plexiform layer of a heterozygous control PCD mouse. The polarity of the synapses is indicated with arrows. In the spine a transversely cut membrane-bound cistern, apparently a spine apparatus, can be seen (sa). Clusters of vesicles (curved arrows) can be seen closely apposed to the presynaptic specialization of both profiles. Calibration bar = 125 nm. (Adapted from Greer and Halasz, 1987.)

tufted cell dendrite other than by the composition of the vesicle pool. In the spine the vesicles are numerous and pleiomorphic, although often elliptical in shape. The postsynaptic specialization in the mitral and tufted dendrite differs dramatically from its counterpart in the spine. In this case, although the area of specialization is marked by an increase in plasmalemmal density, there is no corresponding increase in thickness. As a consequence, the zone of pre- and postsynaptic membrane specialization appears symmetrical. As is well known, correlation of these features with electrophysiologic evidence has unequivocally established that the asymmetric synapse from the mitral and tufted dendrite to the granule cell spine is excitatory but the symmetrical synapse from the spine back to the mitral and tufted cell dendrite is inhibitory (Rall et al., 1966). The two synapses have a frequency that closely approximates a ratio of 1:1 (Greer and Halasz, 1987), and in serial reconstructions they are consistently found to occur in pairs (Greer, 1987a; Greer, 1988).

In homozygous recessive PCD mice tissue examined at high magnification with the electron microscope is difficult to distinguish from control material other than by the absence of the larger pale profiles. The synaptology found in the affected PCD mice does not appear to differ substantively from that of controls (see Figure 8). Of particular importance, there was no evidence of granule cell spines that lacked apposed dendrites at sites of synaptic specialization and the symmetrical and asymmetric synapses occurred at a frequency that closely approximated a ratio of 1:1. This leads to two conclusions. First, as suspected from studying the superficial and deep laminae of the external plexiform layer, there are no apparent synaptologic characteristics that distinguish between the reciprocal dendrodendritic circuits of mitral cells and tufted cells. Second, because there was no evidence of granule cell loss and no evidence of granule cell spines with unapposed zones of synaptic specialization, it seems parsimonious to conclude that following denervation the affected granule cell population establishes new reciprocal circuits, most likely with tufted cell dendrites. Quantitative morphometric procedures confirmed this by showing that the density of synapses increased on tufted cell profiles and, moreover, that the increase in density was greater in the deeper laminae of the external plexiform layer where the preponderance of denervation occurred (see Figure 9). These events, normal organization, denervation, and reorganization, are summarized in Figure 10.

It is interesting to speculate on what factors may contribute to the capacity of reorganization in these circuits and how those factors may influence normal neurobiologic development. First, it is reasonable to suggest that the symmetrical and asymmetric components of the dendrodendritic microcircuit exist in a dynamic equilibrium. If this is not the case, it is more reasonable to expect that a trophic signal from the denervated granule cell initiated reactive dendrodendritic synaptogenesis from neighboring tufted cell dendrites

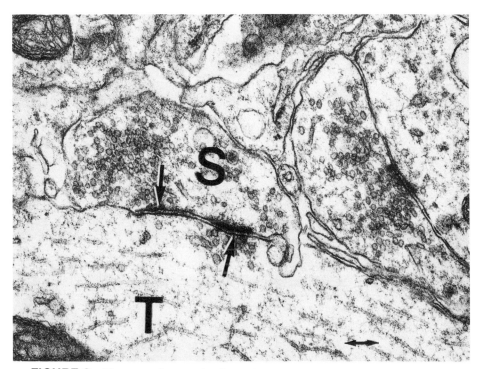

FIGURE 8 Electron micrograph of a typical reciprocal dendrodendritic synapse between a tufted cell dendrite (T) and a granule cell spine (S) from the olfactory bulb of an affected homozygous recessive PCD mouse. The polarity of the synapses is indicated with arrows. Note that the characteristics of this synaptic specialization are indistinguishable from the control shown in Figure 7. Calibration bar = 200 nm. (Adapted from Greer and Halasz, 1987.)

without influencing the formation of the reciprocal inhibitory synapse. A variety of literature attests to the ubiquitous nature of denervation-induced synaptogenesis as a mechanism with which a cell can strive to maintain a minimum level of synaptic input (Raisman, 1985; Devor, 1976; Hamori and Somogyi, 1982). In previous instances in which presynaptic dendrites developed following lesions, the initiating factor appeared to be the presence of an unoccupied postsynaptic specialization (Somogyi et al., 1984). In the case of the PCD mice there is no unoccupied site of membrane specialization so the initiating factor must lie elsewhere. Some support for the notion of a dynamic equilibrium within the reciprocal pair may also come from Benson et al. (1984), who reported a reduction in the frequency of both symmetrical and asymmetric synapses following neonatal nares closure. From a

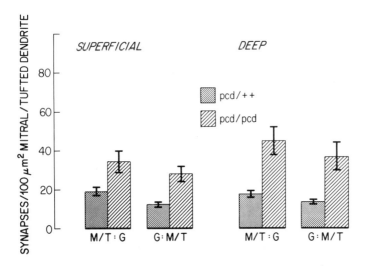

FIGURE 9 Mean (± SEM) asymmetric and symmetrical synapses per 100 μm^2 of mitral and tufted cell dendrites in the superficial and deep external plexiform layer of homozygous recessive mutants and heterozygous control PCD mice. M/T:G, asymmetric mitral and tufted cell synapse onto granule cell spine; G:M/T, symmetrical granule cell synapse onto mitral and tufted cell dendrite. (From Greer and Halasz, 1987.)

different perspective, the significant pruning of granule cell dendritic spines that occurs toward the end of the postnatal developmental period may reflect the reabsorption of spines that do not successfully establish both constituents of the reciprocal pair (Greer, 1984, 1987a; Apfelbach and Weiler, 1985).

In the context of this discussion it is important to consider why some of the granule cell spines were lost rather than all successfully establishing new synaptic circuits. A contributing factor may be the availability of undifferentiated sites on the tufted cell membrane. In the normal olfactory bulb mitral and tufted cells appear to form parallel local circuits with little or no overlap. In light of this it is reasonable to expect that the loss of mitral cells in the PCD mutant has no direct effect on tufted cells or their circuits. Because the density of synapses on the tufted cells increases as a consequence of mitral cell loss we know that sites are available for membrane specialization. This suggests that a limiting factor during normal development is not the availability of sites on the tufted cell. However, a further inference is that the number of sites is limited. In the affected PCD mice those denervated granule cell spines that compete most effectively for the available sites survive

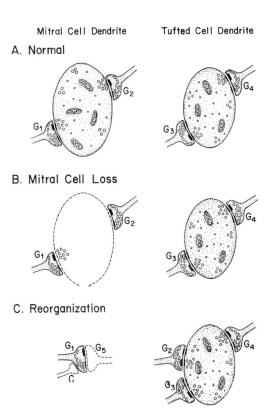

FIGURE 10 Schematic illustration of the sequence of events suggested by the results from the PCD mice. (A) The typical synaptic organization is shown with both mitral cell dendrites (M) and tufted cell dendrites (T) having reciprocal dendrodendritic synapses with granule cell dendritic spines (G_1-G_4). (B) The loss of the mitral cell dendrite is depicted and the corresponding denervation of the granule cell spines G_1 and G_2. (C) Reorganization of the denervated spines is shown. G_2 establishes a new reciprocal pair of synapses with the tufted cell dendrite. G_1 is illustrated in a transitional state. It has not yet established new connections or been lost as depicted by the outline of spine G_5. The unknown effect of granule cell spine reorganization on the distribution of centrifugal axons is illustrated by the outline of an axon (c) terminating on spine G_1. (From Greer and Halasz, 1987.)

and establish in a dynamic equilibrium the reciprocal dendrodendritic synapses. Those spines that do not compete effectively are reabsorbed into the parent dendrite. An interesting outcome of this discussion is why denervated granule cell spines do not form heterologous synapses with each other. Hinds and Hinds (1976a) reported on the transient appearance of spine-to-spine synapses during early olfactory bulb development. However, analyses of the PCD mice did not provide any data suggesting similar synaptic appositions between spines in the adult. It may be that, as in the normal developing organism, synapse formation between spines is a transient phenomenon that is not sustained.

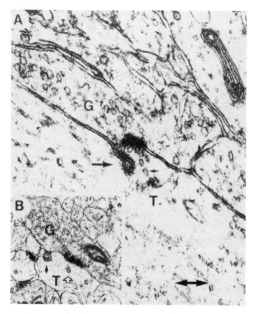

FIGURE 11 Electron micrographs illustrating typical invaginations of granule cell (G) gemmule membranes into tufted cell dendrites (T) in mutant PCD mice. (A) The large arrow denotes a clear double-walled invagination surrounded by a fuzzy clarthin coat. Vesicles (small arrows) associated with the immediately adjacent asymmetric synapse from the tufted cell dendrite to the granule cell gemmule are seen clustered along the invagination. (B) A second example of a double-walled invagination (solid arrow) is seen close to an apparent symmetrical synapse from the gemmule onto the tufted cell dendrite. Note the clusters of polyribosomes (open arrow) adjacent to the postsynaptic specialization in the tufted cell dendrite. Such clusters have previously been reported to occur adjacent to newly developed synaptic specializations. Calibration bar = (A) 100 nm and (B) 200 nm. (From Greer, 1987.)

D. Membrane Remodeling

Transient developmental events are well known in neurobiology. One of interest to development in the olfactory bulb, as well as elsewhere in the nervous system, is the conjugate internalization of apposed membranes. This event was initially recognized by Eckenhoff and Pysh (1983). Because it occurred late in the period of synaptogenesis they speculated that it was a process not intimately related to synaptogenesis but may more simply reflect the remodeling of immature membrane or the recycling of early adhesion sites. This is of interest in the case of the PCD mice because the conjugate exchange of membrane increases significantly, approximately 4.5-fold, following mitral cell loss and during the period of reactive dendrodendritic synaptogenesis (see Figure 11) (Greer, 1987b). Thus it appears that this event may be more directly linked with synaptogenesis than previously thought. Because these processes have frequently been observed in close proximity to areas of synaptic specialization, a tentative suggestion is that they may play a role in the stabilization of synapses, in the context discussed by Changeux and Danchin (1976), rather than the initiation of synaptogenesis.

VI. CONCLUSIONS

As a strategy for unraveling the organization of the olfactory system the instrumental genetic perspective has been productive. Murine neurologic mutants as well as inbred lines of mice provide several models with which we can address provocative questions about basic elements of organization and function within the olfactory system. The mutant reeler is well suited for investigating determinants on long-projection neuron topography and the mutant PCD is more appropriate for studying determinants of local circuit plasticity and organization. As discussed in this chapter, several other models are available that have not been extensively utilized although they provide powerful tools for understanding olfaction. For example, further study of Balb/c mice could provide new insights into the dynamic mechanisms that influence sorting of olfactory receptor cell axons during turnover. In a similar vein, studying the mutant PCD at the glomerular level could provide complementary information about target-axon interactions.

Collectively, these strategies are directed at identifying events that influence neuronal form, position, and connectivity. They do not reflect the notion that the olfactory system is hardwired or constructed on a one-gene, one-neuron basis. To the contrary, as pointed out by Stent (1981), strategies utilizing neurologic mutants are based in part on the belief that neural development is a historical or stochastic process. In the olfactory system this perspective will help us to understand the specificity with which local circuits form, the cellular and molecular basis of odor perception, and the signals influencing the formation of topographic maps within which odor domains are encrypted.

ACKNOWLEDGMENTS

The author wishes to thank Drs. Gordon Shepherd, William Stewart, Patricia Pedersen, Norbert Halasz, and John Kauer for valuable discussions. In addition the technical support of Christine Kaliszewski, Gwen Collins, and Dolores Montoya and the secretarial assistance of Janet Silvestri was invaluable. The author's work has been supported in part by NIH Awards DC00210, NS21563, and NS10174.

REFERENCES

Allen, W. K., and Akeson, R. (1985). Identification of an olfactory receptor neuron subclass:cellular and molecular analysis during development. *Dev. Biol.* **109**:393-401.

Apfelbach, R., and Weiler, F. (1985). Olfactory deprivation enhances normal spine loss in the olfactory bulb of developing ferrets. *Neurosci. Lett.* **62**:169-173.

Barber, P. C. (1981a). Regeneration of vomeronasal nerves into the main olfactory bulb in the mouse. *Brain Res.* **216**:239-251.

Barber, P. C. (1981b). Axonal growth by newly-formed vomeronasal neurosensory cells in the normal adult mouse. *Brain Res.* **216**:229-237.

Benson, T., Ryugo, D., and Hinds, J. (1984). Effects of sensory deprivation on the developing mouse olfactory system. *J. Neurosci.* **4**:638-653.

Caviness, V. L., and Rakic, P. (1978). Mechanisms of cortical development: A view from mutations in mice. *Annu. Rev. Neurosci.* **1**:297-326.

Caviness, V. L., and Sidman, R. L. (1972). Olfactory structures of the forebrain in the Reeler mutant mouse. *J. Comp. Neurol.* **145**:85-104.

Changeux, J. P., and Danchin, A. (1976). Selective stabilization of developing synapses as a mechanism for the specification of neuronal networks. *Nature* **264**:705-712.

Costanzo, R. M., and Graziadei, P. P. C. (1983). A quantitative analysis of changes in the olfactory epithelium following bulbectomy in hamster. *J. Comp. Neurol.* **215**:370-381.

Devor, M. (1976). Neuroplasticity in the rearrangement of the olfactory tract fibers after neonatal transection in hamsters. *J. Comp. Neurol.* **166**:49-72.

Devor, M., Caviness, V. S., Jr., and Derer, P. (1975). A normally laminated afferent projection to an abnormally laminated cortex: Some olfactory connections in the Reeler mouse. *J. Comp. Neurol.* **164**:471-482.

Eckenhoff, M. F., and Pysh, J. J. (1983). Conjugate internalization of apposed plasma membranes in mouse olfactory bulb during postnatal development. *Dev. Brain Res.* **6**:201-207.

Gozzo, S., and Fulop, Z. (1984). Transneuronal degeneration in different inbred strains of mice: A preliminary study of olfactory bulb events after olfactory nerve lesion. *Int. J. Neurosci.* **23**:187-194.

Graziadei, P. P. C., and Monti-Graziadei, G. A. (1978). the olfactory system: A model for the study of neurogenesis and axon regeneration in mammals. In *Neuronal Plasticity*, C. W. Cotman (Ed.). Raven, New York, pp. 131-153.

Graziadei, P. P. C., Levine, R. R., and Monti-Graziadei, G. A. (1978). Regeneration of olfactory axons and synapse formation in the forebrain after bulbectomy in neonatal mice. *Proc. Natl. Acad. Sci. USA* **75**:5230-5234.

Graziadei, P. P. C., Levine, R. R., and Monti-Graziadei, G. A. (1979). Plasticity of connections of the olfactory sensory neuron: Regeneration into the forebrain following bulbectomy in neonatal mouse. *Neurosciences* **4**:713-727.

Greer, C. (1984). A Golgi analysis of granule cell development in the neonatal rat olfactory bulb. *Soc. Neurosci. Abstr.* **10**:531.

Greer, C. A. (1987a). Golgi analyses of dendritic organization among denervated olfactory bulb granule cells. *J. Comp. Neurol.* **257**:442-452.

Greer, C. A. (1987b). Conjugate internalization of apposed dendritic membranes during synaptic reorganization in the olfactory bulbs of adult PCD mice. *Ann. N.Y. Acad. Sci.* **510**:318-320.

Greer, C. (1988). High voltage electronmicroscopic analyses of olfactory bulb granule cell spine geometry. *Assoc. Chemorecep. Sci. Abstr.* **10**:142.

Greer, C. A., and Halasz, N. (1987). Plasticity of dendrodendritic microcircuits following mitral cell loss in the olfactory bulb of the murine mutant Purkinje cell degeneration. *J. Comp. Neurol.* **256**:284-298.

Greer, C. A., and Shepherd, G. M. (1982). Mitral cell degeneration and sensory function in the neurological mutant mouse Purkinje cell degeneration (PCD). *Brain Res.* **235**:156-161.

Greer, C. A., Stewart, W. B., Kauer, J. S., and Shepherd, G. M. (1981). Topographical and laminar localization of 2-deoxyglucose uptake in the rat olfactory bulb induced by electrical stimulation of olfactory nerves. *Brain Res.* **217**:279-293.

Hamori, J., and Somogyi, J. (1982). Presynaptic dendrites and perikarya in deafferented cerebellar cortex. *Proc. Natl. Acad. Sci. USA* **79**:5093-5096.

Hempstead, J., and Morgan, J. (1985). A panel of monoclonal antibodies to the rat olfactory epithelium. *J. Neurosci.* **5**:438-3449.

Hinds, J. W., and Hinds, P. L. (1976a). Synapse formation in the mouse olfactory bulb: I. Quantitative studies. *J. Comp. Neurol.* **169**:15-40.

Hirata, Y. (1964). Some observations on the fine structure of synapses in the olfactory bulb of the mouse, with particular reference to the atypical synaptic configurations. *Arch. Histol. Jpn.* **24**:303-317.

Hudson, R., and Distel, H. (1987). Regional autonomy in the peripheral processing of odor signals in newborn rabbits. *Brain Res.* **421**(1-2):85-94.

Key, B., and Giorgi, P. P. (1986). Soybean aglutinin binding to the olfactory systems of the rat and mouse. *Neurosci. Lett.* **69**:131-136.

Landis, D. M. D., Reese, T. S., and Raviola, E. (1974). Differences in membrane structure between excitatory and inhibitory components of the reciprocal synapse in the olfactory bulb. *J. Comp. Neurol.* **155**:67-92.

Mikoshiba, K., Kohsaka, S., Takamatsu, K., Aoki, E., and Tsukada, Y. (1980). Morphological and biochemical studies on the cerebral cortex from Reeler mutant mice: Development of cortical layers and metabolic mapping by the deoxyglucose method. *J. Neurochem.* **44**:835-844.

Mori, K. (1987). Membrane and synaptic properties of identified neurons in the olfactory bulb. *Prog. Neurobiol.* **29**:275-320.

Mori, K., Kishi, K., and Ojima, H. (1983). Distribution of dendrites of mitral, displaced mitral, tufted and granule cells in the rabbit olfactory bulb. *J. Comp. Neurol.* **219**:339-355.

Mori, K., Fujita, S. C., Imamura, K., and Obata, K. (1985). Immunohistochemical study of subclasses of olfactory nerve fibers and their projection to the olfactory bulb in the rabbit. *J. Comp. Neurol.* **242**:214-229.

Mouradian, L. E., and Scott, J. W. (1988). Cytochrome oxidase staining marks dendritic zones of the rat olfactory bulb external plexiform layer. *J. Comp. Neurol.* **271**:507-518.

Mullen, R., and Herrup, K. (1979). Chimeric analyses of mouse cerebellar mutants. In *Neurogenetics: Genetic Approaches to the Nervous System*, X. Breakfield (Ed.). Elsevier/North Holland, New York, pp. 173-196.

Mullen, R. J., Eicher, E. M., and Sidman, R. L. (1976). Purkinje cell degeneration, a new neurological mutation in the mouse. *Proc. Natl. Acad. Sci. USA* **73**:208-212.

O'Gorman, S., and Sidman, R. (1985a). Degeneration of thalamic neurons in PCD mice. I. Distribution of neuron loss. *J. Comp. Neurol.* **234**:277-297.

O'Gorman, S., and Sidman, R. (1985b). Degeneration of thalamic neurons in PCD mice. II. Cytology of neuron loss. *J. Comp. Neurol.* **234**:298-316.

Onoda, N. (1988a). Monoclonal antibody immunohistochemistry of rabbit olfactory receptor neurons during development. *Neurosciences* **26**:1003-1012.

Onoda, N. (1988b). Monoclonal antibody immunohistochemistry of degenerative and renewal patterns in rabbit olfactory receptor neurons following unilateral olfactory bulbectomy. *Neurosciences* **26**:1013-1022.

Onoda, N., and Fujita, S. C. (1988). Monoclonal antibody immunohistochemistry of adult rabbit olfactory structures. *Neurosciences* **26**:993-1002.

Orona, E., Scott, J., and Rainer, E. (1983). Different granule cell populations innervate superficial and deep regions of the external plexiform layer in rat olfactory bulb. *J. Comp. Neurol.* **217**:227-237.

Orona, E., Rainer, E., and Scott, J. (1984). Dendritic and axonal organization of mitral and tufted cells in the rat olfactory bulb. *J. Comp. Neurol.* **226**:346-356.

Pedersen, P. E., and Blass, E. M. (1982). Prenatal and postnatal determinants of the first suckling episode in the albino rat. *Dev. Psychobiol.* **15**:349-356.

Pourtier, L., and Sicard, G. (1988). Olfactory sensitivity to isovaleric acid in C57BL/6 mice: A behavioral study. *Int. Eur. Chemorecep. Cong. Abstr.* **8**:119.

Price, J. L. (1987). The central olfactory and accessory olfactory systems. In *Neurobiology of Taste and Smell*, T. E. Finger and W. L. Silver (Eds.). Wiley-Interscience, New York, pp. 179-203.

Price, J. L., and Powell, T. P. S. (1970a). The morphology of the granule cells of the olfactory bulb. *J. Cell Sci.* **7**:91-123.

Price, J. L., and Powell, T. P. S. (1970b). The synaptology of the granule cells of the olfactory bulb. *J. Cell Sci.* **7**:125-155.

Price, J. L., and Powell, T. P. S. (1970c). The mitral and short axon cells of the olfactory bulb. *J. Cell Sci.* **7**:631-651.

Price, S. (1977). Specific anosmia to geraniol in mice. *Neurosci. Lett.* **4**:49-50.

Raisman, G. (1985). Synapse formation in the septal nuclei of adult rats. In *Synaptic Plasticity*, C. Cotman (Ed.). Guilford Press, New York, pp. 13-38.

Rall, W., Shepherd, G. M., Reese, T. S., and Brightman, M. W. (1966). Dendrodendritic synaptic pathway for inhibition in the olfactory bulb. *Exp. Neurol.* **14**: 44-56.

Reese, T. S., and Brightman, M. W. (1965). Electron microscopic studies on the rat olfactory bulb. *Anat. Rec.* **151**:492.

Schwob, J. E., and Gottlieb, D. I. (1988). Purification and characterization of an antigen that is spatially segregated in the primary olfactory projection. *J. Neurosci.* **8**(9):3470-3480.

Shepherd, G. M., and Greer, C. A. (1988). The dendritic spine: Adaptations of structure and function for different types of synaptic integration. In *Intrinsic Determinants of Neuronal Form and Function*, R. Lasek (Ed.). Alan R. Liss, New York, pp. 245-262.

Sicard, G., and Pourtier, L. (1988). Specific anosmia in the laboratory mouse: Inferences from various experimental approaches. *Int. Eur. Chemorecep. Cong. Abstr.* **8**:129.

Sicard, G., Royet, J.-P., and Jourdan, F. (1988). A comparative study of 2-deoxyglucose patterns of glomerular activation in the olfactory bulbs of C57 B1/6J and AKR/J mice. *Brain Res.* In Press.

Slotnick, B. M., Graham, S., Laing, D. G., and Bell, G. A. (1987). Detection of propionic acid vapor by rats with lesions of olfactory bulb areas associated with high 2-DG uptake. *Brain Res.* **417**:343-346.

Somogyi, J., Hamori, J., and Silakov, V. (1984). Synaptic reorganization in the lateral geniculate nucleus of the adult cat following chronic decortication. *Exp. Brain Res.* **54**:485-498.

Stent, G. (1981). Strength and weakness of the genetic approach to the development of the nervous system. *Annu. Rev. Neurosci.* **4**:163-194.

White, E. L. (1972). Synaptic organization in the olfactory clomerulus of the mouse. *Brain Res.* **37**:69-80.

White, E. L. (1973). Synaptic organization of the mammalian olfactory glomerulus: New findings including an intraspecific variation. *Brain Res.* **60**:299-313.

Wysocki, C. J., Whitney, G., and Tucker, D. (1977). Specific anosmia in the laboratory mouse. *Behav. Genet.* **7**:171-188.

Wyss, J. M., Stanfield, B. B., and Cowan, W. M. (1980). Structural abnormalities in the olfactory bulb of the Reeler mouse. *Brain Res.* **188**:566-571.

Part V
Human Analyses

22
A Clinical Manifestation and Its Implications

H. N. Wright

*SUNY Health Science Center at Syracuse,
Syracuse, New York*

A clinical manifestation of a disturbance in olfactory ability is frequently found in patients suffering from pseudohypoparathyroidism. This disease frequently has a genetic basis. Patients with type Ia pseudohypoparathyroidism have been described who have an abnormality in the expression of the gene for the α subunit of the guanine nucleotide binding stimulatory (Gs) protein (Carter et al., 1987), and evidence has been provided in studies using frog neuroepithelium that Gs activity is involved in olfaction (Pace et al., 1985). Type Ib pseudohypoparathyroid patients have normal Gs activity. This disease can also be familial and in some families appears to follow an autosomal mode of inheritance. The genetic defects responsible in this instance are unclear but are likely to be heterogeneous.

What I report here is a more detailed analysis of our results obtained earlier from a group of female pseudohypoparathyroid patients, those who were Gs unit normal and Gs unit deficient, in comparison to normal controls. As reported in *Nature* (Weinstock et al., 1986) and somewhat more extensively in the *Annals of the New York Academy of Sciences* (Wright et al., 1987), the guanine nucleotide binding stimulatory protein (Gs unit)-deficient patients presented with a profound deficit in odorant identification, in contrast to pseudohypoparathyroid patients who were not Gs unit deficient and normal controls. Not previously considered, however, was the pattern of error responses for the patient groups in comparison to one another and to normal controls. The results from this type of analysis with multidimensional scaling techniques as a measure of the evidence for similarities, or differences, in the quality coding of odorants directly assesses disturbances in odorant perception that could be related to genetic, hormonal, metabolic, or anatomic abnormalities. It is demonstrated here that even though the pseudohypoparathyroid patients who are Gs unit deficient have a profound deficit in odorant identification, they do not distinguish themselves from pseudohypoparathyroid patients who are not Gs unit deficient in their quality coding of odorants. Further, the quality coding of odorants for all pseudohypoparathyroid patients was found to be different from their normal controls.

A comparison between normal male and female controls indicated they were different from one another with respect to their quality coding for the odorants of the OCM. Since our original complement of pseudohypoparathyroid patients were predominantly female (Weinstock et al., 1986; Wright et al., 1987), our analysis consists of female participants only. Otherwise, differences among the subject groups could be obscured by the differences in quality coding between normal males and females. An additional female patient was added to our pseudohypoparathyroid group, bringing the complement of pseudohypoparathyroid patients to six who were Gs unit deficient and six who were Gs unit normal. Table 1 summarizes their pretreatment characteristics. The details of the laboratory procedures followed have been discussed previously (Weinstock et al., 1986). All the pseudohypoparathyroid patients presented with hypocalcemia and elevated parathyroid hormone levels. In addition, the Gs unit-deficient patients had features of Albright's osteodystrophy (round face, short stature, obesity, short neck, shortened metacarpal and metatarsal bones, and subcutaneous calcifications) and evidence of other hormonal abnormalities, including hypothyroidism and hypogonadism. Table 2 shows their serum values at the time when we measured their sense of smell with the odorant confusion matrix (OCM) (Wright, 1987) and the smell identification test (SIT) (Doty et al., 1984), the results of which are also included. All patients were receiving appropriate medical treatment and monitoring at the time of testing.

Table 3 shows the 10 odorants we now use at our Olfactory Referral Center at the SUNY Health Science Center at Syracuse. Their respective concentrations are such that they represent the perceived counterpart of each odor in nature, as judged by a panel of 30 observers with no complaints of olfactory dysfunction. Propylene glycol is the odorless diluent used to achieve the molarities shown. These 10 odorants are presented successively to the observers with the sniff-bottle technique randomly in blocks of 10, in which each block includes all odorants, with the constraint that when moving from

TABLE 1 Pretreatment Characteristics of Female Patients with Pseudohypoparathyroidism

Patient number[a]	Age (years)	Untreated serum calcium (mg/dl; low limit of normal = 8.4)	Untreated serum phosphorus (mg/dl; upper limit of normal = 4.8)	Peak urinary cAMP response to PTH (nmol/mg Cr; normal 90-315)	Gs activity (% of normal; normal > 75%)
Gs Deficient					
1	37	7.2[b]	4.8[b]	—[c]	55
2	26	8.0	4.9	8.80	66
3	30	8.2	4.6	—[c]	48
4	36	6.6	4.3	4.25	51
5	20	6.0	7.1	7.0	51
*	12	5.7	8.9	5.74	68
Gs Normal					
6	60	5.5	3.3	4.16	87
7	47	6.3	—[c]	15.83	85
8	30	6.4	4.7	5.07	78
9	31	6.8	5.6	4.08	—[c,e]
10	38	7.2[b]	4.3[b]	—[d]	—[c,f]
11	38	6.8	4.4	—[c]	78

[a]Patient numbers refer to those reported previously (Weinstock et al., 1986). The asterisk indicates new female patient.
[b]Values obtained when the patient was receiving treatment with calcium and vitamin D.
[c]Not determined.
[d]Plasma cAMP = 25 pmol/month (normal > 50).
[e]Normal phenotype, a sister of patient 8.
[f]Normal phenotype.

one block of 10 to the next, no one odorant follows itself. The observers are given an alphabetical list of the odors to be presented. Their responses are limited to these 10 odorants, so that even if they are unsure or smell nothing at all, they must make their best guess. A total of 11 blocks are presented, and the first discarded in the data analysis. This results in a 10 × 10 confusion matrix with 100 stimulus-response pairs, in which the rows represent the odorants presented and the columns represent the responses. The cell entires then represent the proportions, or percentage, of responses to each of the odorants. The entire procedure takes about 30 minutes to complete.

Figure 1 shows the composite OCM obtained from 25 normal female controls. The percentage correct is calculated from the main diagonal, whereas the measure of the evidence for quality coding is obtained from those responses

TABLE 2 Serum Values and Percentage Correct Olfactory Ability Test Scores

Patient number[a]	Calcium (mg/dl; low limit of normal = 8.4)	Phosphorus (mg/dl; upper limit of normal = 4.8)	T_4 (μg/dl; lower limit of normal = 5)	TSH (μl/ml; upper limit of normal = 6)	OCM	SIT
Gs deficient						
1	9.0	3.6	4.5	1.6	38	60.0
2	9.0	4.0	7.4	7.8	15	32.5
3	8.8	3.4	7.7	3.8	38	75.0
4	9.8	4.3	8.6	3.3	61	85.0
5	9.6	4.0	9.3	<0.1	20	75.0
*	8.8	5.4	6.6	9.8	61	75.0
Gs normal						
6	8.4	4.3	9.2	3.4	93	95.0
7	9.2	3.7	5.1	5.2	93	90.0
8	8.0	4.0	5.8	3.5	92	90.0
9	9.2	3.6	6.4	5.3	84	85.0
10	7.2	4.3	7.1	13.1[b]	87	92.5
11	9.4	3.9	5.2	1.8	92	92.5

[a]Patient numbers refer to those reported previously (Weinstock et al., 1986). The asterisk indicates new female patient.
[b]Antithyroid antibodies present.

TABLE 3 Odorant Confusion Matrix (OCM) Stimuli

Odor	Odorant	Molarity
Ammonia	Ammonia	3.6015
Cinnamon	*Trans*-cinnamaldehyde	0.0622
Licrocie	Anethole	0.0130
Mint	L-carvone	0.3990
Mothballs	Napthalene	0.0244
Orange	D-limonine	0.0121
Rose	Phenethyl alcohol	1.0467
Rubbing alcohol	Isopropyl alcohol	0.7425
Vanilla	Vanillin	0.0273
Vinegar	Acetic acid	4.3714

CLINICAL MANIFESTATION AND ITS IMPLICATIONS

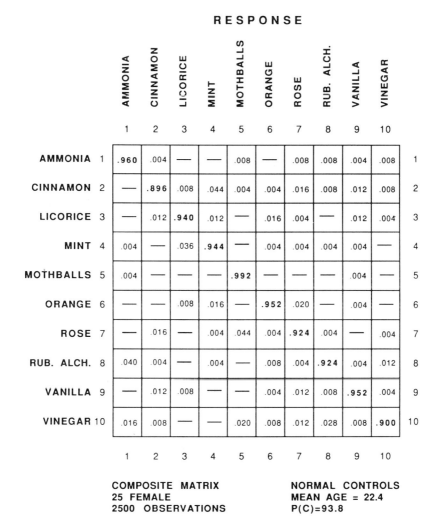

FIGURE 1 Composite OCM obtained from 25 normal female controls with no complaint of olfactory dysfunction.

not on the main diagonal. Interpretation of off-diagonal responses is perhaps best facilitated by examining the results obtained from patients with presumptive absence of input from the olfactory nerve (cranial nerve I). An example of such a patient is shown in Figure 2, who suffered from head trauma. The most plausible explanation for decreased olfactory ability seen in such patients is a shearing of the olfactory nerve from the olfactory bulb at the

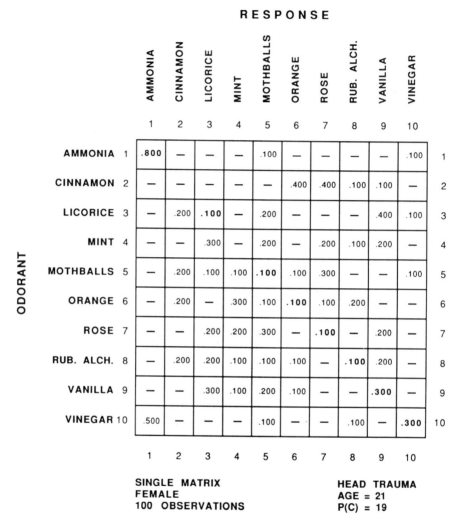

FIGURE 2 OCM obtained from young female patient suffering from head trauma.

level of the cribriform plate. The mechanism for such a shearing action is thought to be the relative movement of the brain structures in contrast to the fixed attachments of the twigs of the olfactory nerve as they pass through the cribriform plate. In the interpretation of these findings we should recognize that the odors of ammonia, mothballs, rubbing alcohol, and vinegar have strong trigeminal (cranial nerve V) components. We have consistently observed that many of these odorants are confused with one another in such patients with presumptive lack of olfactory input when evaluating their sense of smell with the OCM. For this female patient, note that when presented with ammonia, the parosmic substitutions were mothballs and vinegar. Similarly, when presented with vinegar, the error responses were ammonia, mothballs, and rubbing alcohol. Such off-diagonal responses are informative in that they tell us that the trigeminal nerve is functionally intact, but the olfactory nerve is not. Incidentally, and equally important, they further tell us that the correct identification of odorants with strong trigeminal components also requires input from the olfactory nerve.

Further significance of the off-diagonal response patterns is shown in Figure 3 for a female patient suffering from Kallmann syndrome, characterized by agenesis of the olfactory bulb. In this instance, the responses to ammonia are not identical to those found for the previous trauma patient but were nevertheless consistent with our expectations of parosmic substitution among the odorants with strong trigeminal components. The remaining off-diagonal response patterns are more difficult to interpret. It should be recognized, however, that such patients never experienced olfactory input, and in this regard it is somewhat remarkable that the response patterns indeed exhibit some similarity to those patients who previously had the benefit of such input.

The foregoing qualitative analyses assist in the interpretation of findings for the individual patient. More complete quantitative analyses, however, are available to describe differences among patient groups. For this purpose, we turn to multidimensional scaling (MDS) techniques. Such an analysis is based on the premise that odorant identification is a multidimensional process. That is, odorants are perceived to consist of many psychophysical attributes, the most prominent of which is hedonics, and a unique combination of these psychophysical attributes (which are as yet undefined) yields the identification of a particular odorant. Going one step further, it then follows that those odorants close to one another in a multidimensional space will be confused with one another, and those that are distant from one another will not be as readily confused, if at all. Multidimensional scaling offers the opportunity to examine the qualitative differences among the confusions for all odorants with one another simultaneously. More exactly, it describes the relation among all off-diagonal responses, both within and among subject groups.

The composite results for the Gs unit-normal and Gs unit-deficient pseudohypoparathyroid patients are shown in Figures 4 and 5. As reported earlier, there is a vast difference between these patient groups in their ability to correctly identify the odorants. The difference in their mean percentage correct, as calculated from the main diagonal, is 63.5%. The individual results,

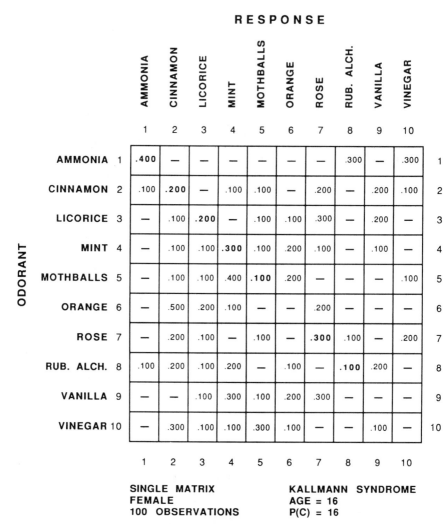

FIGURE 3 OCM obtained from young female patient suffering from Kallmann syndrome.

displayed in Table 2, show no overlap between the subject groups with the OCM. The results from the SIT are similar, but less profound. Such differences between these patient groups are interpreted to indicate differences in the absolute transduction of olfactory information. They do not, however, reveal possible differences in quality coding. For this purpose, we must turn to the descriptive analyses provided by multidimensional scaling.

CLINICAL MANIFESTATION AND ITS IMPLICATIONS

RESPONSE

ODORANT		AMMONIA 1	CINNAMON 2	LICORICE 3	MINT 4	MOTHBALLS 5	ORANGE 6	ROSE 7	RUB. ALCH. 8	VANILLA 9	VINEGAR 10	
AMMONIA	1	.933	—	—	—	—	—	—	.067	—	—	1
CINNAMON	2	—	.917	.017	.033	—	—	.017	—	.017	—	2
LICORICE	3	—	—	.983	—	—	—	.017	—	—	—	3
MINT	4	—	.050	.033	.000	.033	.017	—	—	—	.067	4
MOTHBALLS	5	—	—	—	—	1.00	—	—	—	—	—	5
ORANGE	6	—	—	.017	.117	.100	.733	—	—	.017	.017	6
ROSE	7	—	—	—	.017	.050	—	.950	—	—	—	7
RUB. ALCH.	8	.033	.017	—	—	.017	—	—	.900	.017	.017	8
VANILLA	9	—	.050	.017	.017	—	—	—	—	.917	—	9
VINEGAR	10	—	—	—	—	—	.050	—	.050	—	.900	10

COMPOSITE MATRIX
6 FEMALE
600 OBSERVATIONS

Gs UNIT NORMAL
MEAN AGE = 40.7
P(C) = 90.3

FIGURE 4 Composite OCM for six Gs unit-normal pseudohypoparathyroid patients.

RESPONSE

		AMMONIA	CINNAMON	LICORICE	MINT	MOTHBALLS	ORANGE	ROSE	RUB. ALCH.	VANILLA	VINEGAR	
		1	2	3	4	5	6	7	8	9	10	
AMMONIA	1	**.800**	—	—	—	.017	.017	—	.117	.017	.033	1
CINNAMON	2	.033	**.317**	.133	.100	.017	.150	.100	.033	.083	.033	2
LICORICE	3	—	.183	**.483**	.117	.033	.017	—	.067	.083	.017	3
MINT	4	.017	.167	.200	**.217**	.033	.117	.050	.083	.033	.083	4
MOTHBALLS	5	.083	.017	.017	.033	**.433**	.050	.050	.167	.050	.100	5
ORANGE	6	.050	.117	.133	.100	.033	**.300**	.133	.033	.083	.017	6
ROSE	7	.083	.067	.033	.133	.183	.017	**.167**	.067	.083	.150	7
RUB. ALCH.	8	.083	.050	.017	.050	.033	.017	.033	**.383**	.083	.250	8
VANILLA	9	—	.200	.033	.033	.017	.083	.117	.017	**.450**	.050	9
VINEGAR	10	.100	.133	.083	.067	.017	.067	.017	.050	.133	**.333**	10
		1	2	3	4	5	6	7	8	9	10	

COMPOSITE MATRIX
6 FEMALE
600 OBSERVATIONS

Gs UNIT DEFICIENT
MEAN AGE = 26.8
P(C) = 38.8

FIGURE 5 Composite OCM for six Gs unit-deficient pseudohypoparathyroid patients.

The composite matrices for the normal controls (Figure 1) and the Gs unit-normal (Figure 4) and Gs unit-deficient (Figure 5) pseudohypoparathyroid patient groups were subjected to a weighted (individual differences) multidimensional scaling analysis (ALSCAL) (Young, 1987). The three-dimensional group stimulus configuration for the odorant space obtained from this analysis is shown in Figure 6.

CLINICAL MANIFESTATION AND ITS IMPLICATIONS 327

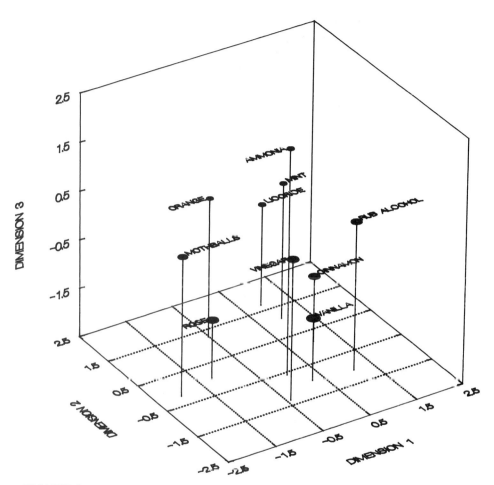

FIGURE 6 Three-dimensional, three-group (normal controls, Gs unit normal, Gs unit deficient), stimulus configuration for the odorant space of the OCM.

Qualitative inspection of the matrices for normal controls and the Gs unit-normal pseudohypoparathyroid groups (Figures 1 and 4) might lead to the suspicion that their patterns of off-diagonal responses were similar to one another, particularly when we consider that the normal controls showed only a modest superiority (3.5%) over the Gs unit-normal patients in their correct identification for the odorants of the OCM. Likewise, inspection of the matrices for the two pseudohypoparathyroid groups (Figures 4 and 5) might lead to the suspicion that their patterns of off-diagonal responses were different from one another, especially when we consider that the Gs unit-normal pseudohypoparathyroid patients showed a marked superiority (63.5%) over the Gs unit-deficient patients in their percentage of correct identifications for the odorants of the OCM. It should be made abundantly clear, however, that the percentage correct as calculated from the main diagonal is completely dissociated from the possible patterns of off-diagonal error responses.

To examine the process of quality coding, the three-dimensional group stimulus space shown in Figure 6 was decomposed into the relative salience for each of the dimensions as perceived by each subject group. More exactly, each subject will stretch, or compress, each of the dimensions in such a way that the relative perceptual values attributed to each of the dimensions by each subject group may be described, and the extent to which they do this is revealed by their group weights (or vectors) in the multidimensional space.

The relative weights ascribed by each of the three subject groups to the three-dimensional stimulus (odorant) space is shown in Figure 7. Here we see that the pseudohypoparathyroid groups do not distinguish themselves from one another with respect to the pattern of their off-diagonal error responses on the OCM and are substantially different from their normal controls. Such results are interpreted to indicate that pseudohypoparathyroid patients, whether Gs unit normal or Gs unit deficient, differ from normal controls in their quality coding of odorants. Further, there appears to be no difference in the quality coding of odorants by pseudohypoparathyroid patients, whether they are Gs unit normal or Gs unit deficient. In the search of an explanation for such findings, we noted that a common feature of the pseudohypoparathyroid patients was that at the time of diagnosis they were hypocalcemic and hyperphosphatemic and presented with elevated parathyroid hormone levels (Table 1). Further, all patients were treated with calcium and vitamin D, and the Gs unit-deficient patients also received thyroid hormone, which has been reported to affect olfactory ability (Henkin, 1968). Any one of these factors, separately or in combination with one another (either pre- or posttreatment), could be associated with olfactory dysfunction (Feldman et al., 1986). The effect of serum parathyroid hormone levels, which were not determined at the time of testing (Table 2), has yet to be determined.

With respect to the difference between the pseudohypoparathyroid groups, our previous conclusion remains unchanged. To explain, even though the quality coding for all pseudohypoparathyroid patients appears different from their normal controls, the Gs unit-deficient patients are shown to be impaired in the transduction of olfactory information in contrast to their Gs unit-normal counterparts. Two separate mechanisms would seem to be involved: first, the absolute amount of information transduced; second, a qualitative distortion of that information, once transduced. It is possible that at least one of these mechanisms could have a genetic basis. It becomes

CLINICAL MANIFESTATION AND ITS IMPLICATIONS

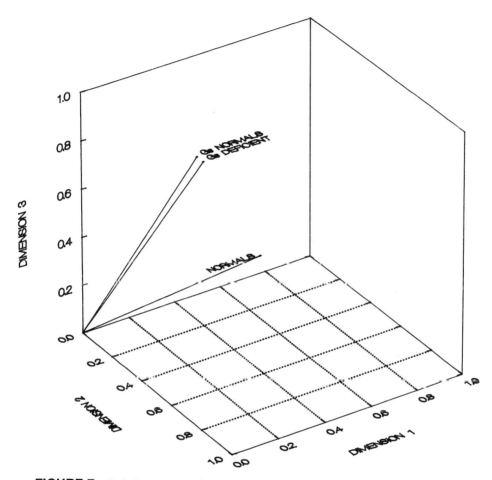

FIGURE 7 Relative group weights (vectors) ascribed by each of the three subject groups (normal controls, Gs unit normal, and Gs unit deficient) to the three-dimensional stimulus (odorant) space of the OCM.

increasingly apparent, therefore, that in the search for those genetic factors that may influence olfactory ability, the impact of consequent endocrine and metabolic abnormalities cannot be ignored.

ACKNOWLEDGMENTS

The author is indebted to Drs. Ruth S. Weinstock for participation in the preparatin of this manuscript and Douglas U. Smith, who performed the multidimensional scaling analyses. This research was supported by grants to the State University of New York, Health Science Center at Syracuse and a Program Project Grant to the Olfactory Referral Center (NS19568) and grants to the Clinical Research Center (RR-229 and RR-35) from the General Clinical Research Center Program of the Division of Research Resources.

REFERENCES

Carter, A., Bardin, C., Collins, R., Simons, C., Bray, P., and Spiegel, A. (1987). Reduced expression of multiple forms of the α subunit of the stimulatory GTP-binding protein in pseudohypoparathyroidism type 1a. *Proc. Nat. Acad. Sci. USA* **84**:7266-7269.

Doty, R., Shaman, P., and Dann, M. (1984). Development of the University of Pennsylvania smell identification test. *Physiol. Behav.* **32**:489-502.

Feldman, J., Wright, H., and Leopold, D. (1986). The initial evaluation of dysosmia. *Am. J. Otolaryngol.* **4**:431-444.

Henkin, R. (1968). Impairment of olfaction and the tastes of sour and bitter in pseudohypoparathyroidism. *J. Clin. Endocrinol. Metab.* **28**:624-628.

Pace, U. E., Hanski, E., Salomon, Y., and Lancet, D. (1985). Odorant-sensitive adenylate cyclase may mediate olfactory reception. *Nature* **316**:255-258.

Weinstock, R. S., Wright, H. N., Speigel, A. M., Levine, M. A., and Moses, A. M. (1986). Olfactory dysfunction in humans with deficient guanine nucleotide-binding protein. *Nature* **322**:635-636.

Wright, H. (1987). Characterization of olfactory dysfunction. *Arch. Otolaryngol. Head Neck Surg.* **113**:163-168.

Wright, H. N., Weinstock, R. S., Spiegel, A. M., and Moses, A. M. (1987). Guanine nucleotide-binding stimulatory protein: A requisite for human odorant perception. *Ann. N.Y. Acad. Sci.* **510**:719-722.

Young, F. W. (1987). In *Multidimensional Scaling: History, Theory, and Applications*, R. M. Hammer (Ed.). Lawrence Erlbaum Associates, New Jersey.

23
A Graphic History of Specific Anosmia

John E. Amoore and Shelton Steinle

Olfacto-Labs, El Cerrito, California

I. INTRODUCTION

Specific anosmia was a term introduced by Amoore (1967) to describe the long recognized but rather neglected phenomenon of "smell blindness" (Blakeslee, 1918). In this condition, a person of otherwise normal smell sensitivity encounters one or more substances that appear odorless or very weak to him or her but that are obviously odorous to the majority of other people. Qualitative summaries of the literature on 76 chemical examples of specific anosmia have been published (Amoore, 1969, 1975). The intention of this chapter is to illustrate graphically the quantitative data that have accumulated in the last 20 years.

The very existence of specific anosmia implies that the olfactory system operates on the basis of selective chemical sensors. Just how selective these sensors can be is illustrated in Figure 1 by four pairs of isomers that differ substantially in their thresholds of detection by the human sense of smell. The chemical differences between the isomers are sufficiently subtle, culminating in the optical enantiomers of carvone, that it is virtually impossible to explain the data except by reference to selective proteinaceous receptor sites. Conversely, it can happen that totally unrelated chemical molecules may exhibit very similar odors (Figure 2) if they have the same external molecular size and shape.

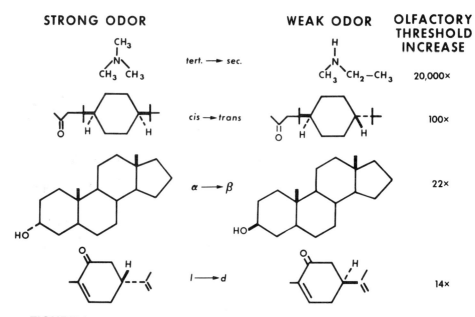

FIGURE 1 Four pairs of isomers that differ substantially in odor threshold (Amoore and Forrester, 1976; Amoore et al., 1977; Pelosi and Viti, 1978). From the top they are trimethylamine and methyl ethyl amine; 4(4'-*cis*- and 4(4'-*trans-tert*-butylcyclohexyl)-4-methyl-pentan-2-one; androstan-3α-ol and 3β-ol; and l- and d-carvone.

Such evidence implies very strongly that receptor proteins are responsible for the initial stage of olfactory detection. A reasonable (but not the only) working hypothesis for the causation of specific anosmia would be an absent or abnormal receptor protein. This would appear to be a fruitful field for classic genetic analysis.

II. METHODS

The procedures for measuring human olfactory thresholds by means of serial dilutions of odorants in aqueous solutions in glass flasks have been described (Amoore et al., 1968). Longer lasting solutions in mineral oil can be presented in plastic squeeze bottles (Amoore and Ollman, 1983). Recently our firm has developed compact, nonspill kits that further improve the practicality of smell testing.

Much of the work on specific anosmia has been reported using concentration scales of binary steps; that is, each consecutive step has double the chemical concentration of the one before. This works well when considering

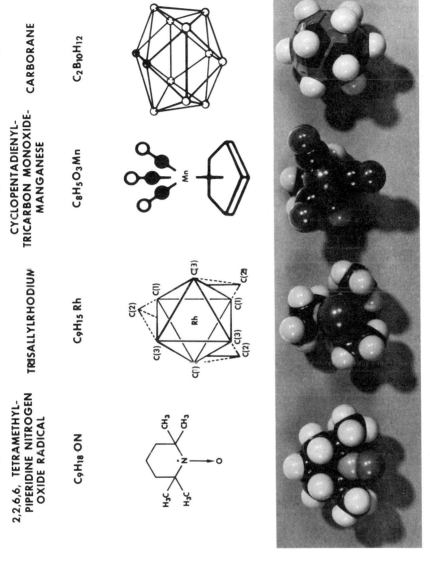

FIGURE 2 Four highly diverse compounds that smell like camphor. (a) Rozantzev and Neiman (1964); (b) Powell and Shaw (1966); (c) Piper et al. (1955); (d) Fein et al. (1963).

only one chemical and one dilution medium. Nevertheless, if we wish to compare the performances of different subjects with different odorants that may have been dispersed in air, water, mineral oil, or other solvents, a more generalized scale is desirable. Amoore and O'Neill (1988) have introduced the "decismel scale," modeled closely on the well-known decibel scale of audiometry. The decismel scale is anchored to the average olfactory detection threshold for healthy 20-year-olds. This is the reference concentration, which leads to the following definition:

TABLE 1 Data Sources for Odor Threshold Distributions in Figures 3-6[a]

Odorant	Persons tested	Percentage anosmics[b]	Odor type	Reference
Trichloroethene	24		Solvent	Olfacto[c]
1,2-Dichloropropane	25		Solvent	Olfacto
Isobutyl isobutyrate	443		Fruity	Amoore, 1968
Naphthalene	25		Mothballs	Olfacto
Pyridine	100		Scallops	Olfacto
4-Ethylphenol	25		Smoky	Olfacto
Phenethyl methyl ethyl carbinol	100	?	Flowery	Olfacto
Tetrahydrothiophene	25	~4	Gassy	Olfacto
Established primary odorants				
Isovaleric acid	443	3	Sweaty	Amoore, 1968
Trimethylamine	177[d]	6	Fishy	Amoore and Forrester, 1976
l-Carvone	87	8	Minty	Pelosi and Viti, 1978
Pentadecalactone	90	12	Musky	Whissell-Buechy and Amoore, 1973
1-Pyrroline[e]	144[d]	16	Spermous	Amoore et al., 1975
1,8-Cineole	85	33	Camphor	Pelosi and Pisanelli, 1981
Isobutyraldehyde	222	36	Malty	Amoore et al., 1976
16-Androsten-3-one[f]	39[d]	47	Urinous	Amoore et al., 1977

[a]Test methods: the first eight odorants (except isobutyl isobutyrate) were tested in mineral oil solution in 2 ounce squeeze-bottles, and the remainder in buffered aqueous solution in 125 ml Erlenmeyer flasks.
[b]Taken from largest available data sets, for example, Table 4.
[c]Olfacto-Labs kits, unpublished calibration data.
[d]Reconstituted distribution, from screening and panel tests.
[e]Data from impure putrescine; its odor is due to 1-pyrroline.
[f]Previously unpublished distribution.

$$\text{Odor level in decismels} = 20 \times \log \frac{\text{test concentration}}{\text{reference concentration}}$$

Individual odor thresholds expressed on this scale become relatively independent of the concentration units, the dilution medium, and even the experimental method for threshold measurement. For comparison with existing scales, quarter log concentration intervals are 5 decismels, and binary dilution steps are approximately 6 decismels. The ratio test concentration to reference concentration is the same as odor units, or olfacties in the older literature.

Olfactory thresholds obtained with our firm's newer test kits are based on the decismel scale. For purposes of comparison in this chapter (Figures 3 through 6), the older data in binary steps have been converted with a fair degree of approximation to the decismel format. The published threshold distributions (Table 1) were obtained from population samples ranging from 24 to 443 persons. These were all recalculated as percentages and displayed on the same vertical scale. When necessary, the data were adjusted to reflect the actual frequencies in the population, not just the sizes of the panels of specific anosmics and normal observers chosen for the experimental work.

For converting binary steps to decismels the factor is 6.021, which was rounded off to 6.0. Since the publications did not report the ages of the subjects, we assumed that their average age was near 40 years and made an arbitrary shift of one binary step in the reported mean to obtain a reference point for the zero of the decismel scale. This is based on the generalization that there is about one binary step increase of threshold concentration for each 22 years of age (Venstrom and Amoore, 1968). The resulting distributions were rounded off to the nearest multiple of 6 decismels. The graphs were constructed from our input data by an IBM AT computer using the Quattro software. Although some of the charts have 5 dS (decismel) increments on the x axis and the remainder have 6 dS increments, the overall scales are virtually identical.

III. RESULTS

The first group of odor threshold distributions are shown in Figure 3. They were selected because they approximate reasonably closely to a normal distribution, that is, a symmetrical bell-shaped curve with no outliers. Ordinary Gaussian statistics may be applied, which yield for each curve a mean value and a standard deviation. The mean is usually at about 5 or 6 dS above the 0 dS reference points for 20-year-olds, because the groups of subjects tested are working-age people, 20-65 years of age and averaging 40-45 years, hence slightly less sensitive to odors. The standard deviations are near ± 12 dS

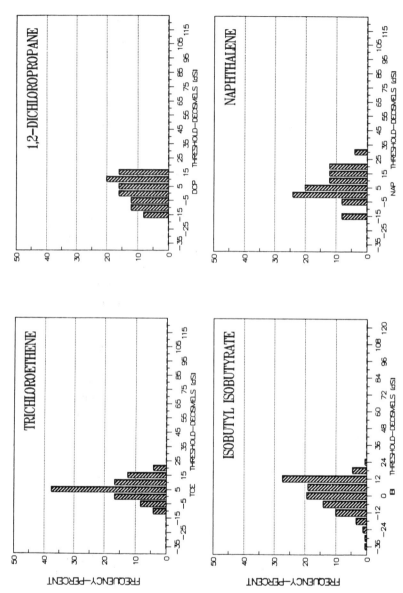

FIGURE 3 Distributions of olfactory thresholds for four odorants that show little or no signs of hyposmia. These are recalculated and adjusted for a standard sample size of 100 subjects. Odor thresholds are expressed on the decismel scale. For references see Table 1.

(±2 binary steps). Strictly speaking, since the x axis is a discontinuous variable (decismel increments of 5 or 6 dS) these are Poisson, not Gaussian, distributions.

It is not clear exactly why these compounds exhibited no outliers, previously defined as beyond ±2 SD (Amoore et al., 1968) or ±24 dS. Reasonable postulates are that the corresponding receptor protein is very rarely defective or that there is no well-fitting receptor site for this compound, which accordingly stimulates several different sites when it reaches a high enough concentration. Isobutyl isobutyrate was selected by Amoore (1967) as a standard odorant for demonstrating "normality" (i.e., lack of general anosmia) in candidate subjects for systematic smell testing, and its fruity odor has been used more or less traditionally for investigating all eight types of specific anosmia that have been established so far (lower half of Table 1). Some other odorant, or mixture of odorants, may prove superior for this purpose.

The next group of four odorants (Figure 4) are a little more interesting. Pyridine has been used for many years (e.g., Sherman et al., 1979) as a standard for clinical smell testing. It shows a fairly symmetrical normal distribution, but this could be illusory. At higher concentrations, 55 dS and above, it is definitely a trigeminal (fifth cranial) nerve irritant that can be detected by persons otherwise anosmic as a result of nonfunctionality of the olfactory (first cranial) nerve. Paradoxically, persons with normal olfactory sensitivity comment that pyridine has a pungent effect above about 30 dS, in addition to its familiar scallop-like odor. Accordingly, persons who are hyposmic to pyridine, and they do exist (Amoore, 1969), may be able to perceive the substance by its mild trigeminal effect.

Phenylethyl methyl ethyl carbinol (PM-carbinol) has been offered by our firm in test kits as a general smell sensitivity test. It reveals about 5% of people who are defined as hyposmic, with thresholds over 24 dS. Further tests are required to determine if this is an example of specific anosmia or the beginnings of general hyposmia. The distribution for tetrahydrothiophene (thiophane) indicated one outlier (at 45 dS) among 25 persons tested. This person was even more deficient with respect to *tert*-butyl mercaptan, but within the normal range for ethylphenol. This is probably an incipient example of true specific anosmia, which awaits detailed stereochemical and olfactometric surveying.

Figures 5 and 6 give the distributions of olfactory thresholds found for each of eight odorants that have been specially selected to demonstrate specific anosmia. Eight different classes of odor are represented, and in each class the compound illustrated is the one that, in its class, showed the greatest average difference of odor threshold between the normal observers and the specifically anosmic persons. The charts are arranged in the sequence of increasing percentages of the population that exhibit the specific anosmia.

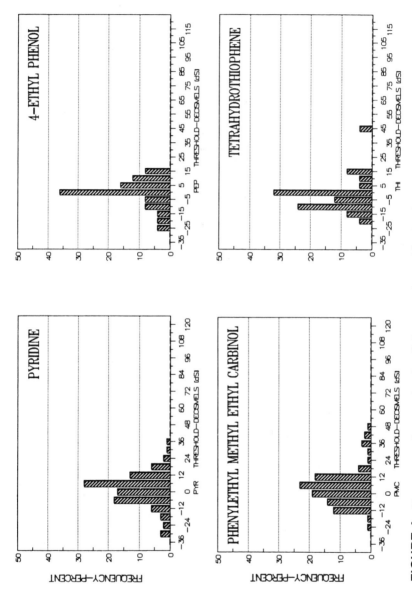

FIGURE 4 Four more odorants showing little or no signs of hyposmia.

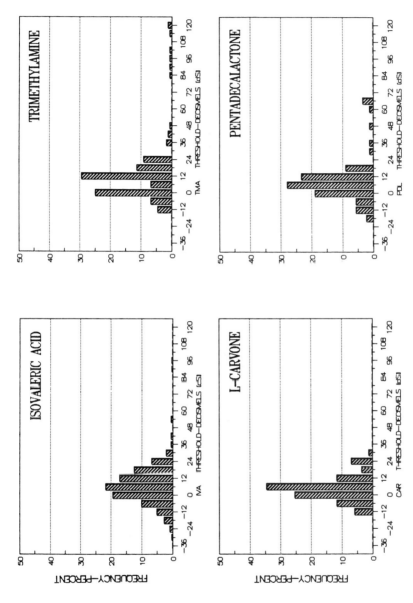

FIGURE 5 Olfactory thresholds for four primary odorants that exhibit, in sequence, increasing proportions of specific anosmics in the population samples. Original threshold data were in binary steps, recalculated as decismels.

In Figure 5, isovaleric acid showed a highly symmetrical normal distribution but with a thin scattering of specific anosmics, about 2 or 3% of the population, who had thresholds over 24 dS (2 SD) up to 96 dS. Trimethylamine exhibited 6% of specific anosmics, with thresholds up to 126 dS (plotted as 120 dS), which is over a million times the normal mean threshold concentration, and an intolerable stench for normal observers. The distribution for l-carvone is perhaps the least convincing as evidence for specific anosmia, with a modest peak at 24 dS, and nobody was found with threshold above 30 dS. This could be on account of overlapping chemical sensitivities of different receptor sites. The pentadecalactone anosmia is clear-cut, with about 12% of persons defined as specific anosmics, one of whom could not smell the saturated solution.

The theme is continued in Figure 6 with even more common anosmias. 1-Pyrroline demonstrated 16% anosmics (Amoore et al., 1975) and 1,8-cineole 33% in the work of Pelosi and Pisanelli (1981) at the University of Pisa. Calibration data by our firm in Berkeley show only about 14% of cineole anosmics. It is not known whether this is a true genetic difference or a consequence of different test methodologies (aqueous dilutions in flasks versus oil dilutions in squeeze bottles).

Isobutyraldehyde is the only substance thus far to have produced a near-classic bimodal curve for the distribution of olfactory thresholds in the population. The antimode is near 24 dS, and the peak for the specific anosmics (36%) is about 55 dS. The most common specific anosmia so far studied in chemical detail is that for androstenone, with 47% unable to detect the odor at the criterion of 24 dS. Indeed, about 23% cannot smell the substance at all, even in its saturated solution (near 72 dS).

Anybody wishing to experiment with these eight odorants may appreciate recipes for preparing aqueous solutions of the discriminatory concentration, or screening test, for each type of specific anosmic (Amoore, 1979).

For each of these eight named specific anosmia compounds, there are more, sometimes many more, that stimulate the same type of odor sensation and belong to the same odor class. This is illustrated in Figure 7 for isobutyraldehyde anosmia. The results are illustrated for representative panels of normal observers and isobutyraldehyde anosmics. For each group, who here resolved themselves into separate modes, the mean threshold was calculated (in binary dilution steps in this illustration). The difference of 9.0 binary steps between the mean thresholds of the two groups measures the anosmic defect.

When these tests were repeated with isobutyl alcohol, the anosmic defect decreased to 4.1 binary steps and, with isobutyl isobutyrate, all but disappeared (0.6 steps). These results indicate that isobutanol belongs to the same anosmia class as isobutyraldehyde, but with a lesser degree of exclusivity.

GRAPHIC HISTORY OF SPECIFIC ANOSMIA 341

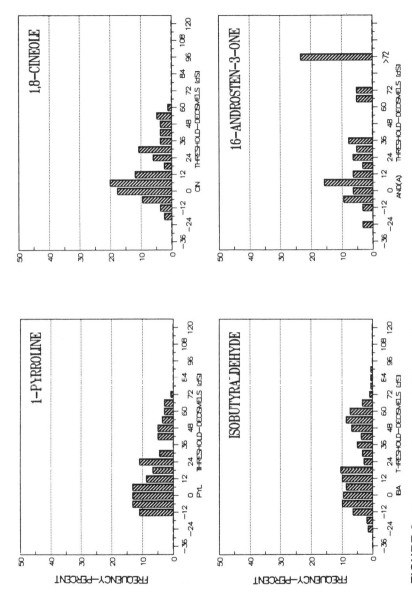

FIGURE 6 Four more odorants exhibiting further increases in the incidence of specific anosmia in the population.

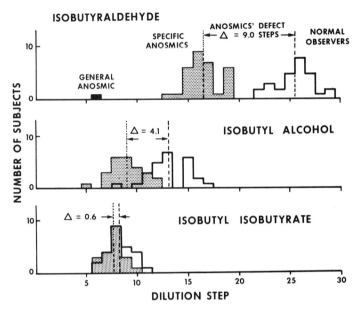

FIGURE 7 Olfactory thresholds of specific anosmics to isobutyraldehyde (stippled histograms) compared with normal observers (heavily outlined histograms). This figure has binary step dilution scales and has the reversed sense on the x axis compared with all other distributions in this paper. (From Amoore, 1982.)

Isobutyl isobutyrate does not belong at all. Among all aldehydes tested, isobutyraldehyde showed the largest anosmic defect and is hence identified as the epitome for this class of odor, which is therefore designated a "primary odor." Isobutyraldehyde itself becomes the corresponding primary odorant. This concept, first put forward by Guillot (1948), has been reviewed by Amoore (1977).

The structural formulas of the eight primary odorants so far clearly established are given in Figure 8, and the corresponding primary odor names in the lower part of Table 1. From the number and nature of the chemicals known to exhibit specific anosmias, it is probable that there are a total of at least 32 primary odors in the human sense of smell (Amoore, 1969, 1977). Any or all of these seem to be good candidates for genetic research.

Two of these odorants, pyridine and androstenone, have been studied in a similar manner by Labows and Wysocki (1984). In Figures 9 and 10 their data have been converted to the decismel scale and compared directly, in composite bar graphs, with our data shown in Figures 4 and 6. There is good agreement for pyridine and fair agreement for androstenone, with some discrepancy at the high sensitivity (low dS) end of the scale.

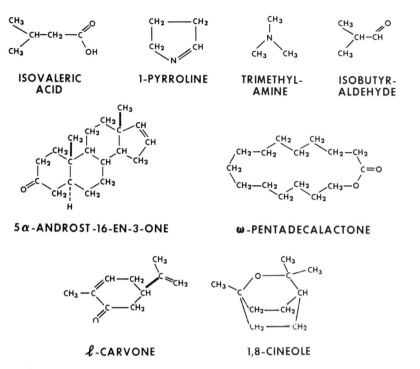

FIGURE 8 Structural formulas of eight primary odorants. (From Amoore, 1982.)

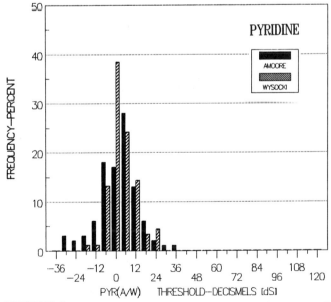

FIGURE 9 Odor thresholds of pyridine: data of Amoore (Figure 4) compared with data of Labows and Wysocki (1984) recalculated on the decismel scale.

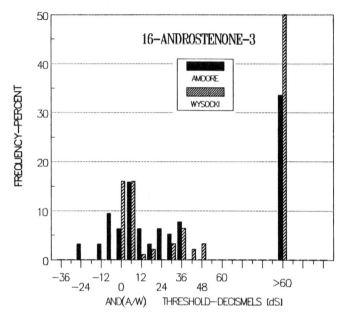

FIGURE 10 Odor thresholds of androstenone: data of Amoore (Figure 6) compared with data of Labows and Wysocki (1984) recalculated on the decismel scale.

Our firm has checked the responses of 20 subjects to chloroacetophenone (Figure 11). This is a lachrymatory (tear gas) and an irritant of the trigeminal nerve. It also has, for younger persons, a noticeable odor like coconut. The chart has been set up as usual with the average odor threshold at 20 years for the 0 dS reference point. On this scale the mean levels for nasal and for eye irritation were about equal, at 17 dS, with little if any decline of sensitivity with age.

During 15 years of research on the specific anosmias at the Western Regional Research Center of the USDA in Albany, California, Amoore and collaborators conducted screening tests for specific anosmia on a total of 764 fellow employees. The main criteria for a specific anosmic were ability to pass a general smell test in the normal sensitivity range (e.g., with isobutyl isobutyrate), associated with failure to detect a specific anosmia compound presented at about 2 SD (4 binary steps or 24 dS) above its normal mean threshold of detection. Although six primary odorants were tested, not all the subjects were tested with all six odorants, since this was not the objective at the time. Nevertheless, the accumulated data, even from an incomplete matrix, permitted an assessment of the incidence of multiple specific anosmias in the same person.

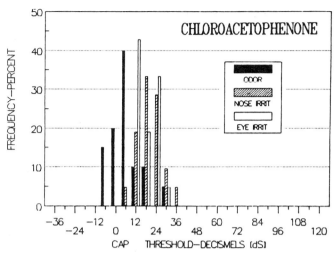

FIGURE 11 Odor and irritation thresholds for chloroacetophenone measured with 20 subjects. The decismel scale is anchored to its odor threshold, and the nose and eye irritation thresholds are expressed on the same scale. (Olfacto-Labs kit calibration data.)

Table 2, as an example, shows the first page of the alphabetical master list, showing the seemingly random occurrence of specific anosmias (indicated by X) and one person exhibiting two specific anosmias. The totals at the foot of the table, consolidated for all the pages in Table 4, provide the basis for the estimates of the frequency of each type of specific anosmia in the population (Table 1). Table 3 provides similar data for all 21 persons in the master list who had isovaleric acid anosmia. These persons, individually and as a group, tend to have a disproportionately large number of specific anosmias among them.

These data on the incidence of multiple specific anosmias are assembled in Table 4. The number of each type of anosmia found (on the diagonal) compared with the number of persons tested (right column) gives the percentage incidence for that type of anosmia. This percentage in turn permits one to calculate the numbers of specific anosmias of the other five types expected among members of the group having a particular specific anosmia. Table 4 provides, for each of the 30 possible permutations, the expected number of anosmias in the subgroup compared with the number actually observed. In most cases there is reasonably good agreement, within a factor of twofold or not much more, but with a general tendency for the observed number of anosmias to be somewhat greater than expected.

TABLE 2 Occurrence of Specific Anosmias[a]

Name	IVA	PYL	TMA	IBA	AND	PDL	Anosmias	Tests
Acco			X	X	O		2	3
Adam	O				O	O	—	2
Ahre	O	X	O	O	O	O	1	6
Aker	O					O	—	2
Alde	O	O	O			O	—	4
Alex	O					O	—	2
Ali					X	O	1	2
Alle	O					O	—	2
Alla	O						—	1
Alli	O	O	O	O		X	1	5
Alme	O	O	O	X	O	O	1	6
Alth	O			O		O	—	3
Amel				O			—	1
Amin	O	O	O	O		O	—	5
Amoo	O	O	O	O	O	O	—	6
Andr			O	O	X	O	1	4
Angl	O					O	—	2
Appl	O						—	1
Aran	O	O	O			O	—	4
Arve					X	O	1	2
Anosmics	—	1	—	2	4	1	8	
Tested	15	7	8	9	7	17		63

[a]First page of the master list of 764 persons tested with one or more primary odorants at the screening concentration (2 SD above mean threshold). The names are abbreviated to four letters. The three-letter odorant designations are IVA, isovaleric acid; PYL, pyrroline; TMA, trimethylamine; IBA, isobutyraldehyde; AND, androstenone; PDL, pentadecalactone. O = test result normal; X = test result specific anosmia; no entry = untested with this odorant.

These results show that, as a first approximation, the incidences of the various types of specific anosmia so far studied are independently distributed in the population. There is, however, one striking exception. As indicated by bold type in Table 4, there is a strong association between the isovaleric acid anosmia and the trimethylamine anosmia. The observed number of persons having both these specific anosmias is about 12 times as many as would be expected on the hypothesis of independent occurrence. It is not known whether the cause of this association is genetic or biochemical.

It has long been supposed that smell blindness is an inheritable defect, probably as a recessive character (Patterson and Lauder, 1948). Figure 12 shows the results of a study on 109 families tested with pentadecalactone at its screening concentration (Whissell-Buechy and Amoore, 1973). As indicated

GRAPHIC HISTORY OF SPECIFIC ANOSMIA

TABLE 3 Specific Anosmia Tests on All the 21 Persons Found to Have the Isovaleric Acid Anosmia[a]

Name	IVA	PYL	TMA	IBA	AND	PDL	Anosmias	Tests
Bart1	X		X	O	O	O	2	5
Bart2	X	O	X	X	X	O	4	6
Burr	X					X	2	2
Chri	X					O	1	2
Colb	X					O	1	2
Free	X		X	X		O	3	4
Gram	X		O	X		O	2	4
Harv	X	O	X			O	2	4
Inge	X	O	X	X	X	X	5	6
Jack	X			X			2	2
Jans	X	O		O		X	2	4
Kara	X						1	1
Kauf	X					O	1	2
Pete	X						1	1
Schu	X	X	O	X	X	O	4	6
Simp	X		X	X			3	3
Simr	X	O		O		O	1	4
Swai	X			X		O	2	3
Thom	X					X	2	2
Vana	X			O		O	1	3
Wrig	X						1	1
Anosmics	21	1	6	8	3	4	43	
Tested	21	6	8	12	4	16		67

[a]Selected from the master list. Abbreviations and symbols as in Table 2.

TABLE 4 Incidence of Multiple Specific Anosmias[a]

Anosmics	IVA, Exp/Obs	PYL, Exp/Obs	TMA, Exp/Obs	IBA, Exp/Obs	AND, Exp/Obs	PDL, Exp/Obs	Total tested
IVA	(21)	1.0/1	**0.4/6**	4.3/8	1.9/3	1.9/4	626
PYL	1.0/1	(36)	1.5/2	6.1/10	6.1/7	3.7/3	221
TMA	**0.5/6**	1.3/2	(16)	4.3/10	3.8/7	1.5/4	284
IBA	3.0/8	9.6/10	4.1/10	(108)	29.3/31	10.7/17	299
AND	2.3/3	9.0/7	4.4/7	28.5/31	(104)	12.0/26	220
PDL	1.4/4	3.1/3	2.4/4	13.7/17	19.9/26	(62)	536

[a]Odorant designations as before. The numerals in parentheses on the diagonal are the number of specific anosmics for the given odorant among the total persons tested shown in the last column. This is the source of the estimated percentages of specific anosmias in the population as given in Table 1. These percentages were then used to calculate the number of persons among, for example, the 21 IVA anosmics who would also be expected to have the PYL anosmia. Each entry of the expected number (Exp) is compared with the number actually observed (Obs).

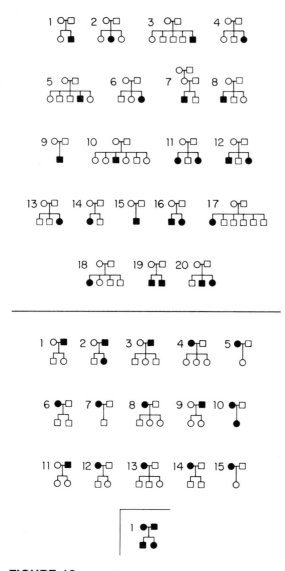

FIGURE 12 Family trees showing the occurrence of specific anosmia to pentadecalactone (solid symbols). (From Whissell-Buechy and Amoore, 1973.)

in the upper half of the chart by solid symbols, children of normal parents can be specific anosmics, and conversely many children of (one) affected parent are normal. In one family with both parents specifically anosmic, both their children were also specific anosmics. A segregation analysis on the offspring from the 20 marriages between "smellers" gave almost exact correspondence between the expected and observed results. These results are consistent with a simple recessive inheritance.

Wysocki and Beauchamp (1984) studied the olfactory thresholds of twins (17 identical pairs and 21 fraternal pairs) to pyridine and to androstenone. The correlations between the thresholds of the members of a twin pair were substantially greater for the identical twins than for the fraternal twins. They concluded that there is a significant genetic component for the variation of sensitivity to androstenone.

IV. DISCUSSION

In this review an attempt has been made to patch together the relevant data from a diverse series of investigations to develop a concordant "history" of specific anosmia. Actually, most of the studies were designed with narrower objectives: to define a single variety of specific anosmia and to understand the stereochemical requirements for the corresponding primary odor. None of the population samples were the same, and the techniques were not fully consistent.

It is our hope that this synthesis will indicate the need for a second round of systematic work in which a coherent technique is applied to test a substantial group of people using all eight primary odorants identified. The resulting data should provide a firm basis for further investigation of possible associations between these and other varieties of specific anosmia and for possible connections with additional genetic or physiologic abnormalities.

V. SUMMARY

Humans exhibit chemically selective defects in the sense of smell, known as smell blindness or specific anosmia. Eight varieties of this defect have been fully characterized in the literature. The reported distributions of olfactory threshold sensitivity in the population are here recalculated and expressed graphically in a consistent manner. These defects are comparatively common (incidences of 3-47% of the population), substantially independent (with one exception), convenient to measure, and where tested have shown a significant degree of heritability as simple recessive characters.

REFERENCES

Amoore, J. E. (1967). Specific anosmia: A clue to the olfactory code. *Nature* **214**: 1095-1098.

Amoore, J. E. (1968). Odor blindness as a problem in odorization. Proc. Operating Sect., Amer. Gas Assoc. Distribution Conf., pp. 242-247.

Amoore, J. E. (1969). A plan to identify most of the primary odors. In *Olfaction and Taste III*, C. Pfaffmann (Ed.). Rockefeller University Press, New York, pp. 158-171.

Amoore, J. E. (1975). Four primary odor modalities of man: Experimental evidence and possible significance. In *Olfaction and Taste*, Vol. V, D. A. Denton and J. P. Coghlan (Eds.). Academic Press, New York, pp. 283-289.

Amoore, J. E. (1977). Specific anosmia and the concept of primary odors. *Chem. Senses Flav.* **2**:267-281.

Amoore, J. E. (1979). Directions for preparing aqueous solutions of primary odorants to diagnose eight types of specific anosmia. *Chem. Senses Flav.* **4**:153-161.

Amoore, J. E. (1982). Odor theory and odor classification. In *Fragrance Chemistry*, E. T. Theimer (Ed.). Academic Press, New York, pp. 27-76.

Amoore, J. E., and Forrester, L. J. (1976). Specific anosmia to trimethylamine: The fishy primary odor. *J. Chem. Ecol.* **2**:49-56.

Amoore, J. E., and Ollman, B. G. (1983). Practical test kits for quantitatively evaluating the sense of smell. *Rhinology* **21**:49-54.

Amoore, J. E., and O'Neill, R. S. (1988). Proposal for a unifying scale to express olfactory thresholds and odor levels: The "decismel scale." *Proceedings of the 1988 Air Pollution Control Association Annual Meeting*, Paper No. 78.5. Air and Waste Management Association, Pittsburgh.

Amoore, J. E., Venstrom, D., and Davis, A. R. (1968). Measurement of specific anosmia. *Percept. Motor Skills* **26**:143-164.

Amoore, J. E., Forrester, L. J., and Buttery, R. G. (1975). Specific anosmia to 1-pyrroline: The spermous primary odor. *J. Chem. Ecol.* **1**:299-310.

Amoore, J. E., Forrester, L. J., and Pelosi, P. (1976). Specific anosmia to isobutyraldehyde: the malty primary odor. *Chem. Senses Flav.* **2**:17-25.

Amoore, J. E., Pelosi, P., and Forrester, L. J. (1977). Specific anosmias to 5α-androst-16-en-3-one and ω-pentadecalactone: The urinous and musky primary odors. *Chem. Senses Flav.* **2**:401-425.

Blakeslee, A. F. (1918). Unlike reaction of different individuals to fragrance in Verbena flowers. *Science* **48**:298-299.

Fein, M. M., Bobinski, J., Mayes, N., Schwartz, N., and Cohen, M. (1963). Carboranes. I. The preparation and chemistry of 1-isopropenylcarborane and its derivatives (a new family of stable clovoboranes). *Inorg. Chem.* **2**:1111-1115.

Guillot, M. (1948). Anosmies partielles et odeurs fondamentales. *C.R. Acad. Sci.* **226**:1307-1309.

Labows, J. N., and Wysocki, C. J. (1984). Individual differences in odor perception. *Perfumer Flavorist* **9**:21-26.

Patterson, P. M., and Lauder, B. A. (1948). The incidence and probable inheritance of "smell blindness." *J. Hered.* **39**:295-297.

Pelosi, P., and Pisanelli, A. M. (1981). Specific anosmia to 1,8-cineole: The camphor primary odor. *Chem. Senses* **6**:87-93.

Pelosi, P., and Viti, R. (1978). Specific anosmia to l-carvone: The minty primary odor. *Chem. Senses Flav.* **3**:331-337.

Piper, T. S., Cotton, F. A., and Wilkinson, G. (1955). Cyclopentadienyl-carbon monoxide and related compounds of some transitional metals. *J. Inorg. Nuclear Chem.* **1**:165-174.

Powell, J., and Shaw, B. L. (1966). Triallylrhodium (III) and a crotyl-butadiene complex of rhodium. *Chem. Commun.* 323-325.

Rozantzev, E. G., and Neiman, M. B. (1964). Organic radical reactions involving no free valence. *Tetrahedron* **20**:131-137.

Sherman, A. H., Amoore, J. E., and Weigel, V. (1979). The pyridine scale for clinical measurement of olfactory threshold: A quantitative reevaluation. *Otolaryngol. Head Neck Surg.* **87**:717-733.

Venstrom, D., and Amoore, J. E. (1968). Olfactory threshold in relation to age, sex or smoking. *J. Food Sci.* **33**:264-265.

Whissell-Buechy, D., and Amoore, J. E. (1973). Odour-blindness to musk: Simple recessive inheritance. *Nature* **242**:271-273.

Wysocki, C. J., and Beauchamp, G. K. (1984). Ability to smell androstenone is genetically determined. *Proc. Natl. Acad. Sci. USA* **81**:4899-4902.

24
Individual Differences in Human Olfaction

Charles J. Wysocki and Gary K. Beauchamp

Monell Chemical Senses Center, Philadelphia, Pennsylvania

I. INTRODUCTION

Genetic methods of analysis have made significant contributions to the study of neurobiology and sensory systems, chromatic vision in particular (see contributions in Breakefield, 1979; Gershon et al., 1981). Moreover, genetic methods have provided fundamental insights into invertebrate chemoreceptive mechanisms. However, with few exceptions (Amoore, 1971), genetic methodologies have not been exploited in the study of mammalian nasal chemoreception. In summarizing research directions in the chemical senses, Beidler (1987) devoted a major section to a discussion of the use of genetic analyses of chemosensory function in the pursuit of molecular receptors and an understanding of cell differentiation and receptor protein regulation. Indeed, he stated that "the marriage of genetics and molecular biology offers the rewards of decisive information concerning chemosensory transduction that can be obtained in no other manner."

The genetic contribution to variation in normal olfaction remains to be defined. Reports by Hubert et al. (1980, 1981) suggest that little genetic variation exists within the normal range of olfactory acuity. Estimates of heritability are, however, limited by the reliability of the test, and the procedures utilized by Hubert and colleagues were coarse. Hence the issue of a genetic component of variation within normal ranges remains unresolved. Rabin and

Cain (1986) suggested that little variation exists among individuals within the normal range of human olfactory perception. Stevens et al. (1988), however, present additional data suggesting that variation can be extreme; they noted that intraindividual variation was at times as great as interindividual variation (sometimes 10,000-fold).

Although considerable progress has been made in the field of olfaction, knowledge of this sensory system lags considerably behind that of the visual and auditory systems. Our laboratories have pursued psychophysical and genetic strategies in an attempt to reveal the mechanisms underlying olfaction. Our approach is founded on the premise that studies of the extremes of natural variation in olfactory perception can lead to an understanding of the normal underlying processes. Key to this approach are our investigations of the well-documented individual differences in the ability to smell 5α-androst-16-en-3-one (specific anosmia to androstenone). In the past few years we have learned that sensitivity to this odor is regulated by genetic, developmental, and environmental factors.

II. SPECIFIC ANOSMIA: BACKGROUND

In some instances there are very large individual differences in sensitivity to specific odors among individuals with otherwise normal or average olfactory acuity. Such deficits in sensitivity to one odor or class of odors have been called specific anosmias (Amoore, 1971). Examples are shown in Table 1. Some specific anosmias have been studied in pedigree or sex-linkage analyses (e.g., Whissel-Buechy and Amoore, 1976), but most have not. Genes may play a role in controlling specific anosmias; however, drawing conclusions regarding the inheritance of specific anosmias can be difficult, due in part to disagreements among investigators who use different methods for evaluating olfactory sensitivity and for classifying individuals into categories.

TABLE 1 Some Compounds for Which Specific Anosmia Exists and Percentage of Adults Having the Anosmia

Compound	% Anosmic
Androstenone	47
Isobutyraldehyde	36
1,8-Cineole	33
1-Pyrroline	16
Exaltolide	12
1-Carvone	8
Trimethylamine	6
Isovaleric acid	3

Anosmia to hydrogen cyanide (HCN) provides an example. Estimates of the frequency of HCN anosmia vary from 17 to 54% for males and from 0 to 53% for females. Some studies report HCN anosmia to be sex-linked (i.e., affected by genes on the X-chromosome) (Srivastava, 1961); others do not (Hauser et al., 1958). There may also be differences among people who smell HCN. Some people barely detect the smell of 20% potassium cyanide solutions; others say it is strong (Brown and Robinette, 1967). As we have found in studies of sensitivity to androstenone (see later), developmental changes may also alter sensitivity to this odor, thereby compromising genetic analysis. A significant gender difference in HCN sensitivity was found for adults, but not for children (Brown and Robinette, 1967). Furthermore, the distribution of children's thresholds to HCN changes with age (Brown and Robinette, 1967).

Other specific anosmias are less well studied, and only two have received any form of genetic analysis: specific anosmia to Exaltolide, having a musky smell to those who detect it, and to androstenone. Familial pedigrees of Exaltolide anosmia are consistent with the hypothesis that the inability to detect an odor is inherited as an autosomal recessive (Whissel-Buechy and Amoore, 1976). The model to account for genetic variation in the sensitivity to androstenone is likely to be more complex (see later).

III. SPECIFIC ANOSMIA: ANDROSTENONE

Androstenone is a steroid of gonadal origin that is plentiful in the saliva of male pigs. When detected, it facilitates the lordosis response made by receptive sows (Reed et al., 1974). Not surprisingly, androstenone is found in pork products. It is also present in celery and truffles. Human secretions and excretions contain androstenone (Brooksbank et al., 1974; Bird and Gower, 1983; Claus and Asling, 1976); it is more abundant in males than in females. There are a number of reports that this or related odors alter human behavior, act as a sexual attractant, or have special effects on emotions and sexual responses (Cowley et al., 1977; Kirk-Smith et al., 1978; Benton, 1982; Van Toller et al., 1983; Filsinger et al., 1984, 1985), but these remain controversial and tentative.

Our particular interest in the compound stemmed from the extreme variation in response to it among adults. Previous studies have shown that many otherwise normal adults cannot detect this odor at vapor saturation, and others are exquisitely sensitive to it; estimates of a specific anosmia center around 40-50% (Labows and Wysocki, 1984). Not only is there vast variation in threshold (Figure 1), but different individuals appear to experience the odor as having different qualities. The odor of androstenone is described as urinous, sweaty, musky, like sandalwood, or, importantly, having no smell.

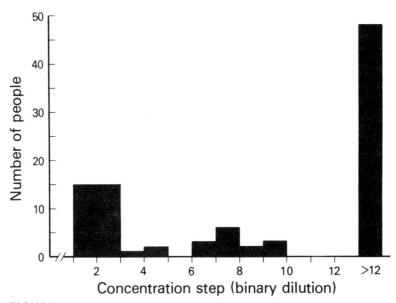

FIGURE 1 Androstenone detection thresholds were determined in approximately 100 individuals. People in the group on the left are the most sensitive and generally dislike the smell androstenone, which typically is labeled as urinelike. People in the middle group, although capable of smelling androstenone, generally perceive a much different odor quality; they typically label androstenone as sandalwood, perfume, musky, or sweet and are not usually offended by the smell. People in the group on the right are anosmic to androstenone, but they could smell pyridine. (Adapted from Labows and Wysocki, 1984.)

Here we describe the results of a series of studies that serve to illustrate the role of genetic variation in the determination of individual differences in sensitivity to the odor of androstenone. Importantly we noted that developmental changes also affect sensitivity to this odor. Additionally we observed that exposure to androstenone can affect perception of its odor.

A. Methods of Sensory Evaluation

Typically, we utilize one or two odors, pyridine (the odor of spoiled musk) and amyl acetate (having a banana- or pearlike odor), to screen for general anosmia, but our primary focus is on responses to androstenone. Odorless light white mineral oil is used as the diluent. Each concentration step of each odor is paired with a dedicated oil blank of equivalent volume. Stimulus and blanks are presented in cleaned, 300 ml polypropylene squeeze bottles with plastic fliptop caps. Eight or ten binary dilutions of pyridine are used; each contains a final volume of 30 ml. The highest concentration is 3.0 ×

10^{-3}% v/v, which is below the concentration necessary to elicit trigeminal responses; the lowest concentration is 5.85×10^{-6}% v/v. Pyridine was chosen as a general olfactory stimulus because it is frequently used as a stimulus in smell testing and few people fail to detect its odor (Sherman et al., 1979). Twelve binary dilutions of amyl acetate are used, ranging from 4.0×10^{-3} to 2.0×10^{-6}% v/v. Twelve binary dilutions of androstenone are also used, ranging from 1.0×10^{-1}% w/v (step 12) to 4.88×10^{-5}% w/v (step 1).

Sensitivity is determined in a two-sample (odor versus blank), forced-choice, ascending concentration series. After an individual successfully identifies the odor in four (pyridine and amyl acetate) or five (androstenone) successive trials, the series ceases and the lowest of the four or five correct trials is entered as the threshold for that series. Four series are conducted for each odor. The median value for the four series for each odor is the "threshold" for the individual. Individuals who are anosmic to an odor have a numeric threshold value (usually an integer greater than the highest of the series) even though the individual failed to meet the criterion and could not identify the odor-containing squeeze bottle at the highest concentration. Some anosmic individuals may have a threshold slightly lower than the highest concentration because they may have correctly guessed the odor-containing bottle on one or more of the last trials.

B. Genetic Studies

To evaluate the possibility that individual differences in perception of androstenone may be due to genetic differences between people, detection thresholds for androstenone and pyridine were determined in 17 identical and 21 fraternal twin pairs (Wysocki and Beauchamp, 1984). Threshold determinations were accomplished with an ascending concentration two-sample (odor versus blank) forced-choice procedure (Figure 2). Details concerning odor purity, exact testing procedures, determination of zygosity, and statistical procedures can be found in the original publication. The results were straightforward: 100% of the identical twin pairs were either both sensitive to the smell of this odor or both insensitive to the odor (sensitivity was defined as the antimode of the bimodal threshold-concentration histogram); only 61% of the fraternal twins were concordant. There was also no evidence for a genetic effect on pyridine sensitivity.

The results of this study seemed to suggest a rather simple and elegant model system. Sensitivity to androstenone has a major genetic component; perhaps genetic differences between people can explain all the variance in this phenotype. Recent studies, however, indicate a complex mode of inheritance (Pollack et al., 1982; Wysocki et al., 1989). Moreover, results from studies of developmental factors and the effects of exposure on perception now demonstrate that this apparent simplicity was an illusion.

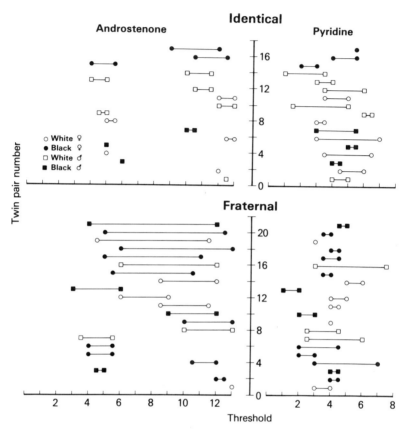

FIGURE 2 Androstenone and pyridine detection thresholds for each member of 17 identical and 21 fraternal twin pairs. Pair-mates are connected by horizontal lines. A single entry for a twin pair indicates identical thresholds for each. (Adapted from Wysocki and Beauchamp, 1984.)

To investigate the mode of inheritance we determined sensitivity to androstenone and pyridine in both parents and at least two biologic offspring from 67 families in the Philadelphia area. Although strong evidence for genetic modulation of sensitivity to androstenone (but not pyridine) was observed again, patterns of inheritance and associations among relatives were complex (Table 2).

The pattern of relationships found among family members is often generated when sex chromosomes are involved. Fathers possess only one X-chromosome and pass it to their daughters, which inflates the phenotypic relationship between father and daughter for X-linked traits; fathers cannot,

TABLE 2 Correlation and Regression Coefficients for Threshold Sensitivity

	Androstenone			Pyridine		
	Mother	Son	Daughter	Mother	Son	Daughter
Father	0.20	0.14	0.32	0.20	0.01	0.14
Mothere	—	0.40	0.27	—	0.15	0.09

however, pass X-linked genes to their sons. Mothers possess two X-chromosomes and provide one to both sons and daughters. Hence our results, coupled with an observed sex difference in anosmia to androstenone (42% of fathers and 32% of mothers were anosmic) suggest X-linkage of one or more genes involved in regulating sensitivity to the compound. Sensitive and insensitive males would thus be coded AY and aY, respectively. The proportion of androstenone-anosmic females [aa], however, exceeds expectation for an X-linked trait. Random inactivation of the X-chromosome (Lewin, 1980; Lyon, 1962, 1988), which occurs in the tissues of females, coupled with variation in the proportion of sensory cells containing an active recessive allele in heterozygous [Aa] females, could account for this discrepancy. In these Aa females, receptor cells that express the dominant A gene (androstenone sensitive) could comingle in the olfactory epithelium with receptor cells that express the recessive a gene (androstenone insensitive). In some females, the frequent occurrence of a-containing cells could far outnumber A-containing cells. This could result in functional anosmia to androstenone.

For heuristic purposes only we assumed that X inactivation occurred in the female olfactory neuroepithelium (a likely but as yet unsubstantiated outcome). Further, we assumed that the olfactory neuroepithelium of females exists as an X-chromosome mosaic; some sensory cells express the genes of the paternal X but others express the genes of the maternal X. We then determined the most likely genotype of each individual by inspecting family dendrograms. Under these above assumptions, members from all 67 families were assigned tentative genotypes. After this, the mean sensitivity of each presumptive genotypic group was calculated. The results are contained in Figure 3. Note that presumptive heterozygous females who express products of the "sensitive" gene, that is $A(a)$ females, are more sensitive to androstenone than are presumptive heterozygous females who express products of the "insensitive" gene, that is, $a(A)$ females. Also note that $A(a)$ females are less sensitive than AA females. In the former case, only a subsample of receptor cells possess gene products that result in sensitivity to androstenone, but in the latter case all sensory cells possess such gene products. Although no solid evidence exists for a positive correlation between numbers of an-

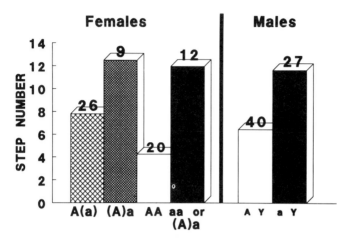

FIGURE 3 Androstenone thresholds were determined for members of 67 families having both parents and at least two offspring (high numbers along the y axis indicate insensitive individuals). After assuming X-linkage of one or more genes and female X-chromosome inactivation (including the presumed genes affecting sensitivity to androstenone), the presumptive genotype of individuals was determined by inspection of family dendrograms. An A gene confers sensitivity to androstenone; a confers insensitivity. Among heterozygous females, either A or a could be deactivated, as indicated by (A) or (a).

drostenone-sensitive receptor cells and general sensitivity to androstenone, the premise has face validity.

Regardless of its mode of expression, the gene(s) for sensitivity to androstenone may reside within a family of genes that may function to regulate olfaction in general. The concept of gene families has been verified in other systems (e.g., color vision). Locating one of these genes could result in substantial progress in understanding the biochemistry and neurobiology of olfaction.

C. Developmental Changes

Our initial developmental study using androstenone (Schmidt and Beauchamp, 1988) was incidental to a more general interest in olfaction in infancy and childhood, a neglected area. The few published experiments that have focused on this issue suggest that children are sensitive to odors (Cernoch and Porter, 1985; Engen and Lipsitt, 1965; Engen et al., 1963; MacFarlane, 1975; Rovee, 1969) but that their hedonic experience of odors is quite different from that of adults. In several paradigms children younger than 5 years have failed to respond differentially to odors that are judged by adults to have different hedonic values (e.g., Stein et al., 1958; Peto, 1936; Lipsitt et al., described in Engen, 1982).

Based on these and other observations some investigators have concluded that young children do not have aversions to odors that adults find offensive (Peto, 1936; Stein et al., 1958), that they are more tolerant of odors than adults (Moncrieff, 1966), or that the range of hedonic experience is more limited for children than for adults (Engen, 1974). Furthermore, the apparent absence of odor aversions in young children has led to the suggestions that there are no inherently unpleasant odors and that all hedonic reactions are acquired through learning (Engen, 1974; see Schmidt and Beauchamp, 1988).

These conclusions may be premature. Several recent studies are consistent with the proposition that offensive odor reactions begin in early infancy (e.g. Cernoch and Porter, 1985; Self et al., 1972; Steiner, 1977, 1979). Furthermore, methodologic difficulties in evaluating odor hedonics in young children could be involved in the earlier failures to elicit offensive responses. In light of these considerations we (Schmidt and Beauchamp, 1988) investigated the hedonic reactions of 3-year-olds using a method designed to minimize potential methodologic artifacts and increase olfactory sampling by the child. The test was a forced-choice categorization procedure (good odor versus bad odor). A testing session was introduced as a smell game with one simple rule: good things should be given to a stuffed toy of Big Bird and bad or yucky things were to be given to a stuffed toy of Oscar the Grouch. Both characters are from *Sesame Street*, a popular children's TV program with which all subjects were familiar. Children were told that they would be asked to smell something and if they liked the smell they should point to Big Bird, but if they did not like the smell they should point to Oscar the Grouch (to put in his trash can). Adults also were evaluated in the same way.

Subjects were tested with nine odors selected to represent a wide range of qualities and hedonic values. To minimize intensity differences, which could influence judgments, intensity of the following odors were matched by adults: C-16 aldehyde (strawberry), phenylethylmethylethyl carbinol (PEMEC-floral), *l*-carvone (spearmint), methylsalicylate (wintergreen), eugenol (cloves), amyl acetate (banana), butyric acid (strong cheese or vomit), pyridine, and androstenone.

The overall results of this study can be summarized quite simply. With few exceptions children and adults exhibited a similar pattern of preferences and aversions. Odors adults judged as good were generally judged good by children, and vice versa. Thus even children as young as 3 years have distinctive, reliable preferences and aversions for odors. The most interesting exception was androstenone; children and adults differed in their affective judgments. More (13 of 14) children reported this odor as bad relative to adults (10 of 17). The proportion of adults reporting a bad odor is consistent with the rate of anosmia found in other studies. This is clearly not the case for the children. That this age difference in response to androstenone was not due to a general response bias is shown by the different pattern of responses to pyridine. All the adults judged pyridine as bad but this was true for only 71% of the 3-year-olds.

We tentatively conclude that most 3-year-olds detect androstenone. This implies that before adulthood some proportion of individuals lose their ability to detect androstenone. We explored this hypothesized developmental change in older children.

In this study (Dorries et al., 1989) androstenone and pyridine thresholds were measured in subjects aged 9-20 years and in a group of older adults. Descriptive and hedonic responses to these two odors plus PEMEC and a no-odor control (mineral oil) were also obtained from these subjects plus an additional group of 6- to 8-year-olds.

A total of 247 individuals were tested. For threshold determination a forced-choice ascending series of paired trials very similar to that just described were used. Following threshold testing subjects were given the highest concentration of each odor and the blank and, using a standard face-scale, were asked to provide hedonic judgments. Further details can be found in Dorries et al. (1989).

Consistent with the work already described, we found a significant increase with age in the proportion of subjects, particularly males, insensitive to androstenone. The proportion of males insensitive to androstenone tripled between the ages of 9-14 and 15-20 years. Furthermore, among those males who could smell androstenone, the average sensitivity declined. Among females the pattern was somewhat different. Although less extreme, there was a tendency for there to be a greater proportion of insensitive females at the older ages, suggesting that some females also lose their ability to detect androstenone. In contrast to the males, however, among those females who could smell androstenone greater sensitivities were apparent in older individuals.

Hedonic ratings also indicated a developmental shift in perception of androstenone (Figure 4). Consistent with our work with 3-year-olds, no child in the youngest group rated androstenone as good and almost two-thirds rated it as bad. Since these children tended to rate the blank as good, we concluded that a large proportion of them detected androstenone and perceived it as bad. Furthermore, this percentage declined with age. These data, plus those of the National Geographic Smell Survey (Wysocki and Gilbert, 1989), support the hypothesis that the probability that an individual will detect the odor of androstenone declines with age between childhood and young adult-

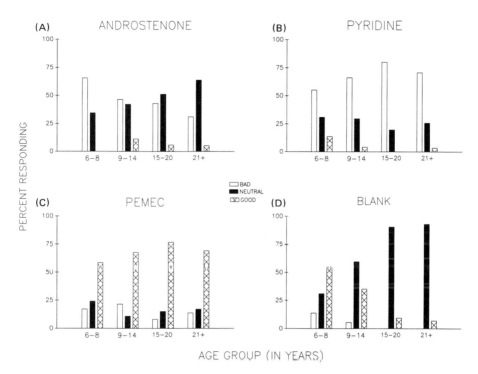

FIGURE 4 Percentage of subjects by age group rating the odorant as bad, neutral, or good when presented with (A) a high concentration of androstenone; (B) a high concentration of pyridine; (C) neat PEMEC; and (D) a mineral oil blank. (From Dorries et al., 1989.)

hood. They further suggest that this decline may be the most apparent, at least for males, at or near adolescence. Thus individual differences in perception of androstenone must involve developmental factors interacting with genetic variation. Environmental factors have recently been shown to influence sensitivity, further complicating the story.

D. Environmental Effects

During the course of research one of us (Wysocki), who could not detect androstenone, apparently experienced induced sensitivity; after a period of intermittent contact with the compound a distinct odor could be perceived. To explore this further androstenone-anosmic individuals were given repeated measures of sensitivity to pyridine, amyl acetate, and androstenone.

Long-Term Experience

Subjects, all of whom were anosmic to androstenone, were given threshold tests for androstenone, pyridine, and amyl acetate once a week for 7 weeks. After the first week of baseline tests the subjects were assigned to an experimental or a control group. Subjects in the experimental group sniffed androstenone and amyl acetate continuously for 3 min each, three times per day, for 6 weeks. Control subjects were tested weekly but did not sniff odor samples between sessions.

An ability to perceive androstenone became apparent in 10 of 20 experimental subjects as demonstrated by a substantial decrease in threshold (see Figure 5); none of the 18 control subjects exhibited a comparable shift in sensitivity (Wysocki et al., 1989). The reason that 10 of the 20 experimental subjects did not develop an ability to detect androstenone is not known.

What changed in people who became sensitive to the odor of androstenone remains in question. One hypothesis presumes a peripheral locus. Here, minimal androstenone receptors exist in some individuals who are insensitive to it and extended exposure induces clonal expansion or selection of such receptors in a manner analogous with immune response to antigen. Alternatively, exposures to androstenone may affect changes in enzyme levels or activity in the epithelial and/or mucous layers. The enzymes may be necessary to activate or alter the molecule, to facilitate odor-receptor binding, to protect androstenone from degradation before reaching the receptor, or to act as a carrier molecule. This explanation relies upon the necessity of an inducible enzyme system, again a well-established phenomenon.

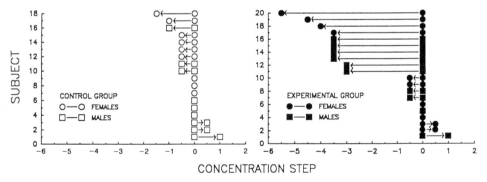

FIGURE 5 Individual changes in androstenone threshold from session 1 to session 7 for all subjects tested. Zero represents the individual's score at session 1; all subjects were insensitive to androstenone at the beginning of the experiment. A shift in the negative direction indicates an increase in sensitivity to androstenone. (From Wysocki et al., 1989.)

Consequent changes in receptor numbers or types or in enzyme systems could thus raise odor stimulation to the level of conscious perception. This is an attractive idea, but the data cannot exclude central changes following exposure, for example in the olfactory bulb or elsewhere in the central nervous system (Wilson et al., 1987).

Short-Term Experience

Results from the experiment designed to assess the effects of long-term exposure to androstenone suggest that exposing androstenone-anosmic individuals to the compound sensitizes some individuals. During the experiment the duration of sniffing androstenone amounted to slightly over 6 hr across 6 weeks. Indeed, the majority of changes in sensitivity occurred after 1-3 weeks of sniffing. Hence, the total sniff time required to induce sensitivity appears to be as little as 1-2 hr. If our hypothesis regarding a shift in receptor types or enzyme system is correct, then the effects of exposure should require a substantial time to be observed. Under normal conditions the olfactory epithelium is estimated to turn over in 25-35 days (Moulton, 1978; Graziadei and Monti-Graziadei, 1980). In 2 weeks roughly 40-56% of the epithelium is replaced.

As a first attempt to determine whether passage of time during exposure is important, we condensed the equivalent of a 30 day exposure to androstenone into three, 90 min sessions and tested for changes in sensitivity to the compound following each of the daily exposures. Eight androstenone-anosmic adults were tested for sensitivity to pyridine, amyl acetate, and androstenone in five test sessions. Session 1 was a pretest to determine eligibility; in sessions 2, 3, and 4 sensitivity was determined before each of the 90 min exposures to androstenone (the odor was placed on surgical masks worn by the subjects); session 5 was conducted the day following the final 90 min exposure. In sessions 2 and 5 subjects also provided magnitude estimates for the loudness of five, 1 kHz tones, five concentrations of amyl acetate, four concentrations of androstenone, and a blank. The results were straightforward; no shift in sensitivity to androstenone was observed during or after massed exposure to the compound in any subject (Figures 6 and 7).

The protocol used during this experiment (passive exposure to androstenone on a surgical mask) differed from that employed in the previous long-term experience experiment (active sniffing of androstenone). It is possible that this subtle difference in exposure regime could account for the lack of sensitivity to androstenone noted subsequent to passive inhalation of the compound. Perhaps it is necessary to actively sniff androstenone to induce sensitivity. This possibility necessitates further investigation.

Whether a single exposure to androstenone can affect long-term changes in sensitivity to the compound was also unknown; hence we conducted an

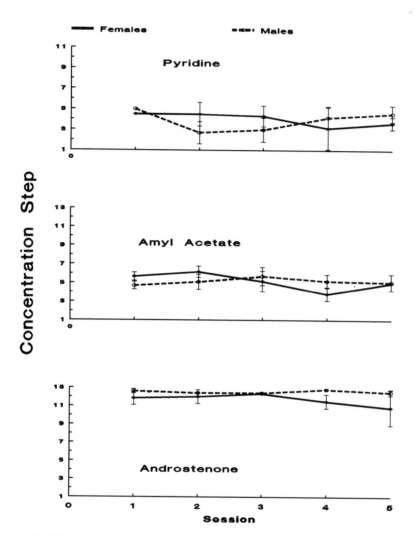

FIGURE 6 Thresholds for pyridine, amyl acetate, and androstenone were determined before massed exposure (session 1), before each of three, 90 min exposures to androstenone (sessions 2-4), and on the day following the final exposure. The y axis represents the binary dilution step for each odor.

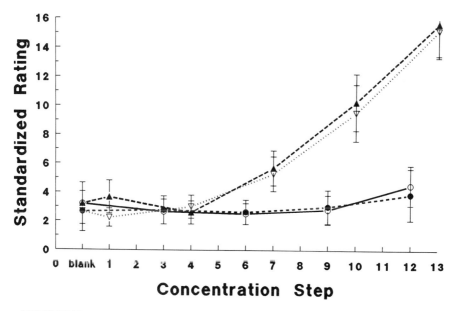

FIGURE 7 Magnitude estimates for amyl acetate (triangles) and androstenone (circles) were obtained from androstenone-anosmic adults before (open symbols) and after (blackened symbols) three, 90 min sessions of exposure to androstenone. Raw estimates were standardized using each individual's estimates of loudness of auditory stimuli. Individual- and session-specific constants were determined by equating auditory estimates to a mean of 10. These constants were then used to adjust odor magnitude estimates. Note the flat functions for androstenone, both before and after exposures, obtained from androstenone anosmics. When similarly tested, individuals who are sensitive to androstenone generate a function with a positive slope (unpublished observations).

experiment to determine whether a brief exposure to androstenone would alter sensitivity. Individuals in this experiment experienced androstenone during an initial exposure session (a single-threshold test conducted in our standard format). They returned 6 weeks later for a repeat evaluation (again, a single-threshold test session). No systematic exposures to androstenone occurred during the interim. No change in sensitivity to androstenone occurred. None of the nine subjects could detect androstenone during the follow-up test. We conclude that time and repeated exposures both appear to be critical in determining whether an individual becomes sensitized to androstenone. The total amount and duration of exposure and, importantly, whether active sniffing of androstenone is crucial remain to be determined.

Feedback During Testing

Testing was previously performed without subjects being told after each trial whether they were right or wrong (feedback). If androstenone was inherently difficult to perceive, then during testing feedback may be expected to improve performance as it does in many learning situations (Deese and Hulse, 1967). If cognitive factors were responsible for the variation in the ability to detect androstenone, then providing the subject with assessments of accuracy after each two-choice trial may be expected to improve detection of the odor; the tester would cue the subject after each trial, which should presumably provide reinforcement and facilitate detection by the subject. We selected 20 androstenone-sensitive subjects and performed the experiment.

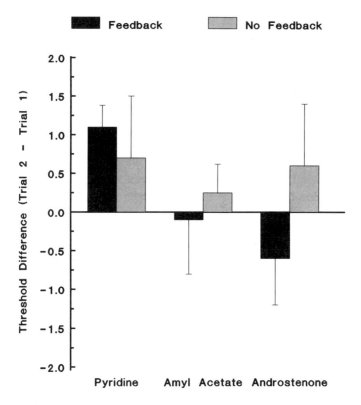

FIGURE 8 Sensitivity to pyridine, amyl acetate, and androstenone was determined without feedback regarding accuracy of choice in trial 1. Sensitivity was reassessed in trial 2; however, half the subjects received feedback but the remainder did not. The difference between these scores is indicated. For each of the odors, the groups did not differ on the first trial.

During trial 1 no feedback was given. During trial 2 half the subjects were told whether they were right or wrong after each response. Three odors were used: androstenone, amyl acetate, and pyridine.

All subjects could smell the odors; however, during the second trial those receiving feedback neither improved performance relative to the first trial nor scored better than the no-feedback group (see Figure 8). Importantly, feedback did not affect performance in tests of androstenone. Furthermore, the test-retest correlations for androstenone were high: feedback $r = 0.89$; no-feedback $r = 0.72$. Overall test-retest correlations were lower for pyridine ($r = 0.50$) and amyl acetate ($r = 0.61$). We conclude that trial-by-trial feedback about accuracy does not improve the perception of androstenone in androstenone-sensitive individuals. This experiment was repeated with androstenone-anosmic individuals serving as subjects. Analogous results were obtained; that is, feedback did not result in lower "thresholds."

IV. CONCLUSIONS

Variation in response to the odor of androstenone is affected by genes, by development, and by individual experience. It is not yet possible to comprehend the patterns underlying what appear to be almost contradictory data. How can experience influence a trait that appears to have such a large genetic component? Is induced sensitivity a permanent change? Why do only half the insensitive individuals become sensitive following exposure? What characterizes individuals who appear to lose sensitivity to the odor as they age? Why do they tend to be adolescent males? At present we do not have answers for these questions.

Is the induction of sensitivity to an odor by exposure to it restricted to androstenone-like substances, or is it a much more common phenomenon? There is little in the literature to answer this question. Modest increases in sensitivity to odor have been found with repeated threshold testing (Rabin and Cain, 1986), but these have been attributed to cognitive and performance factors rather than changes in sensitivity. Much needs to be done in this area with odors for which specific anosmias exist, as well as with odors that are readily detected. In this regard one can speculate that expert tasters and smellers, such as perfumers or wine tasters, may increase their sensitivity to the substances of their expertise as a consequence of repeated exposure. Furthermore, perhaps resistance to change in sensitivity in some older people, in the face of a general decline in odor sensitivity with age (see contributions in Murphy et al., 1989), is due in part to their exercising of their facilities.

Finally, among those individuals who could detect androstenone odor, there were differences in its quality and affective tone. Engen (1974) has emphasized the importance of learning in the determination of odor pleasantness. The results of recent work suggest that developmental changes and individual differences in the perceived pleasantness of odors may also derive from nonexperiential causes. Generally, individuals who are more sensitive to androstenone are more likely to label it as unpleasant (Dorries et al., 1989), and anecdotal evidence indicates that very sensitive individuals find it unpleasant at their very lowest level of detection.

The odor of androstenone provides an interesting subject for study. Significant individual differences in perception of this odor are influenced by genes, by factors associated with age, perhaps endocrine changes, and by exposure history. Our work has illustrated the complexity of the human response to odors and has highlighted the singularity of individual experience. Future work with androstenone could provide fundamental insights into the nature of chemosensory perception.

REFERENCES

Amoore, J. E. (1971). Olfactory genetics and anosmia. In *Handbook of Sensory Physiology*, L. M. Beidler (Ed.). Springer-Verlag, Berlin, pp. 245-256.

Beidler, L. (1987). Research directions in the chemical senses. In *Neurobiology of Taste and Smell*, T. E. Finger and W. L. Silver (Eds.). Wiley Neuroscience, New York, pp. 423-437.

Benton, D. (1982). The influence of androstenol—a putative human pheromone—on mood throughout the menstrual cycle. *Biol. Psychol.* **15**:249-256.

Bird, S., and Gower, D. B. (1983). Estimation of the odorous steroid, 5α-androst-16-en-3-one, in human saliva. *Experientia* **39**:790-792.

Breakefield, X. O. (Ed.) (1979). *Neurogenetics: Genetic Approaches to the Nervous System*. Elsevier, New York.

Brooksbank, B. W. L., Brown, R., and Gustafsson, J. A. (1974). The detection of 5α-androst-16-en-3α-ol in human axillary sweat. *Experientia* **30**:864-865.

Brown, K. S., and Robinette, R. R. (1967). No simple pattern of inheritance in ability to smell solutions of potassium cyanide. *Nature* **215**:406-408.

Cernoch, J. M., and Porter, R. H. (1985). Recognition of maternal axillary odors by infants. *Child Dev.* **56**:1593-1598.

Claus, R., and Asling, W. (1976). Occurrence of 5α-androst-16-en-3-one, a boar pheromone, in man and its relationship to testosterone. *J. Endocrinol.* **68**:483-484.

Cowley, J. J., Johnson, A. L., and Brooksbank, B. W. L. (1977). The effect of two odorous compounds on performance in an assessment-of-people test. *Psychoneuroendocrinology* **2**:159-172.

Deese, J., and Hulse, S. H. (1967). *The Psychology of Learning*. McGraw-Hill, New York.

Dorries, K., Wysocki, C. J., Beauchamp, G. K., and Schmidt, H. J. (1989). Changes in sensitivity to the odor of androstenone during adolescence. *Dev. Psychobiol.* **22**:423-436.

Engen, T. (1974). Method and theory in the study of odor preferences. In *Human Responses to Environmental Odors*, J. Johnston, D. Moulton, and A. Turk (Eds.). Academic Press, New York, pp. 121-141.
Engen, T. (1982). *The Perception of Odors*. Academic Press, New York.
Engen, T., and Lipsitt, L. P. (1965). Decrement and recovery of responses to olfactory stimuli in the human neonate. *J. Comp. Physiol. Psychol.* **59**:312-316.
Engen, T., Lipsitt, L. P., and Kaye, H. (1963). Olfactory responses and adaptation in the human neonate. *J. Comp. Physiol. Psychol.* **56**:73-77.
Filsinger, E. E., Braun, J. J., Monte, W. C., and Linder, D. E. (1984). Human responses to the pig sex pheromone 5 alpha-androst-16-en-3-one. *J. Comp. Psychol.* **98**:220-223.
Filsinger, E. E., Braun, J. J., and Monte, W. C. (1985). An examination of the effects of putative pheromones on human judgements. *Ethol. Sociobiol.* **6**:227-236.
Gershon, E. S., Matthysse, S., Breakefield, X. O., and Ciaranello, R. D. (Eds.) (1981). *Genetic Research Strategies in Psychobiology and Psychiatry*. Boxwood Press, Pacific Grove, California.
Graziadei, P. P. C., and Monti-Graziadei, G. A. (1980). Neurogenesis and neuron regeneration in the olfactory system of mammals. III. Deafferentation and reinnervation of the olfactory bulb following section of the fila olfactoria in rat. *J. Neurocytol.* **9**:713-728.
Hauser, H. J., Moor-Jankowski, J. K., Truog, G., and Geiger, M. (1958). Klinische, genetische und gerinnungs-physiologische Aspekte der Hamophilie B bei den Bluten von Tenna, mit einem Beitrag zur Genetik der Gerinnungsfaktoren. *Acta Genet. (Roma)* **8**:25-35.
Hubert, H. G., Fabsitz, R. R., Feinleib, M., and Brown, K. S. (1980). Olfactory sensitivity in humans: Genetic versus environmental control. *Science* **208**:607-609.
Hubert, H. G., Fabsitz, R. R., Brown, K. S., and Feinleib, M. (1981). Olfactory sensitivity in twins. In *Twin Research. III. Proceedings of the Third International Congress on Twin Studies. Progress in Clinical and Biological Research*, Vol. 69, L. Gedda, P. Parise, and W. E. Nance (Eds.). Alan R. Liss, New York, pp. 97-103.
Kirk-Smith, M., Booth, D. A., Carroll, D., and Davies, P. (1978). Human social attitudes affected by androstenol. *Res. Commun. Psychol. Psychol. Behav.* **3**: 379-384.
Labows, J. N., and Wysocki, C. J. (1984). Individual differences in odor perception. *Perf. Flav.* **9**:21-26.
Lewin, B. (1980). *Gene Expression*. Wiley Interscience, New York, pp. 441-454.
Lyon, M. (1962). Sex chromatin and gene action in the mammalian X-chromosome. *Am. J. Hum. Genet.* **14**:135-148.
Lyon, M. (1988). The William Allan Memorial Award Address: X-chromosome inactivation and the location and expression of X-linked genes. *Am. J. Hum. Genet.* **42**:8-16.
MacFarlane, A. (1975). Olfaction in the development of social preferences in the human neonate. In *The Human Neonate in Parent-Infant Interaction*, Ciba Foundation Symposium, Amsterdam, pp. 103-117.

Moncrieff, R. W. (1966). *Odour Preferences.* Leonard Hill, London.
Moulton, D. G. (1978). Olfaction. In *Handbook of Behavioral Neurobiology.* Volume 1. *Sensory Integration,* R. B. Masterton (Ed.). Plenum Press, New York, pp. 91-118.
Murphy, C. L., Cain, W. S., and Hegsted, M. (Eds.) (1989). *Nutrition and the Chemical Senses in Aging: Recent Advances and Current Research Needs.* Annals of the New York Academy of Sciences, Vol. 561, New York Academy of Sciences, New York.
Peto, E. (1936). Contribution to the development of smell feeling. *Br. J. Med. Psychol.* **15**:314-320.
Pollack, M. S., Wysocki, C. J., Beauchamp, G. K., Braun, D., Jr., Calloway, C., and Dupont, B. (1982). Absence of HLA association or linkage for variation in sensitivity to the odor of androstenone. *Immunogenetics* **15**:579-589.
Rabin, M. D., and Cain, W. S. (1986). Determinants of measured olfactory sensitivity. *Percept. Psychophys.* **39**:281-286.
Reed, H. B. C., Melrose, D. R., and Patterson, R. L. S. (1974). Androgen steroids as an aid to the detection of oestrus in pig artificial insemination. *Br. Vet. J.* **130**: 61-67.
Rovee, C. K. (1969). Psychophysical scaling of olfactory response to the aliphatic alcohols in human neonates. *J. Exp. Child Psychol.* **7**:245-254.
Schmidt, H. J., and Beauchamp, G. K. (1988). Adult-like preferences and aversions in three-year-old children. *Child Dev.* **59**:1136-1143.
Self, P. A., Horowitz, F. D., and Paden, L. Y. (1972). Olfaction in newborn infants. *Dev. Psychol.* **7**:349-363.
Sherman, H. H., Amoore, J. E., and Weigel, V. (1979). The pyridine scale for clinical measurement of olfactory threshold: A quantitative re-evaluation. *Otolaryngol. Head Neck Surg.* **87**:717-725.
Srivastava, R. P. (1961). Ability to smell solutions of sodium cyanide. *East. Anthropol.* **14**:189-195.
Stein, M., Ottenberg, P., and Roulet, N. (1958). A study of the development of olfactory preference. *Am. Med. Assoc. Arch. Neurol. Psychol.* **80**:264-266.
Steiner, J. E. (1977). Facial expressions of the neonatal infant indicating the hedonics of food-related chemical stimuli. In *Taste and Development: The Ontogeny of Sweet Preference,* J. M. Weiffenback (Ed.). U. S. Government Printing Office, Publication Number NIH 77-1068, Washington, D.C., pp. 173-189.
Steiner, J. E. (1979). Oral and facial innate motor responses to gustatory and some olfactory stimuli. In *Preference Behavior and Chemoreception,* J. H. A. Kroese (Ed.). Human Information Retrieval, London, pp. 247-262.
Stevens, J. C., Cain, W. S., and Burke, R. J. (1988). Variability of olfactory thresholds. *Chem. Senses* **13**:643-653.
Van Toller, C., Kirk-Smith, M., Wood, N., Lombard, J., and Dodd, G. H. (1983). Skin conductance and subjective assessments associated with the odour of 5-alpha-androstan-3-one. *Biol. Psychol.* **16**:85-107.
Whissel-Buechy, D., and Amoore, J. E. (1976). Odor-blindness to musk: Simple recessive inheritance. *Nature* **242**:271-273.

Wilson, D. A., Sullivan, R. M., and Leon, M. (1987). Single-unit analysis of postnatal olfactory learning modified olfactory bulb output response patterns to learned attractive odors. *J. Neurosci.* **7**:3154-3162.

Wysocki, C. J., and Beauchamp, G. K. (1984). Ability to smell androstenone is genetically determined. *Proc. Natl. Acad. Sci. USA* **81**:4899-4902.

Wysocki, C. J., and Gilbert, A. N. (1989). The National Geographic Smell Survey: The effects of age are heterogeneous. In *Nutrition and the Chemical Senses in Aging*: *Recent Advances and Current Research Needs*, Ann. N.Y. Acad. Sci., Vol. 561, C. L. Murphy, W. S. Cain and D. M. Hegsted (Eds.). New York Academy of Sciences, New York, pp. 12-28.

Wysocki, C. J., Dorries, K., and Beauchamp, G. K. (1989). Ability to perceive androstenone can be acquired by ostensibly anosmic people. *Proc. Natl. Acad. Sci. USA* **86**:7976-7978.

Index

A

Acrocentrics, 7
Adenylate cyclase, receptor-mediated stimulation of, 15, 16
ADP ribosylation, 15
Amino acid substitution, mutation-induced, 216-218
Androstenone, effects on anosmia
 age-related developmental changes, 356, 360-363
 background, 354, 355
 characteristics, 355
 duration of exposure and total amount, 364-367
 environmental effects, 363
 enzyme activity, 364
 genetic studies, 357-360
 methods of sensory evaluation, 355-357

[Androstenone]
 role of feedback in enhancement of odor perception, 368-369
Anosmia
 androstenone effects, 354-369
 background, 331, 332, 354, 355
 heritable defect, 346-349
 human olfactory threshold, 332-347, 356-370
 hyposmia, 338
 incidence of multiple specific anosmia, 347
 irritation threshold distribution, 338
 multiple specific anosmias, 344-346
 odor-specific, 79, 94
 percentage of adults presenting it, 354

375

[Anosmia]
  results of strain differences in odor perception, 292, 293
  structural formula of eight primary odorants, 342, 343
Aphrodisin
  amino acid sequence of, 174, 176
  homology of aphrodisin with odorant-binding protein, 178
  isolation and characterization of, 170-179
  pheromonal action mechanisms, 169, 179-183
Autophosphorylation, intracellular signaling, 42

B

Backcrosses, 5, 6
  protocol for congenic strain development, transfer of dominant taster allele from donor strain to inbred partner strain, 249, 250
Bacterial chemotaxis
  intracellular signaling, 42-44
    communication modules, 42-44
    protein phosphorylation, 42
    role of MCP, 42-44
  sensory system, 29-33
    flagellar signaling, 30, 31
    locomotor response, 30
    stimulus detection, 30
    transduction pathways and components, 31-33
  structure-function studies of MCP transducers, 33-40
    covalent modifications, 33-35

[Bacterial chemotaxis]
    domain organization and membrane topology, 33
    mutant studies, 35-40
  transmembrane signaling
    conformation changes in receptor domain, 41
    cytoplasmic signaling domain, 40-42
    methylation segments, 40, 41
Bitterness, 227-233, 235, 240
  linkage studies for PRPs and bitter taste in mouse and human, 281-288
  recombinants between bitter taste and PRP genes, 284, 286
  PRP genes and proteins
    human, 280, 281
    mouse, 281
Bruce effect (*see* Pregnancy block)

C

Calcium pump, ionic mechanisms of chemoreception, 52
cAMP, receptor proteins in chemoreception, 16, 54-56
cDNA libraries, 1
Chemoreception
  chemoresponse behavior, 47-49
  adaptation, 53
  evidence for receptors, 52-54
  chemosensory transduction pathways, 52
  Mendelian mutants, 53
  evolution of pheromonal specificity, 61-66, 68-72
  neurological and behavioral comparisons with other moths, 68

INDEX 377

[Chemoreception]
  response specificities of receptor neurons, 66, 67
  ionic mechanisms, 51, 52
    calcium pump, 52
    membrane potential hypothesis, 49-51
      hyper/depolarization of action potentials, 49
    second messengers, 52
  pheromonal specificity evolution, 61-73
  receptor proteins, 54-56
    cAMP, 54-56
  role of genetics, 47-57
  specificity
    contact, 66-73
    evolution of, 68-73
    pheromonal, 66-73
Chemosensation, 61
Chloramphenicol, taste determination, 236
1,8-Cineole, identification of odorant-evoked current transients, 26
Class I antigens, 187-207
  cellular origins, 194-196
  excretion of, 196-199
  in mouse MHC, 212-222
  olfactory responses to MHC-associated urinary odors, 199, 200
Class II antigens, 187
Complement component 5, 2
Congenic mice, 211-222
  development of, 243, 248-256
  use of *Soa* gene, 244-248
  taste profile of, 256-261
Conspecific-associated chemical cues
  evidence for heritable variation, 155-157

[Conspecific-associated chemical cues]
  fitness, 161, 162
  feeding and oviposition, 147-163
    adult aggregation, 148-151
    male peptide-stimulated oviposition, 154, 155
    nonmating effects of males on oviposition, 152, 153
    olfactory attraction to larval cues, 153
    oviposition in sites with larva or eggs, 153, 154
    responses to field resources, 157-160
  fitness and responses to, 160, 161
Cyclic AMP (*see* cAMP)

D

Decismel scale, use in obtaining olfactory thresholds, 334-337, 339, 343, 347
Denatonium benzoate, taste determination, 236
Diacylglycerol, 18
Diphtheria toxin A polypeptide, 3
Donor strain, congenic strain development in mice, 249-252
D-phe
  murine taste sensitivity to, 267-276
    environmental communication, 276
    receptor mechanisms of transduction sites, multiple and single, 267-273
  sweet taste response

[D-phe]
  murine strain differences and response to D-phe, 270
  single-locus control for, 270-273
D-phenylalanine (*see* D-phe)
Dystrophin, 2

E

Embryonic stem cell (*see* ES cells)
ENU, 7
Enzyme I, PTS transport system, 31
ES cells, 4, 5
  homologous recombination, 5
Escherichia coli, sensory signaling in bacterial chemotaxis, 29-44
Ethylnitrosourea (*see* ENU)

F

Flush-crash-founder cycle, 122
Food-associated chemical cues, 147-163
  attractants, 147
  fermentation products, 147-163
Forskolin, 16, 26
Furosemide, identification of odorant-evoked current transients, 25, 26

G

Gene
  analysis, overview, 1
  cloning, 1, 2
  mapping
    backcrosses, 5, 6, 249, 250
    extensive linkage homology, 8, 281-286

[Gene]
  in situ hybridization to metaphase chromosomes, 1-8
  overcoming limitations to, 6
  PRP and bitter taste in mouse and humans, 281-286
  retroviral insertion, 3-5
  RI strains, 5, 6, 235, 236
  somatic cell hybrids, 6
  taste receptors, 125-134
G_i, adenylate cyclase inhibition, 15, 16
α_{2u}-Globulin superfamily, extracellular proteins
  description of, 175-177
  transport functions, 169, 175, 176
Glycine, taste determination, 233, 235
G_o, 15
Gonadotropin releasing hormone, 2
G proteins, 17, 18
  function coupling with olfactory amino acid receptors, 14, 15
  receptor-mediated stimulation of adenylate cyclase, 15, 16
  as a second messenger, 16
Gs, quality coding of odorants, 317-329
GTP binding proteins (*see* G proteins)
Guanine nucleotide binding stimulatory protein (*see* Gs)

H

Homologous recombination, gene replacement, 5
HPr, PTS transport system, 31

HPRT, 5
H-2 (see MHC)
Hyposmia, 337, 338 (see also Anosmia)
Hypoxanthine phosphoribosyltransferase, see HPRT

I

Immune function and chemosensory identity, MHC of mice, 211-222
Inbred partner strain, 263
  congenic strain development in mice, 249-252
  mechanisms influencing sorting of olfactory receptor cell axons during turnover, 292, 294, 295
Inositol triphosphate, 18
Insertional mutations, 5
  creation of, 3, 4
Isovaleric acid, mechanism behind encoding of olfactory information, selective anosmia in mice, 292, 293

K

Kallmann syndrome, agenesis of olfactory bulb, 323
Kineses, types of, 48
Klinokinesis, 48

L

Lectins, 14
Lesch-Nyhan syndrome, 5
Lithium chloride, genetics of taste in mice, 228
[L-^{3}H]Alanine, 15
[L-^{3}H]Arginine, 15

Lyon, Mary
  t complex on chromosome 17, 7

M

Major histocompatibility complex (see MHC)
Mate recognition system, 68-73, 109-122
MCP
  chemotactic response mediation, 31-33
  mediation of behavioral effects, 42, 43
  transducer mutants, 35-40
    biased output signals, 38-40
    chemoreception defects, 36, 37
    locked output signals, 37, 38
    structure-function studies, 33-40
Metallothioneine promoter, 3
Methyl-accepting chemotaxis proteins (see MCP)
MEV, 6
MHC, 187-207
  chemosensory identity and immune function, 212-222
  determination of mating preference, 221, 222
  genetic map of mouse, 212
  mutants, 216-218
  olfactory discrimination, 213-216
  role as an odorant, 222
Mitral cell loss, Purkinje cell degeneration, 295-300
Moloney murine leukemia virus, insertion mutation, 4
Multiple ecotropic virus (see MEV)

Murine genetics
  current methods, 1-8
  c-myc oncogene, 3
  homologous genes in mice and humans, 279, 288
  human growth hormone, 3
  link to fur color gene b, 273
  MHC-associated chemosensory communication, 211-222
    initial trials, 213
    mating preference determination, 221-222
    mutations, 216-218
    olfactory discrimination, 213-216
    pregnancy block, 219-220
    radiation chimera mice, 218-219
    use of congenic mice, 212-213
    Y maze, 213, 214, 219, 222
  neurological mutants, 311
    mechanisms of cortical organization, 291-312
  salivery proline-rich chromosome 6 proteins, 269, 270
  sensitivity to D-phe, 267-276
    multiple receptor site, 268-275
    single receptor site, 275, 276
    strain differences, 270-276
  taste determination techniques, 227, 228
    acetate attachment to ring Carbon 2, 229, 230
    bitterness, 229-233, 269, 270
    saltiness, 238-239
    sweetness, 233-238
    taste bud distribution, 263-265
  tissue-specific expression, 3
  transgenic mice
    production of, 2
    use of, 3, 4

Mutant reeler, gene determinants of long-projection neuron topography, 291, 293, 294
Mutations, 1, 2
  amino acid substitution, 216-218
  as an animal model for human disease, 2
  bacterial chemotaxis, 35-40
  chemoreception, 53-56
  detection tests, 7
  insertional, 3-5
  MCP, 35-40
  murine neurological mutants, 291, 293, 294, 311
  odor perception, 91-94, 97-106
  olfaction, 79-94, 97-106, 293-295
    antennal morphology, 98-101
  retrovirus-induced, 4, 5
  taste, 133
Myelin proteolipid protein, 2

N

Na^+ conductance, odorant-activated, 25, 26
Neuroendocrine response to genetic identity, 220

O

OBP, mechanisms of pheromonal action, 169-183
  homology of aphrodisin with odorant binding protein, 178
OCM, clinical manifestations and implications, 320-329
Odorant binding proteins (see OBP)
Odorant confusion matrix (see OCM)

INDEX 381

Odor perception
  anosmia, 79, 94, 292, 293, 368, 369
    effects of androstenone, 368, 369
  commensal bacteria in urine, 203-207
  discrimination mechanisms, 79-94
  effects of androstenone, 368, 369
  Gs activity, 317
  MHC-associated communication system, 211-222
  mutant antennal morphology, 97-106
    identification of dysgenic homoeotic mutant, 101-106
    isolation of mutants, 99-101
  neurogenetics, mutants, 91-94
  OCM, 320-329
  odorant binding proteins, 169-183
  odor-mapping, 88-91
  quality coding of odorants, pseudohypoparathyroidism, 317-329
  salient features in normal flies, 80-91
  sensory physiology, 85-88
  transplantation antigen excretion, 187-207
Odor threshold distribution, in assessing specific anosmia, 331-351
Olfaction
  amiloride-sensitive odorant-activated Na$^+$ conductance, 25, 26
  mRNA translation system, 26
  antennal mutants, precedent for use of, 97-106

[Olfaction]
  behavioral response genetic factors, 120, 121
  cytogenetic localization of olfactory genes on X chromosome, 92
  food-associated chemical cues, 147-163
  gene identification, 94, 95
  G proteins, 13-20
    amino acid receptors, 14, 15
    G protein-linked receptor-effector systems, 14-18
    ion channel activity, 15, 17
    second messenger, 16, 17
  in humans
    individual differences, 353-370
    specific anosmia, 355-369
  MHC-associated urinary odors, 199, 200
    nature of odorants, 200-203
  neurologic mutant reeler, 293, 294
  discovery, 291, 293
  odor perception
    androstenone, 353-370
    central projections of odors, 88-91
    class I antigens, 187-207
    contribution of glomerular circuitry to odor processing, 294, 295
    Drosophila, 79-94
    effect of background odor on larval taxis, 83, 84
    hamster, 169-183
    larval responses of olfactory mutants to odors, 93
    mice, 211-222
    mutant antennal morphology, 97-106

[Olfaction]
    odorant-evoked current transients by RNA, 25-27
    odorant-specific gene products, 26
    olfactory neurogenetics, 91-94
    role of inbred Balb/c, 294, 295
    sensory physiology, 85-88
pheromones
    aphrodisin effects, 169, 179-183
    genetics of moth, 109-122
    odorant-binding proteins, 169-183
    production, 115, 116
    transport proteins, 169-183
Purkinje cell degeneration, 292, 295
    granule cell morphology, 300-303
    membrane remodelling, 311
    mitral cell loss, 296-300
    synaptic organization, 303-310
response to MHC-associated urinary odors, 199-200
selectivity of chemical sensors, 331, 332
signal transduction
    cAMP, 16, 54-56
    genetic/immunological probes, 18-20
    olfaction-specific gene products, 25-27
    site-directed antibodies to ion channels, 19, 20
specific anosmia
    abnormal or absent receptor proteins, 331-349, 369
    genetic contribution to olfaction, 353-370
    receptor cell subsets, 292, 293

Ornithine transcarbamylase, 2
Orthokinesis, 48, 56
Ouabain, 25
    identification of odorant-evoked current transients, 26

P

Phenylethyl methyl ethyl carbinol (*see* PM-carbinol)
Phenylthiourea (*see* PTC)
Pheromone systems, 109-122, 169-183
    action mechanisms, 179-183
    aggregation pheromones, (*D. melanogaster*), 148-151
    aphrodisin (hamster), 169-183
    background, 110-111
    biosynthetic pathways for pheromone components, moth, 110
    communication, 68
    evolution
        in chemoreceptors, 61-73
        with CNS, 71, 73
    European corn borer pheromone system, 112-121
    females (moths)
        data summaries for female pheromone blend production for paternal races and crosses, 113, 114
        sex pheromone production, 110, 113
    males (moths)
        antennal receptors, 113-115
        data summaries of male antennal olfactory cell response phenotypes for ECB races and crosses, 115, 116
        upwind flight response, 117-119, 121

[Pheromone systems]
  mate recognition system, 61-73, 109-122, 147-163
    sex-linked marker, 119, 120
  neurophysiologic and behavioral comparisons, 68
  odorant-binding proteins, 169-183
  olfactory cells
    detection of, and pheromone production, 115, 116
    odorant perception, 169-183
    recombinations of, and behavioral response genetic factors, 120-121
  receptor neurons
    evolution of, 68-73
    response specificities, 66, 67
  sex-linked gene, 119-120, 121
    allozyme differences as genetic markers, 119-120
  speciation debate, 109, 122
  stability of, 111-112
Phosphatidylinositol-4,5-biphosphate, 18
Phospholipase C, role of G proteins, 17, 18
PM-carbinol in identifying hyposmia, 337, 338
PMF, transduction pathways for chemotaxis, 31
Pregnancy block, MHC-associated chemosensory communication, 219, 220
Promoter-enhancer regions, 3
Protein-encoding sequences, 3
Protein phosphorylation, bacterial chemotaxis, 42
Proton-motive force (*see* PMF)

PRP, description, 280
  genes and proteins, humans
    mechanisms of function, 280
    role of tannins, 280, 281
    PRP polymorphisms, 280, 281
  linkage
    to bitter taste genes in mice and humans, 269, 270, 279, 282, 284-288
    studies assigning *PRP* genes to mouse chromosome 6, 281-284
Pseudohypoparathyroidism, 317-329
  quality coding of odorants
    Gs unit normal and deficient, 317-329
    odorants used, 318, 320
    patient pretreatment characteristics, 318, 319
    type Ia and b, 317
PTC
  genetics of tasting in mice, 228
  homologous *PRP* and bitter taste genes, 279, 284, 286-288
  polymorphic sensitivity, 269
PTS, transduction pathways for chemotaxis, 31
Purkinje cell degeneration, gene determinants of local circuit plasticity and organization, 292, 295-311

Q

Quinine, taste genetics, 230, 231, 235

R

Radiation chimera mice, 218-219

Radioligand binding techniques in olfaction, 14, 15
Recombinant inbred strains, 5, 6 (*see also* RI strains)
Resource response variation, *Drosophila melanogaster*
  adult aggregation, 148-151, 154-156, 158-161
  chemical modification of food resources, 153, 157-160
  conspecific-associated chemical cues, 147-163
    responses to field resources, 157-160
  density changes
    attraction to larval cues, 153
    fitness-related traits, 160
    oviposition, 154, 160
  heritable variation
    ecologic correlates, 161
    genetic correlation between emission of and response to chemicals, 155-157, 162
  hydrocarbon-associated pheromones, 150-153, 155
  oviposition
    nonmating effects of males, 152, 153
    peptide-induced, 154, 155
    stimulation and inhibition, 148-155, 160, 161
    surface texture-dependent, 153, 154
Restriction fragment-length polymorphisms (*see* RFLP)
RFLP, 1, 6
RI strains, 5, 6
  bitter taste, 235, 236
    linkage studies of PRP and bitter taste in mouse and human, 281-286

Robertsonian fusion chromosome, 7
RNA, odorant-evoked current transients, 25-27

S

Salivary proline-rich proteins (*see* PRP)
Salmonella typhimurium, sensory signaling in bacteria chemotaxis, 29-44
Saltiness, 228, 229, 238, 239, 259, 275
Second messengers
  G proteins as, 16, 17
  taste sensitivity, 128, 129
Smell blindness (*see* Anosmia)
Soa gene, 227, 244, 249, 256, 261, 269
  bitter taste, 236, 257, 258
  salty taste, 228, 229, 259
  sour taste, 259, 260
  sweet taste, 232-234
SOA
  congenic mice development, 243-246, 248-256
    donor strain and inbred partner, 249-252
    two-bottle preference tests, 244, 245, 248
  dimorphic taste sensitivity to bitterness, 269
  interstrain response of glossopharyngeal nerve to, 284
  taste profile of mice, 227, 244-261
    bitterness, 229, 230, 232, 233
    factors influencing sensitivity and intake behavior, 263-265
    taste bud distribution, 263-265

Specific anosmia (*see* Anosmia)
α-Spectrin, 2
Strychnine, taste genetics, 230
Sucrose octaacetate (*see* SOA)
Sugars, genetic alteration of multiple taste receptor sites, 125-134
Sweetness, 233-238, 240
  difference in ability of congenic lines to taste SOA, 243-261
  receptor transduction
    mechanisms of, 267-276
    multiple sites, 125-134, 267-275
    single sites, 275-276

T

Tannins, 280
Taste
  acesulfame, 236, 237
  bitterness perception, 229-233, 235, 240
    acetate group attachment to ring carbon 2, 229-230, 232
    chromosome 6
    cycloheximide, 233, 235
    Cyx, 233, 235, 269, 279, 284, 287
    Glb, 235, 279, 284, 287
    linkage studies of PRP genes and bitter taste in mouse and human, 279-288
    Qui, 233, 235, 269, 279, 284, 287
    Rua, 233, 235, 269, 279, 284, 287
    SOA gene, 230, 232, 269
  cyclohexamide, 238

[Taste]
  discriminability
    CNS integration of taste receptor responses, 143-145
    factors affecting it, 139-145
  multiple receptor sites, 126, 127, 143-145
  mutations affecting the pyranose site, 133
  receptor transduction, 137-145
  sweetness perception, 267-276
  saccharin, 236
  saltiness perception, 275
  sodium chloride, 238, 239
  sensitivity of
    alteration due to gene dosage, 129-133
    genetic variation of, 127-129
    possible mechanisms, 137-145
    role of second messengers, 128, 129
    strain distribution pattern of taste sensitivity, 284, 285
  Tre gene control of, 129
  sweetness perception, 125-134, 240
    chromosome 4 gene influence on taste receptor mechanisms, 267-276
    glycine, 233, 234, 236, 268
    gymnemic acid, 268, 269
    interspecies differences, 267-269
    link to fur color gene *b*, 273
    multiple receptor sites, 143-145, 267-276
    responses to D-amino acids, 274, 275
    single receptor sites, 275, 276
Thyroglobulin, 2
TPI enzyme, sex-linked marker, 119

[TPI enzyme, sex-linked marker]
conformation of males in flight tunnel-response profiles, 120
Transgenes, 4
Transgenic mice
production of, 2
uses of, 3, 4
Transplantation antigens, physiology of
class I antigens
association with commensal bacteria in urine, 203-207
cellular origin of, 194-196
responses to MHC-associated odors, 187-200
role as odorant, 200-207
quantitation and turnover, 192-194
target cell lysis inhibition, 191
urinary cell excretion, 196-199
description, 187
α,α-Trehalose octaacetate, murine taste genetics, 229
Triose phosphate isomerase (see TPI)
Two-bottle preference test
SOA aversion versus nonaversion, 248-256

[Two-bottle preference test]
sweetness, 270-273
taste mediation in SOA avoidance, 243-245
Tyrosinase, 2

U

Ussing method, 25

V

Vertebrate reproduction, behavior and selection
gene-associated, 213, 219-220
Vomeronasal organ, detection of macromolecular pheromone, 169, 171
role of odorant binding proteins, 169, 171

W

Wild-type revertants, 4

X

Xenopus oocyte, olfaction transduction, 25, 26
in ovo mRNA translational system, 25-27